Machine Learning and Data Mining for Emerging
Trend in Cyber Dynamics

Haruna Chiroma · Shafi'i M. Abdulhamid ·
Philippe Fournier-Viger · Nuno M. Garcia
Editors

Machine Learning and Data Mining for Emerging Trend in Cyber Dynamics

Theories and Applications

 Springer

Editors
Haruna Chiroma ⓘD
Mathematical Sciences
Abubakar Tafawa Balewa University
Bauchi, Nigeria

Shafi'i M. Abdulhamid ⓘD
Information Technology & Cyber Security
Community College Qatar
Doha, Qatar

Philippe Fournier-Viger ⓘD
School of Computer Science
Harbin Institute of Technology
Shenzhen, China

Nuno M. Garcia ⓘD
Instituto de Telecomunicações
University of Beira Interior
Covilha, Portugal

ISBN 978-3-030-66287-5   ISBN 978-3-030-66288-2 (eBook)
https://doi.org/10.1007/978-3-030-66288-2

This Springer imprint is published by the registered company Springer Nature Switzerland AG
The registered company address is: Gewerbestrasse 11, 6330 Cham, Switzerland

Contents

A Survey of Machine Learning for Network Fault Management

Mourad Nouioua, Philippe Fournier-Viger, Ganghuan He, Farid Nouioua, and Zhou Min

Abstract Telecommunication networks play a major role in today's society as they support the transmission of information between businesses, governments, and individuals. Hence, ensuring excellent service quality and avoiding service disruptions are important. For this purpose, fault management is critical. It consists of detecting, isolating, and fixing network problems, a task that is complex for large networks, and typically requires considerable resources. As a result, an emerging research area is to develop machine learning and data mining-based techniques to improve various aspects of the fault management process. This chapter provides a survey of data mining and machine learning-based techniques for fault management, including a description of their characteristics, similarities, differences, and shortcomings.

1 Introduction

Computer networks are crucial to today's society as they support communication between individuals, governments, and businesses. They are used not only to connect

M. Nouioua · P. Fournier-Viger (✉) · G. He
Harbin Institute of Technology (Shenzhen), Shenzhen, China
e-mail: philfv8@yahoo.com

M. Nouioua
e-mail: mouradnouioua@gmail.com

G. He
e-mail: heganghuan@gmail.com

M. Nouioua · F. Nouioua
University of Bordj Bou Arreridj, El Anceur, Algeria
e-mail: faridnouioua@gmail.com

F. Nouioua
Aix-Marseille Université, LIS, UMR-CNRS 7020, Marseille, France

Z. Min
Huawei Noah's Ark Lab, Shenzhen, China
e-mail: zhoumin27@huawei.com

© The Author(s), under exclusive license to Springer Nature Switzerland AG 2021
H. Chiroma et al. (eds.), *Machine Learning and Data Mining for Emerging Trend in Cyber Dynamics*, https://doi.org/10.1007/978-3-030-66288-2_1

1

desktop computers, but also all kinds of electronic devices such as smartphones, wearable devices, sensors, and industrial machines. They also play a key role in emerging domains such as sensor networks [66, 87], vehicular networks [86], cloud computing [8], big data analysis [79], and the Internet of Things [81].

To ensure effective and efficient communication between devices, a network must be carefully designed in terms of physical and logical topology, and software must be properly configured. This requires considering various aspects such as budget, facilities, performance, and security requirements. Then, during a network's lifetime, various maintenance tasks must be carried out such as to replace, install, and upgrade equipment and software. Moreover, a key activity is *fault management*, which is carried out to ensure a network's security, availability, reliability, and optimize its performance [21, 71].

Fault management aims at solving problems that are occurring in a network. It consists of four main tasks, which are (1) Detecting, (2) Diagnosing, (3) Isolating, and (4) Fixing network faults [84, 87]. Fault management is not an easy task because faults may be caused by complex interactions between network devices and sometimes only appear for a short time. A good fault management process may consist of (1) Preventive measures and logging solutions that may raise alarms indicating potential problems, (2) A method for prioritizing the most important alarms or faults to analyze, and (3) Appropriate methods to isolate and fix the issues. Fault management can be quite time-consuming and costly, especially for large and heterogeneous networks. Hence, it has become critical to develop improved fault management techniques [84, 87].

Since more then two decades, some attempts at developing computer systems for fault management were made. For instance, in the 1990s, some expert systems were designed that relied on a knowledge base of rules to diagnose network problems. But a drawback of such systems was that specifying rules by hand requires expert knowledge, these rules would not be noise tolerant, and that writing these rules is time consuming and prone to errors [42, 60].

To build computer systems for fault management that do not rely heavily on domain experts, a promising fault management approach has been to apply data mining and machine learning-based techniques [5, 17, 19, 48, 56, 69, 74, 77]. These techniques allow to semi-automatically extract knowledge and learn models from data. Though there has been several studies in this direction, no survey has been published on this topic.

This chapter fills this gap by reviewing key studies on data mining and machine learning for fault management. The chapter provides a brief description of each study, their key ideas, and the advantages and limitations of the proposed techniques. The reviewed studies are categorized into two main categories based on the type of algorithms that they used: (1) Pattern mining-based approaches (e.g., itemset mining, association rule mining, and clustering) and (2) Machine learning-based approaches (e.g., neural networks, decision trees, Bayesian networks, and dependency graphs).

The rest of this chapter is organized as follows: Sect. 2 reviews some important concepts related to telecommunication network fault management. Then, Sects. 3

and 4 survey data mining-based approaches and machine learning-based approaches, respectively. Finally, a conclusion is drawn.

2 Network Fault Management

This section introduces important concepts related to telecommunication networks and fault management.

A computer network connects a set of devices that can exchange information and share resources. Typical devices found on a network are end-user computers (e.g., workstations), servers, mobile devices, and networking devices (e.g., routers and switches). Because networks can be used in a wide range of contexts and to address different use cases, numerous hardware and software technologies have been proposed for networking. Choosing a set of technologies requires considering various criteria such as cost, security, and performance. For example, some key requirements for building a sensor network may be to preserve battery life and perform distributed calculations. This is quite different from the requirements for large-scale computer networks such as the Internet, or those of a mobile GSM or UMTS network.

The goal of fault management is to detect, identify, and correct malfunctions in telecommunication networks [21, 71, 84, 87]. A *fault* is a malfunction that occurs on a network and may cause errors. A fault can have various consequences such as making a network device unavailable or degrading its performance. For example, a router hardware malfunction may cause the device to reboot and to be temporarily unavailable. Though a fault may cause several undesirable events, a fault is said to not be caused by any other events. In other words, a fault is the root cause of some error(s). An *error* is defined as a discrepancy between some observed values and some expected values or the violation of some conditions [71]. An error is caused by a fault and can propagate inside a network causing other errors.

Two main types of faults may occur [84, 87]: *hard faults* (a device cannot communicate with others) and *soft faults* (a device continues to operate but with an abnormal behavior such as sending corrupted data or incorrectly routing data). Moreover, from the perspective of time, faults can be categorized as *permanent* (an action must be taken to fix the issue), *temporary/transient* (the fault may appear only for a short time) and *intermittent* (the fault may periodically re-appear if no action is taken) [87].

Fault are sometimes not observable for various reasons such that no evidences have been collected. To more easily detect faults, alarms may be raised by network devices. An *alarm* is a symptom that can be observed of a potential fault. Alarms can be generated by devices or network management systems to provide information about potential faults that may cause errors. Alarms are very important as they allow to infer the existence of faults so that remediation steps can be taken [71]. Many alarms may appear in a network because a single fault may cause multiple alarms and because some alarms may be triggered in situations where no fault has occurred.

To facilitate alarm management, alarms are often categorized into different *severity levels* such as cleared, low, medium, high, and critical. Based on severity levels,

| 1) Fault detection | 2) Fault localization | 3) Fault isolation | 4) Fault recovery |

Fig. 1 A generic network fault management process

alarms may be treated differently [74]. For instance, while an alarm of the cleared level may be ignored, an alarm considered as critical may trigger an emergency message sent by SMS to a network administrator so that he can take quick action. Moreover, alarms can be recorded in logs on each device or can be collected from all devices by some fault management software to facilitate alarm analysis. In *passive fault management*, each device is responsible of communicating its alarms to other devices or the user, while in *active fault management*, a device may be periodically probed by other devices to verify its state. Because device clocks may not be synchronized, it is sometimes not possible to know which alarms occurred first. A solution to this issue can be to use distributed algorithms to synchronize logical clocks [50]. A human can investigate an alarm and fix a fault. Besides, it is also possible to put recovery procedures into place (scripts) to automatically handle faults or errors detected by some specific alarm types [71].

Finding the reason(s) why some error occurs in a network can sometimes be quite difficult because of the complex and dynamic interaction between devices, and because some faults are not permanent, errors can propagate or are influenced by other faults or events. For example, some faults may only appear in some circumstances such as when the battery level of a sensor is low.

Managing faults in a telecommunication network is generally done by the following steps [71]: (1) Automatically collecting data about alarms generated by devices, (2) Preparing the data and enriching the data with additional information (if needed), (3) Identifying alarms that should be investigated with higher priority and inferring the root cause of alarms, (4) Applying recovery procedures or dispatching technicians to specific locations (physically or virtually) to isolate and fix the issues. This process is illustrated in Fig. 1. Step 1 is called *fault detection*, Step 2 is named *fault localization*, while Step 3 is sometimes called *fault localization*, *fault isolation* or *root cause analysis*. Finally, Step 4 is named *fault recovery* [71].

Another reason why fault management is challenging is that while thousands of alarms may be generated in network devices, the number of technicians or budget for maintaining a network is limited. Hence, it is easy for technicians to be overloaded with thousands of alarms and being unable to investigate all of them. Accordingly, it is critical to be able to prioritize some alarms and their relationships to investigate the most important alarms first. However, identifying the most important alarms is not easy. In some case, some fault in a device may cause many other alarms. Moreover, some alarms may only be triggered in some very specific and complex situations involving multiple devices.

The following Sects. 3 and 4 give a survey of the different approaches for fault management using pattern mining and machine learning techniques, respectively.

3 Pattern Mining-Based Approaches

To develop innovative fault management approaches, several studies have used pattern mining techniques. Pattern mining is a key task in the field of data mining, which consists of analyzing data to identify interesting patterns that may help to understand the data or support decision-making [3, 28, 57, 59, 64, 80]. Several pattern mining algorithms have been designed to search for patterns. To apply an algorithm, a user typically has to set constraints on patterns to be found. For example, some algorithms are designed to find frequent patterns (patterns that appear at least a minimum number of times in a database) [3, 28, 59, 64], while others for clustering can find groups of similar data records [80]. Most of pattern mining algorithms are unsupervised as the goal is to discover new knowledge.

Pattern mining techniques have been used to analyze telecommunication data and find various types of patterns for fault management. The next three subsections review studies that have used three main types of pattern mining techniques: (1) Episode and association rule mining, (2) Sequential pattern mining and (3) Clustering algorithms.

3.1 Episode and Association Rules Mining-Based Approaches

Hatonen et al. [38] and Klemettinen et al. [48] first proposed using pattern mining techniques to analyze telecommunication network data. They designed a system called Telecommunication Alarm Sequence Analyser (TASA) to discover correlations between alarms.

Let A be a set of alarm types and T be a set of timestamps. The input of TASA is a sequence of alarms $S = \langle (a_1, t_1), (a_2, t_2), \ldots (a_n, t_n) \rangle$ that has been recorded from a network, where each alarm $a_i \in A (1 \leq i \leq n)$ is annotated with a timestamp $t_i \in T$. From such sequence of alarms, TASA applies a frequent episode mining algorithm [33, 34, 59] to extract *episode rules* [48].

An episode E is a tuple that has the form $\langle a_1, a_2, \ldots a_k \rangle$ where $a_i \in A$, for all $i \in \{1, \ldots, k\}$. An episode rule has the form $E_1, E_2, \ldots E_m \Rightarrow_{(x,y)} a$, where $E_j \subseteq A (1 \leq j \leq m)$ and $a \in A$, and x and y are two amounts of time where $y > x$. This rule is interpreted as if all alarms of E_1 appears in any order, and then are followed by alarms of E_2 in any order, ..., and then are followed by alarms of E_m in any order, and all of this happens in no more than x seconds, then the alarm a will appear no more than $z = (y - x)$ seconds later.

To find interesting episode rules, two measures are used called the support and the confidence. Let there be an episode rule $\alpha = (X \Rightarrow_{(x,y)} Y)$ and a sequence S, where X and Y are the rule antecedent and consequent, respectively. The support of the episode rule α in TASA is the absolute number of occurrences of the episode $\langle X, Y \rangle$ in the sequence. Formally, given an episode rule $\alpha = X \Rightarrow_{(x,y)} Y$ and a sequence S, $Support(\alpha) = |Occ(\langle X, Y \rangle, S)|$, where $Occ(\langle X, Y \rangle, S)$ is the set of time intervals where the episode $\langle X, Y \rangle$ occurs in sequence S.

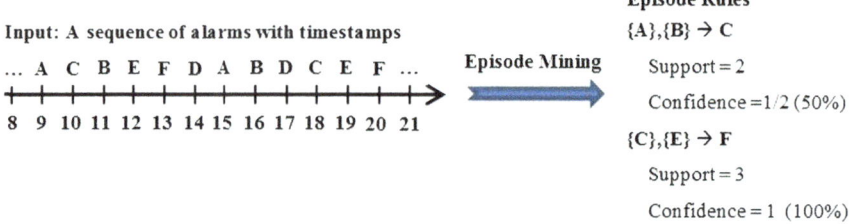

Fig. 2 Example of discovering episode rules between alarms with TASA

The confidence measure can be interpreted as the conditional probability that the whole episode occurs within time y given that the left side of the episode rule has already occurred within time x. Formally, we refer to the set of occurrences of an episode $\langle X, Y \rangle$ in a sequence S within time t as $Occ(\langle X, Y \rangle, S, t)$. Based on this, the confidence of an episode rule $\alpha = X \Rightarrow_{(x,y)} Y$ is $\text{Confidence}(\alpha) = \frac{|Occ(\langle X,Y \rangle,S,y)|}{|Occ(\langle X \rangle,S,x)|}$.

The TASA system outputs all rules having at least a minimum support and a minimum confidence [48] defined by the user. Moreover, the user can set constraints on the maximum amount of time for the duration x and y of a rule.

Figure 2 shows an example of episode rule extraction from a sequence of alarms using TASA. For this example, only timestamps from the 9th to the 20th s of that sequence are considered. Alarms are denoted as A, B, C, D, E, and F. We can see from Fig. 2 that $Occ(\langle A, B, C \rangle, S) = \{[9, 18], [15, 18]\}$. Thus, Support$(A, B \Rightarrow_{(x,y)} C = 2)$. If we consider the time durations x and y as $x = 3$ and $y = 5$ s, then $Occ(\langle A, B, C \rangle, S, 5) = \{[15, 18]\}$ and $Occ(\langle A, B \rangle, S, 3) = \{[9, 11], [15, 16]\}$. Accordingly, Confidence$(A, B \Rightarrow_{(3,5)} C) = 1/2$. Taking another episode rule $(C, E \Rightarrow_{(3,5)} F)$, Since $Occ(\langle C, E, F \rangle, S) = \{[10, 13], [10, 20], [18, 20]\}$, Support$(C, E \Rightarrow_{(x,y)} F) = 3$. Moreover, since $Occ(\langle C, E, F \rangle, S, 5) = \{[10, 13], [18, 20]\}$ and $Occ(\langle C, E \rangle, S, 3) = \{[10, 12], [18, 19]\}$, Confidence$(C, E \Rightarrow_{(3,5)} F) = 2/2 = 1$.

Besides finding episode rules, TASA can also discover interesting associations between alarm properties by applying an association rule mining algorithm [3, 63]. Let there be a set of property values P to describe alarms. Furthermore, let $DE = \{D_1, D_2, \ldots D_l\}$ be the descriptions of l alarm occurrences, where $D_h \in P (1 \leq h \leq l)$. An association rule has the form $X \Rightarrow Y$, where $X, Y \subseteq P$ and $X \cap Y = \emptyset$, and is interpreted as if an alarm occurrence has property values X, then it also has property values Y. For example, a rule $linkAlarm \Rightarrow BF$ may indicate that if an alarm of type $linkAlarm$ occurs, it is associated to a network device called BF. Two measures are used to select rules. The support of an association rule $X \Rightarrow Y$ is calculated as $sup(X \Rightarrow Y) = |\{D | X \cup Y \subseteq D \in DE\}|$, and the confidence of a rule $X \Rightarrow Y$ is calculated as $conf(X \Rightarrow Y) = sup(X \Rightarrow Y)/|\{D | X \subseteq D \in DE\}|$. An example of association rule extraction using TASA is shown in Fig. 3, where properties are denoted as A, B, C, and D. Contrarily to episode rules, association rules do not consider time.

Description of four alarm occurrences

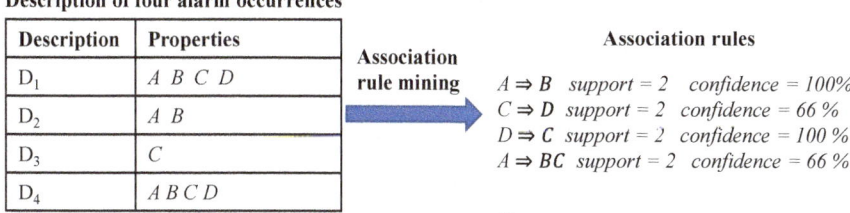

Description	Properties
D_1	$A \ B \ C \ D$
D_2	$A \ B$
D_3	C
D_4	$A \ B \ C \ D$

Association rule mining

Association rules

$A \Rightarrow B \quad support = 2 \quad confidence = 100\%$
$C \Rightarrow D \quad support = 2 \quad confidence = 66\%$
$D \Rightarrow C \quad support = 2 \quad confidence = 100\%$
$A \Rightarrow BC \quad support = 2 \quad confidence = 66\%$

...

Fig. 3 Example of discovering association rules between alarm properties with TASA

The TASA system also offers various tools to visualize rules, sort rules, and group rules having similar properties. A drawback of the TASA system is that it is not suitable for processing long alarm sequences.

3.2 Sequential Pattern Mining-Based Approaches

Another pattern mining technique that has been applied for fault management is sequential pattern mining [28, 64].

Lozonavu et al. [56] proposed a method to discover alarm correlation patterns using sequential pattern mining [64]. This method not only discovers sequential patterns indicating interesting relationships between alarms instances, but also studies correlation and the relationships between different network elements such as network nodes and network problems [56]. The input is a sequence of alarms with their timestamps $\langle (a_1, t_1), (a_2, t_2), \ldots (a_n, t_n) \rangle$, as defined in the previous subsection. The approach of Lozonavu et al. first partitions the input sequence into several sequences based on the timestamps of alarms. Two consecutive alarms (a_v, t_v) and (a_{v+1}, t_{v+1}) of the input sequence are put in the same sequence partition if $t_{v+1} - t_v < maxGap$, where $maxGap$ is a user-defined parameter. The result is a sequence database containing multiple input sequence partitions. In that sequence database, the timestamps of events are discarded. Then, the PrefixSpan [64] algorithm is applied on that database to find subsequences that appear in at least $minsup$ sequence partitions, where $minsup$ is a parameter set by the user. The result is a set of *sequential patterns* (frequent subsequences of alarms). A sequential pattern has the form $\langle b_1, b_2, b_3, \ldots, b_y \rangle$ indicating that some alarm b_1 appeared before another alarm b_2, was followed by b_3 and so on.

Then, the approach of Lozonavu et al. constructs a relationship graph, which shows the relationships between different alarms. This is done by transforming the discovered sequential patterns into relations. More precisely, the relationship graph is a directed weighted graph where each node represents a distinct alarm, a directed edge between two alarms indicates that an alarm occurred before the other in at least one discovered sequential pattern, and the weight of a relation reflects the strength

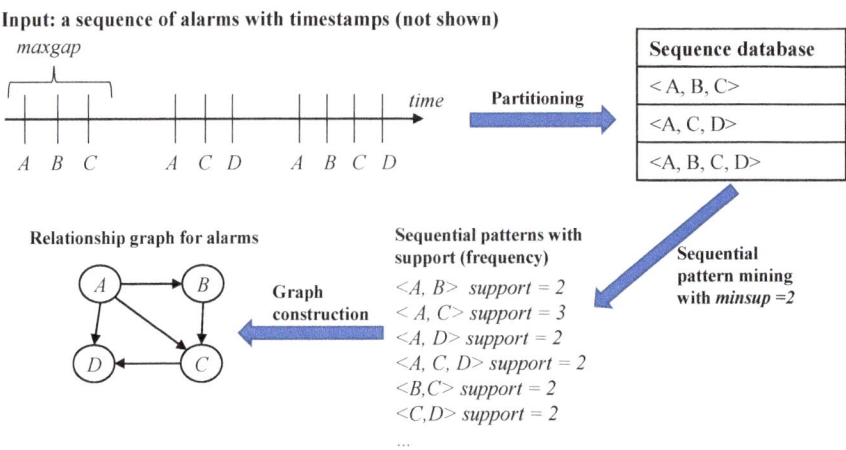

Fig. 4 Example of applying Lozonavu et al.'s approach to study temporal relationships between alarms

of this relation which is calculated by the confidence measure. The confidence of a pattern $\langle b_1, b_2 \rangle$ is defined as how many sequence partitions contains the pattern $\langle b_1, b_2 \rangle$ divided by how many sequence partitions contain b_1. An example of using the approach of Lozonavu et al. is illustrated in Fig. 4, where alarms are denoted as A, B, C and D.

Lozonavu et al. also presented a second type of relationship graph where the relationships between network devices are studied instead of the alarms. This graph is also derived from the sequential patterns because each alarm instance is associated with a device. It has been demonstrated that these graphs can then be used by network experts to better understand the network behavior and to discover hidden relations between network elements that were not known by network experts.

In another study, Wang et al. [74] have presented a system called Automatic Alarm Behavior Discovery (AABD) to study the behavior of alarms and select important alarms that should be brought to the attention of network operators from the thousands of alarms that may occur in a network. The AABD system takes as input a sequence of alarms, where each alarm is described using several fields such as the alarm type, the time that the alarm was produced and that it was cleared, the name of the device that has sent the alarm, and the network domain of the device.

AABD first preprocesses the input sequence to filter out invalid alarm instances. An alarm is said to be invalid if some of its fields contain invalid or missing values. For example, an alarm having no timestamp will be discarded as it will not be useful for the subsequent analysis performed by AABD. During the pre-processing step, alarm instances are also grouped by network domains as the behavior of an alarm may not be the same for devices of different domains.

Then, AABD applies an algorithm called Transient Flapping Determination (FTD) to detect alarms that are transient, i.e., alarms that usually only appear for a short time before they are cleared. A simple approach to determine if an alarm is transient is to calculate the duration of each of its instances. Then, given two parameters Ω and CT, an alarm is called transient (flapping) if the proportion of its instances that have a duration no greater than CT is greater than Ω. Then, for each transient alarm, all alarm instances having a duration that is less than CT are discarded to reduce the number of instances. It was shown that this approach based on transient alarm detection can reduce the number of alarms presented to operators by more than 84% [74]. However, a drawback of this definition is that setting the CT parameter for different alarm types in different situations requires expert knowledge and is time-consuming. Thus, Wang et al. also proposed a method to automatically set these parameters for each alarm type based on the distribution of each alarm's duration [74]. The settings of the CT parameter for an alarm type is called a *flapping rule*.

Then, AABD divides the sequence of alarms into several sequences to obtain a sequence database and applies the PrefixSpan [64] algorithm to extract frequent sequential patterns representing temporal correlations between alarm instances. This process is similar to the approach of Lozonavu et al. but the transformation from the input sequence to a sequence database is done differently. First, AABD calculates the support (occurrence frequency) of each alarm to find the N most frequent alarms, called main alarms, where N must be set by the user. Then, AABD creates a sequence for each main alarm. The created sequence contains that main alarm and all the alarms that occurred in the same time window width (w) with that main alarm. In other words, the created sequence will contain that main alarm and all the alarms that occurred no more than $w/2$ s before and $w/2$ s after. In the experiments, the value of the time window width w was set to 5 min because it was empirically observed that 90% of related alarms occurred within 5 min. Sequential patterns are then extracted from the resulting sequence database. Note that each network domain is treated individually because the behavior of the same alarm type may not be the same for different domains.

Then, the AABD system creates rules called P-C Rules (Parent-Child rules) to reduce the number of alarms presented to users. A PC-rule indicates that an alarm (called *parent*) may cause several other alarms (called *child*). A PC rule generation is done using an algorithm named PCRG, which takes as input the sequential patterns and also a knowledge-base called the P^2 lookup table. This latter is created by network administrators for each network domain based on historical trouble tickets as well as based on their rich experience. The P^2 table specifies two possible relationships between alarm pairs, that is an alarm may cause another alarm or two alarms are mutually exclusive. In the P^2 lookup table, each potential parent alarm is represented by its name and a serial number which is an ordered list of integers that reveal the ranking property of this potential parent alarm. The PCRG algorithm utilizes the serial numbers to determine the relationship between alarms, to test the possibility of generating a PC rule from each discovered sequential pattern. It was found that PC rules can greatly reduce the number of alarms presented to users.

Costa et al. [19] proposed a complete alarm management system. The system's architecture has two main modules. The first one is a rule management system, which discovers rules in alarm data using a modified sequential pattern mining algorithm (GSP) [70], which reveals alarm correlation and can be used to perform root cause analysis. Rules can be edited by hand. Moreover, a reinforcement learning algorithm is applied to evaluate how these rules are used for network malfunction resolution to refine the rule database. The second module performs alarm prioritization using a neural network to select alarms that should be treated with higher priority.

The system first pre-processes raw alarm data to keep only the relevant alarm attributes and sorts alarm instances by their start time. Moreover, if several alarm instances of the same type appear within a small amount of type, they are replaced by a single alarm with a counter so that patterns found only contain different alarms. The result is a sequence of alarms with timestamps. Then, sequential pattern extraction is performed by applying a modified GSP algorithm that extracts frequent sequential patterns using a sliding-window. From these patterns, rules are generated, which are evaluated using the lift and confidence measures to filter out spurious rules. To identify root cause problems, each extracted sequence is divided into two parts: The first part is the first alarm instance of the sequence and the second part is the following alarm instances with their occurrence counts. This method has some similarity to the approach of Lozonavu et al. [56] but it can be seen as more elaborated. The approach of Raúl is quite flexible as the sliding-window can be automatically enlarged in specific situations. The system was applied to data from a large Portugese telecommunication company and reduced the number of alarms presented to the user by up to 70%.

3.3   Clustering-Based Approaches

Clustering algorithms are another type of unsupervised data mining techniques that have been used to perform alarm correlation and root cause analysis in telecommunication networks. Generally, the goal of clustering is to extract *clusters* (groups of instances) from a dataset, such that similar instances are grouped together while dissimilar instances are put in different clusters [40]. Instances are described using attributes and similarity between instances can be measured using various measures.

Clustering techniques, as other data mining techniques, have been studied for detecting and diagnosing faults in alarms data of telecommunication networks. In the context of alarm correlation analysis, clustering is used to automatically group alarms that occurred within a same short period of time and may have been triggered by the same cause. Clustering is interesting for fault management because it does not require training data and it can group alarms that are related into clusters based on various criteria such as their occurrence times. Then, experts can further analyze these clusters to discover the root cause of problems. For example, it has been suggested that the earliest alarm of a cluster may be considered as its root cause.

Sozuer et al. [69] proposed a method to discover clusters in alarm event data. Let A be a set of alarm types. The proposed approach first transforms raw alarm event data into a sequence database, defined as a set of n sequences $SDB = \{s_1, s_2, \ldots, s_n\}$. Each sequence s_i $(1 \leq i \leq n)$ is an ordered list of alarm sets $\langle (A_1, t_1), (A_2, t_2), \ldots (A_m, t_m) \rangle$ where each alarm set A_j $(1 \leq j \leq m)$ has a timestamp t_j and $A_j \subseteq A$. Each alarm instance has a start time and clearance time, which defines a time interval called its life cycle. An alarm set is created by picking an alarm instance and adding all other alarm instances that have a start time within its life cycle. Moreover, alarm instances within an alarm set are sorted by time. This approach allows capturing that an alarm may trigger other alarms until it is cleared. After the sequence database creation, two weight metrics are calculated for each item (alarm type) in the sequence database. These metrics are the term frequency and inverse item frequency, which are typically used for document classification. Then, item vectors are formed based on these weight vectors. Thereafter, the K means clustering algorithm is applied on the normalized item vectors to obtain a set of clusters where each cluster contains alarms that are highly related in view of the selected weight measure.

An experimental evaluation with two weeks of data from a radio network was performed to evaluate the approach. It was found that clusters could be used to perform prediction more accurately than using the RuleGrowth [32] sequential rule mining algorithm. A limitation of the approach of Sozuer et al. is that clustering is applied separately on each network node. Thus, generalizations for multiple nodes cannot be found.

In another study, Hashmi et al. [37] used various clustering and outlier detection techniques to analyze one year of network failure data collected from a national broadband network. Each failure was analyzed with respect to five attributes: fault occurrence date, time of the day, geographical region, fault cause (from 92 possible causes), and resolution time. Using clustering (k-means, fuzzy c-means or self-organizing maps), interesting spatio-temporal clusters of faults were found. For example, one cluster indicated that fault resolution typically takes a long time between 1 and 4 PM in a specific region for some fault types. Such insights can be useful to improve service in that area. Moreover, anomaly detection techniques were applied (either local outlier factor or local outlier probabilities) to identify abnormal data points. For example, some detected anomalies were faults occurring during the night [37]. In experiments, the sum of squared errors (SSE) and Davies-Bouldin index (DBI) values were used to evaluate the performance of different techniques. Note that the DBI is the average ratio of intra cluster variance and inter cluster distance of all clusters. Obtaining a low DBI value is desirable because it indicates that there is a high separation between clusters. The results indicated that the k-means clustering method outperformed fuzzy c-means clustering. In fact, the k-means clustering algorithm was able to create a larger separation between different clusters which improved the accuracy by obtaining clusters that are very different from each other. Moreover, experiments have shown that clustered self-organizing maps outperformed k-means and fuzzy c-means in terms of both SSE and DBI.

3.4 Summary and Perspective

This section has reviewed some key approaches for fault management using pattern discovery algorithms. The reader can refer to Table 1 for a summary of these approaches. The main benefit of using a pattern mining approach is that patterns

Table 1 Summary of reviewed pattern mining-based approaches

Study	Category	Input	Output	Observation	Measure	Dataset
[38, 48]	Episode mining	Event sequence extracted from alarms data records	Episode rules, association rules	(1) Not suitable for long event sequences (2) The number of generated rules is large	Support, confidence	Several real-world datasets from telecommunication companies
[56]	Sequential pattern mining (PrefixSpan)	Alarm sequence database extracted using silent periods as delimiters	Relationship graph is extracted from frequent sequential patterns	The graph relationship is difficult to extract and to use by experts in case of large alarm datasets	Support, confidence	Dataset triggered in a 3G mobile network
[74]	Sequential pattern mining (PrefixSpan)	Alarm sequence database extracted using occurrence number of alarms with time window	Flapping rules, Parent–Child rules	a P^2 lookup table is required for each network domain which is not easy to create	Support	6 alarms datasets with several network domains: 2G,3G,4G, CS and PS
[19]	Sequential pattern mining (GSP)	Alarm sequence database extracted using a sliding window	Association rules extracted from frequent sequential patterns	Reinforcement learning is performed to filter unimportant rules	Support, confidence, lift	Real-world dataset containing alarms from a Portuguese telecommunications company
[69]	Clustering (K-means)	Alarm sequence database extracted using the life cycle of alarms	Clusters of alarms	The clustering method is applied only on data of each node separately.	Term frequency (TF), inverse item frequency (ITF)	Alarms data extracted from a Nokia radio access network logs
[37]	Clustering (various)	Network failure log	Clusters of failures and anomalies	Can discover spatio-temporal clusters of failures and anomalies	Various	Data from a national broadband provider

found can be easily understood by humans. There are many interesting possibilities for carrying further research on pattern mining for fault management. The next paragraphs discusses some of these possibilities.

First, it is observed that most of the above approaches are designed to handle rather simple data types (mostly discrete sequences) where alarms are viewed as events that have some attribute values. It would be interesting to consider more complex data representation to extract richer patterns. For instance, none of the above studies consider the spatial dimension (the network topology) in the pattern mining process. An interesting possibility is to view the network as a dynamic graph where alarms are spreading along edges (communication links) between vertices (network devices) to find spatio-temporal patterns. To extract such patterns, graph-based pattern mining algorithms could be considered or extended. For example, there exists several algorithms for discovering patterns in dynamic attributed graphs. A dynamic attributed graph is a graph where vertices may be described using multiple attributes, and the graph evolves over time (edges and vertices may be added or deleted, and attribute values may change) [20, 25]. A network can be modeled as such a graph, where node attributes can be used to represent alarms. Then, various types of patterns could be discovered to reveal relationship between alarms such as (1) *Cohesive co-evolution pattern* [20] (a set of vertices that are similar and display the same trends for some attribute(s) during a time interval), (2) *Triggering patterns* [45] (a rule of the form $L \rightarrow R$ where L is a sequence of attribute variations followed by a single topological change, R), and *significant trend sequences* [25] (correlated attribute variations for connected nodes).

Another possibility is to use other time representations for the input data as well as for patterns found. For example, while most of reviewed studies consider a strict sequential ordering between events in patterns, some algorithms have been designed to extract partial orders (patterns where events are partially ordered) [23, 65]. It is possible to consider richer relationships between events by explicitly representing each event as a time interval [18].

Another possibility is to consider extensions of traditional frequent pattern mining algorithms. For example, some extensions were proposed to handle items (events) having weights indicating their relative importance [53], weights and quantities [29, 72, 73], and cost values [27], as well as to use fuzzy functions [54] and taxonomies of items [12, 30, 82]. Using such algorithms would allow to consider richer information.

Another research direction is to explore the use or development of appropriate techniques for visualizing alarm patterns. For example, Jentner and Keim presented a detailed survey of many visualization techniques for patterns [41].

An important issue that could be also studied is how to reduce the number of patterns presented to the user by selecting the most important patterns or summarizing these patterns by reducing redundancy. There are several studies on this direction in the field of pattern mining such as to mine concise summary of patterns such as closed patterns [35, 51], maximal patterns [31, 58], and generator patterns [26, 83].

Another research direction is to go beyond the classical measures to select patterns such as the *support* and the *confidence*. A well-known problem with the support measure is that frequent alarms are sometimes unimportant and a limitation of the *confidence* is that it is very sensitive to the frequency of a rule's consequent in a database. Thus, other measures could be considered such as the *lift* [11], and application-specific measures could be designed.

Lastly, it is possible to build upon the research on stream pattern mining to adapt the current approaches for real-time processing [15, 61].

4   Machine Learning-Based Approaches

Apart from pattern mining-based techniques, several studies have applied machine learning techniques for network fault management. The next subsections review studies that have used different types of machine learning techniques: (1) artificial neural networks, (2) decision tree learning, (3) Bayesian networks, (4) Support-Vector Machines, (5) dependency graphs, and (6) other approaches.

4.1   *Artificial Neural Networks-Based Approaches*

Artificial Neural networks (ANN) are one of the most popular types of machine learning models [52, 55]. Their structure and learning mechanisms are loosely inspired by the human brain. Over the years, multiple ANN architectures have been proposed for different needs. A traditional feed-forward ANN contains multiple layers of neurons that are interconnected, where some neurons receive some numeric values as input, while others output numeric values. An ANN is typically trained in a supervised manner using some examples of inputs and corresponding desired outputs. After training, a neural network can predict output values for new input values. Some key properties of artificial neural networks is that they can approximate various nonlinear functions (mapping inputs to outputs), and that they are resilient to noise [21, 71]. Because of these properties, artificial neural networks are suitable for fault management.

Wietgrefe et al. [77] presented an artificial neural network-based alarm correlation system for correlating alarms in a GSM network. The proposed system is called Cascade Correlation Alarm Correlator (CCAC). It relies on an ANN to find the causes of alarms. Each input neuron of the ANN represents an alarm type and takes a binary value (the alarm is active or inactive), while each output layer neuron correspond to a problem's cause. The neural network is trained using sets of alarms with their known causes. During the training, weights of an hidden layer of neurons are adjusted. Then, after training, the neural network can be fed with alarm data to obtain the likely alarm causes. It was demonstrated that CCAC works well even in the presence of noisy data, where noise is defined as some missing alarms or some additional irrelevant alarms. In a comparative study done by Wietgrefe [76], it was found that the trained ANN is more accurate at finding alarm causes than several other approaches such as case-based reasoning and rule-based reasoning approaches. A limitation of this study is that it does not consider the temporal relationships between alarms, and how devices are interconnected.

To take the time ordering of alarms into account, Marilly et al. [60] proposed an hybrid approach for fault management that combines a multi-layer feed-forward ANN with signal-processing techniques. This method receives as input an alarm log and first removes redundant alarms. Then, each alarm is feed to the input layer of the ANN. Each input neuron represents either an alarm type or a logical network entity. These inputs then pass through two hidden layers to produce an output (an alarm class). After repeating this process for each alarm, a sequence of alarm classes is obtained ordered by time, which forms a signal. Then, the signal is treated using signal-processing methods such as a time-frequency visualization to extract pertinent information that could help identifying the cause of the malfunction (e.g., peaks of alarms that could indicate an equipment breakdown). This approach is robust to noise but still requires that an expert intervenes to identify malfunction in signals.

In another study, Arhouma and Amaitik [5] have compared different types of ANN models for alarm correlation in the ALMADAR GSM network by changing parameters of a neural network such as the network type, number of hidden neurons, and the learning algorithm. As in the work of Wietgrefe et al. [77], input neurons represent alarms while output neurons represent an initial cause of failure. It was found that the cascade-forward network type with Levenberg–Marquardt back-propagation training function gave the best diagnoses.

Differently, from the above studies, Barreto et al. [7] designed a fault detection approach to monitor the condition of a cellular network and detect abnormal behaviors. In this approach, the current state of a network is described as a vector of KPI (key performance indicator) values. Then, an ANN is trained with normal and abnormal states to then be able to classify a novel network state as abnormal or not. In that study, various neural network types were compared such as Self-Organizing Maps (SOM) and Neural Gas using a simulator.

In summary, although neural network-based approaches have the ability to extract patterns from complex and noisy data, these methods sometimes have a long training time and have difficulty to predict correctly for data that is largely different from the training data [21, 71]. Another issue with neural network-based approaches is that it is sometimes difficult to understand their inner-working.

4.2 Decision Tree-Based Approaches

Another machine learning technique that has been used for fault management is *decision tree learning*. It is a supervised learning method, which requires to provide a set of training instances described using some attributes and where an attribute is selected as the target attribute to be predicted. Each possible value for the target attribute is called a *class*. From the training data, a decision tree is built using a learning algorithm. Then, the model (tree) can be used to classify (predict the class) new instances (data records).

A *decision tree* is a tree-like structure that is designed to support the classification of data instances based on their attribute values. In a decision tree, each leaf represents a decision (a class), while internal nodes represent a test of an attribute and each outgoing branch from an internal node represents a possible value of this attribute. An instance can be classified using a decision tree by traversing it from the root node and following the branches based on the instance's attribute values until a leaf node is reached [14].

Chen et al. [17] proposed a decision tree based method for failure diagnosis in large Internet systems. A decision tree was trained to classify the failed and successful requests occurring during faulty periods. In the tree, the leaf nodes indicate whether a request is successful or has failed while internal nodes represent different system features such as a Machine's name and the name of a software program running on that machine. To identify causes of failures, post-processing is performed on paths of the generated decision tree by ranking them based on their correlation with failures. At this point, important features that correlate with the largest number of failures are selected.

Kiciman and Fox [47] have proposed a system for automatic fault detection in Internet services in which runtime paths of different requests served by the system are studied to identify the system's faulty components. A runtime path is a sequence of components, resources, and control-flow used to service a request. Besides, after detecting and recording the anomalous or successful requests with their corresponding paths, a decision tree is used to identify the components that may cause the failures. More precisely, a decision tree classifies the different paths into anomalous or successful paths based on their properties. Then, from the decision tree, a set of rules are extracted and used to extract the software and hardware components that lead to the failures.

Apart from fault localization in computer networks, decision trees have also been used for fault localization in other domains such as speech recognition [14] and Enterprise software systems (ESS) [67]. A drawback of decision tree based approaches for fault localization is that they may suffer from a degraded accuracy when dealing with noisy data [21, 71].

4.3   Bayesian Networks-Based Approaches

Bayesian networks are a type of Probabilistic Graphical Models that can be used to build models from data or from expert opinions. Bayesian networks are directed acyclic graphs (DAG) in which nodes correspond to random variables over a multi-valued domain while edges represent the causal relationship between nodes which is measured by the conditional probability [21]. Bayesian networks can be used to deal with a wide range of tasks such as prediction, decision under uncertainty, and diagnosis [13].

Barco et al. [6] proposed a model based on discrete Bayesian networks, namely smooth Bayesian network, for automatic fault identification in cellular networks. The main purpose of this model is to improve the fault identification process by decreasing the sensitivity of diagnosis accuracy which happens principally due to the imprecision of the model's parameters. In other words, the model was designed to improve the diagnosis accuracy by overcoming the inaccuracy of a model's parameters. The proposed approach considers alarms and key performance indicators registered daily by the network management system for fault identification. Experimental results on data from GSM/GPRS networks have shown that the proposed smooth Bayesian network outperforms traditional Bayesian networks in case of inaccuracy of the model's parameters.

In another study, Ruiz et al. [68] used a Bayesian Network to identify the causes of network failures at the optical layer, and give them probabilities. The input of the Bayesian network is monitoring data represented as two time series about bit error rates and received power. The data was discretized to train the Bayesian network. Then, it can predict two types of failures, namely inter-channel interference and tight filtering. The approach was found to provide high accuracy in a simulated environment.

Khanafer et al. [46] developed a Bayesian network-based fault isolation approach for Universal Mobile Telecommunications System (UMTS) networks. Given some symptoms (KPIs and alarms), the system can predict the cause. Because data about symptoms is continuous, Khanafer et al. first discretized the data (using two methods, namely percentile-based dicretization and entropy minimization discretization). This allowed to automatically find thresholds for symptoms that may indicate faults. Then, a Naive Bayes network was built in which the conditional probabilities linking causes to symptoms were learnt from training data based on the thresholds. Experiments on data from a real UMTS network have shown that the proposed approach identified the correct cause of problems 88% of the time.

To handle time, Ding et al. [42] proposed an approach for fault management in IP networks, which relies on a dynamic Bayesian network. In that network, dependencies between network objects (e.g., devices, software processes) are modeled, and how they evolve over time with conditional probabilities. The dynamic Bayesian network can be used to predict the state of a network object and the evolution of dependencies between two objects. Moreover, it can also be used to infer the likely causes of some observed symptoms using backward inference. This study focused on faults caused by *soft changes* (changes that gradually occur over time) in a network, and their causes. The approach was only tested with simulated data.

Besides fault isolation, Bayesian networks were also used to predict faults in communication networks [2, 39, 49] to perform pro-active maintenance.

The above studies have shown promising results but Bayesian networks have some limitations. One of them is that several Bayesian models rely on some assumptions of independence between some events. Another is that a considerable amount of training data is required to correctly estimate conditional probabilities. A third one is that complex temporal relationships between alarms and interactions between devices are difficult to model using Bayesian Networks.

4.4 Support-Vector Machine-Based Approaches

Another popular machine learning technique that is used for fault management is support-vector machine (SVM). It is a supervised technique that is generally used for classification or regression analysis. SVM is a linear classifier which is based on the margin maximization principle [1].

Given labeled data as input, the main idea of SVM is to find an optimal separating hyperplane that divides the input data into two classes. Note that to deal with nonlinear problems, the data can be mapped to higher dimensions to find a separating hyperplane. This mapping is performed using kernel methods and this process is called nonlinear SVM [4].

Wang et al. [75] have combined SVM with double-exponential smoothing (DES) to predict optical network equipment failures. To perform this task, some selected indicators are used as features to predict equipment failures. If an indicator is related to equipment failure, a change of this indicator's value will directly affect the equipment's state. Besides, since the relation between different indicators and equipment failures is not linear, a kernel function with punishment vector needs to be selected. This selection is performed by trying multiple kernel functions with different punishment factor values on SVM, using tenfold cross-validation to calculate their accuracy. Then, the combination that yield the highest accuracy is selected. Before applying SVM, DES is used to predict the value of each indicator at time $(t + T)$ taking the historical data from time $(t - n)$ to $(t - 1)$. At this point, the SVM method is applied to predict equipment failures at time $(t + T)$. Note that $(t - n)$ to $(t - 1)$ is a period of time that is selected in terms of days in the experiment and $(t + T)$ is the next period of time from the end of the previous time period to the end of the whole observation period. It was found that this improved SVM method can achieve an average of 95% accuracy (it can predict 95% of equipment failures).

In another study, Yuan et al. [85] proposed a system to automatically identify the root causes of problems in computer systems based on low-level traces of their behavior. This approach is different from that of utilizing a text-based search to find solutions to problems, which has been used in other systems. The proposed system has two main components: the tracer and the classifier.

The tracer collects the list of events that occur in the system when a problem's symptom is reproduced. Besides, the tracer records most system calls, which have several attributes such as Sequence number, Process ID, and Thread ID. After collecting all system call sequences related to symptoms, the system extracts *n-grams* from them. An n-gram can be viewed as sequential pattern where events must be consecutive. Then, the system encodes each log sequence as a bit vector where each dimension indicates the presence or absence of an n-gram.

At this point, the SVM classifier is applied on the set of bit vectors generated in the previous step for predicting the root cause of new traces from the previous registered traces with their known root causes. Besides, to prevent over-fitting due to the limited data, k-fold cross-validation is applied. It divides the training data into k partitions and then repeat selecting one partition to test it with the classifier trained with the remaining data until all data has been used for testing. The proposed approach was evaluated using four case examples containing diverse root causes. A prediction accuracy of nearly 90% was obtained.

Zidi et al. studied fault management for Wireless Sensor Networks (WSNs). A WSN is a set of autonomous devices that collaborate together through a wireless channel. WSNs are used to collect, process, and send data in various situations. However, WSNs may suffer from numerous failures [88]. Hence, Zidi et al. [88] proposed a new SVM-based technique for failure detection in WSNs. The proposed approach has two phases: The first phase is performed on anticipated time; whereas, the second one is performed on real time. The first phase is the learning phase where the main objective is to obtain a decision function from learning data using SVM. Besides, data is composed of a set of normal data as well as a set of faulty data. Then, the decision function is further used in the second phase to detect in real time whether a new observation is normal or belongs to faulty cases. To evaluate the performance of the proposed method, 21 datasets were formed from a previously published database. The collected datasets are composed of a set of sensor measurements where different types of faults are injected with certain degrees. The experimental results show that the proposed method achieves high detection rate (99% in most cases).

4.5   Dependency Graph-Based Approaches

These approaches are based on the construction of a dependency graph to represent different network elements. In [44], a graph-based method is developed for the fault localization problem. Besides, a dependency graph is constructed for the different network objects. Each vertex in the graph dependency is assigned with a weight which represents the probability that this object fails (triggers alarms) independently of other dependent network elements. On the other hand, each directed edge between two vertices is assigned with a weight that represents the strength of dependency between these vertices. In other words, the assigned weight is the conditional probability that the failure of one object is due to the failure of the other object. These weights can be estimated from the system specification information or from the history of previous failures [44].

After the graph construction, the system finds the domain of each alarm in the system which is defined as the set of objects that can cause this alarm. Note that, the problem of finding alarm domains is formulated as a variant of single source problem. At this point, a set of localization algorithms are used to discover alarm correlation patterns and identify fault locations.

Bouillard et al. [9, 10] have proposed another graph dependency based approach to correlate alarms in a network. This method is based on the assumption that frequently occurring alarms just refer to general information about the system, while rare alarms are viewed as more important since they may reveal a critical problem in the system. Accordingly, this method focuses on observing rare alarms.

First, this approach calculates the most frequent alarms. Then, based on these alarms, the alarm sequence is cut into set of small patterns (set-patterns) where frequent alarms are used as separators. Then, set-patterns are reduced using some transformation rules to facilitate their analysis.

At this point, the dependency graph from the reduced sets pattern is constructed and is divided into a set of subgraphs where each subgraph focuses on one rare alarm or a set of rare alarms. Finally, these subgraphs are further analyzed by network experts to discover the root cause of these alarms.

4.6 Other Approaches

Several of the reviewed approaches are passive. Johnsson and Meirosu [43] proposed an active fault management approach to perform fault localization in a packet-switched network. In that study, a network is viewed as an undirected graph connecting devices, and a fault is a performance degradation such as a large packet loss rate or transmission delay. Periodically, packets are sent between pairs of device to evaluate the network performance. They collected information about delays, jitters, and errors for the different edges of this graph and then used to calculate the probability that an edge is the source of a fault. The proposed approach utilizes probabilistic inference with a discrete state-space particle filter (also called histogram filter) to calculate the most probable location of a fault. This approach is lightweight and is applied in real time. The approach was evaluated using a simulator.

4.7 Summary and Perspective

This section has reviewed several studies, which have applied machine learning for fault management. Table 2 provides a summary of the main reviewed approaches discussed in this section. Most of the approaches surveyed in this section can be viewed as supervised approaches (requiring training data).

There are many possibilities for future work about using machine learning techniques. First, only a handful of machine learning techniques have been used for fault management, and there has been many advances in this field in recent years. Thus, newer techniques may be considered such as deep learning models [52], which may provide better results.

Second, it would be interesting to explore using models that consider richer information. For example, some models designed to handle temporal information could be used such as LSTM (Long Short Term Memory) [36]. Finding a way of also considering the network topology could lead to interesting results.

Third, the use of larger datasets with more features could be helpful to train better models. Semi-supervised machine learning techniques could also be used to reduce the need for human intervention. Synthetic but realistic data could also be generated to increase the size of the training data.

Fourth, an interesting alternative to passive monitoring (where a network management system waits passively for alarms sent by network nodes) is active monitoring (where the management system probes each nodes to verify its state). The advantage of active monitoring is that it may help recovering from faults more quickly. However, designing active monitoring techniques brings more challenges for network management [21].

Fifth, we have noticed that there are some recent studies on the use of machine learning in general and neural networks in particular for LTE and 5G networks [16, 22]. However, to the best of our knowledge, few studies have applied neural

Table 2 Summary of machine learning-based approaches

Study	Category	Input	Output	Observation	Algorithm	Dataset
[77]	Neural network	The input layer is a set of binary alarm vectors where each alarm is represented by a neuron	Each neuron of the output layer represents an initial cause of failure	The algorithm was found to tolerate well noisy data but only a few alarm vectors were used	Cascade Correlation Algorithm	Subset of 94 alarms chosen from real network
[60]	Neural network	Each input neuron represents a network's logical entity or alarm type from alarm log	An output neuron represents an alarm class	The neural network's output classes must be predefined by a network operator (not an easy task)	Multi-layer feed-forward ANN	Small set of alarm logs of an SDH network
[5]	Neural network	Each input neuron represents an alarm	Each output neuron represents an initial cause of failure	The neural network is applied separately to each network subarea. Small training set (55 patterns)	Cascade-forward ANN	Subparts of real GSM mobile network
[7]	Neural network	A vector of KPIs representing a network state	The state is abnormal or not		SOM, Net-Gas and others	Data from simulator
[17]	Decision tree	A set of vectors corresponding to request paths with their associated features	A set of paths extracted from a decision tree	Leaf nodes that contain less failure requests will be ignored in the diagnosis process	C4.5	Months of data from eBay's production website
[47]	Decision tree	A set of runtime paths with their associated features	A set of rules extracted from a decision tree's paths	The decision tree is a part of a pinpoint approach that is applied without requiring any apriori information from experts	ID3	A testbed is deployed with three Internet services that include a wide variety of failures
[6]	Bayesian network	Causes and symptoms (KPIs, alarms)	The conditional probabilities of fault causes for the observed symptoms	In experiments, the diagnostic model does not take alarms into consideration	Naive Bayes model	3 months of data from a GSM/GPRS network
[68]	Bayesian network	Time series about received power and error rates	One of two cause of failure	The model has good accuracy but consider only two causes	Bayesian network	Simulated data
[46]	Bayesian network	KPIs and alarms	A cause of failure	Discretization methods were employed to find thresholds for continuous variables	Naive Bayes model	UMTS network data

(continued)

Table 2 (continued)

Study	Category	Input	Output	Observation	Algorithm	Dataset
[75]	SVM	A set of vectors, each vector is composed of set of indicator values	Predict if the equipment will fail	Tenfold cross-validation is used twice for selecting relevant indicators to failures as well as for selecting the best failure diagnostic model	double-exponential smoothing with Support-Vector Machine	44 d of real data collected in a WDM network from a telecommunications operator
[85]	SVM	System call sequences	Predict the class failure of new traces	To perform a prediction, it is necessary to reproduce the problem's symptoms, which is not always convenient	Linear SVM classifier	Collect event traces of four computer problems occurring in the Windows XP SP2 system
[88]	SVM	A set of observation vectors where each vector has 12 dimensions. Each dimension represent a sensor measurement and these sensor measurements are taken in 3 consecutive instances	Detection of faulty data	The faulty data are injected by authors with different rates of faults and different types of faults. This method allows to detect faults in real time. However, it cannot predict faults before its occurrence to prevent it	Non-linear SVM classifier	A set of 21 labeled WSNs datasets, each dataset consists of a set of sensor measurements where different type of faults are injected
[9]	Dependency graph	An event sequence divided into small pattern-sets using frequent alarms as delimiters	The dependency graph is divided into small subgraphs. Conditional probabilities are used to find root causes	This method is based on the assumption that rare alarms can give information about failure sources	Dependency graph	1 year alarm log from two element networks of the Alcatel-Lucent operator
[9]	Dependency graph	An event sequence divided into small pattern-sets using frequent alarms as delimiters	The dependency graph is divided into small subgraphs. Conditional probabilities are used to find root causes	This method is based on the assumption that rare alarms can give information about failure sources	Dependency graph	1 year alarm log from two element networks of the Alcatel-Lucent operator
[44]	Dependency graph	Network elements are modeled as a dependency graph	Discover the number of fault hypotheses and assign a confidence measure to each one	This method was not tested on a real network	Dependency graph	Simulated network with 50 nodes, where each node has a failure probability
[43]	Other	KPIs about packet transmission on different links	The network link that is likely faulty	A fault is a performance degradation. Considers the network topology. Can be applied in real-time	Probabilistic inference	Simulated data

network in LTE and 5G networks for fault management [24, 62]. Thus, exploiting recent advances in machine learning in general and neural network in particular for fault management in 5G and LTE networks is a promising direction for future work.

Another research direction is to develop hybrid machine-learning based systems for fault management, which is to combine different machine learning techniques in a complementary way to efficiently perform fault management, while overcoming the disadvantages of using only one machine learning technique [13].

Moreover, with large and complex networks, both time and complexity become very large which makes the computation intractable [21]. As a result, developing techniques to reduce runtime and memory requirements of models while preserving their accuracy is crucial.

Finally, it is worth noticing that fault management in computer networks is a very active research topic. The reason is that, even with the development of many techniques based on pattern mining and machine learning, novel networking technologies and applications raise new challenges. Hence, existing systems become insufficient and must be improved [78].

5 Conclusion

This chapter has presented a survey of the main studies on using data mining and machine learnin- based techniques for network fault management, including a description of their characteristics, similarities, differences, and shortcomings. This is an active research area with many research opportunities.

References

1. Adankon, M.M., Cheriet, M., et al.: Support vector machine. In: Encyclopedia of Biometrics. Springer, Boston, MA (2009)
2. Agbinya, J.I., Omlin, C.W., Kogeda, O.P.: A probabilistic approach to faults prediction in cellular networks. In: International Conference on Mobile Communications and Learning Technologies, Conference on Networking, Conference on Systems, p. 130 (2006)
3. Agrawal, R., Srikant, R., et al.: Fast algorithms for mining association rules. In: Proceedings of the 20th International Conference on Very Large Data Bases, VLDB, vol. 1215, pp. 487–499 (1994)
4. Amirabadi, M.: A survey on machine learning for optical communication (machine learning view). arXiv preprint arXiv:1909.05148 (2019)
5. Arhouma, A.K., Amaitik, S.M.: Decision support system for alarm correlation in GSM networks based on artificial neural networks. In: Conference Papers in Science, vol. 2013, Hindawi (2013)
6. Barco, R., Díez, L., Wille, V., Lázaro, P.: Automatic diagnosis of mobile communication networks under imprecise parameters. Exp. Syst. Appl. 36(1), 489–500 (2009)
7. Barreto, G.A., Mota, J.C.M., Souza, L.G.M., Frota, R.A., Aguayo, L.: Condition monitoring of 3G cellular networks through competitive neural models. IEEE Trans. Neural Netw. 16(5), 1064–1075 (2005)

8. Botta, A., de Donato, W., Persico, V., Pescapè, A.: Integration of cloud computing and internet of things: a survey. Fut. Gen. Comp. Syst. **56**, 684–700 (2016)
9. Bouillard, A., Junier, A., Ronot, B.: Alarms correlation in telecommunication networks. Research Report RR-8321, INRIA (2013). https://hal.inria.fr/hal-00838969
10. Bouillard, A., Junier, A., Ronot, B.: Impact of rare alarms on event correlation. In: Proceedings of the 9th International Conference on Network and Service Management (CNSM 2013), pp. 126–129. IEEE (2013)
11. Brin, S., Motwani, R., Ullman, J.D., Tsur, S.: Dynamic itemset counting and implication rules for market basket data. ACM Sigmod Record **26**(2), 255–264 (1997)
12. Cagliero, L., Chiusano, S., Garza, P., Ricupero, G.: Discovering high-utility itemsets at multiple abstraction levels. In: Proceedings of 21st European Conference on Advances in Databases and Information Systems, pp. 224–234 (2017)
13. Cai, B., Huang, L., Xie, M.: Bayesian networks in fault diagnosis. IEEE Trans. Indus. Inform. **13**(5), 2227–2240 (2017)
14. Cerňak, M.: A comparison of decision tree classifiers for automatic diagnosis of speech recognition errors. Comput. Inform. **29**(3), 489–501 (2012)
15. Chen, M., Zheng, A.X., Lloyd, J., Jordan, M.I., Brewer, E.: Failure diagnosis using decision trees. In: International Conference on Autonomic Computing, Proceedings, pp. 36–43. IEEE (2004)
16. Chen, Y.C., Peng, W.C., Lee, S.Y.: Mining temporal patterns in time interval-based data. IEEE Trans. Knowl. Data Eng. **27**(12), 3318–3331 (2015)
17. Chen, C.C., Shuai, H.H., Chen, M.S.: Distributed and scalable sequential pattern mining through stream processing. Knowl. Inform. Syst. **53**(2), 365–390 (2017)
18. Chen, M., Challita, U., Saad, W., Yin, C., Debbah, M.: Artificial neural networks-based machine learning for wireless networks: a tutorial. IEEE Commun. Surv. Tutor. **21**(4), 3039–3071 (2019)
19. Costa, R., Cachulo, N., Cortez, P.: An intelligent alarm management system for large-scale telecommunication companies. In: Portuguese Conference on Artificial Intelligence, pp. 386–399. Springer (2009)
20. Desmier, E., Plantevit, M., Robardet, C., Boulicaut, J.F.: Cohesive co-evolution patterns in dynamic attributed graphs. In: International Conference on Discovery Science, pp. 110–124. Springer (2012)
21. Ding, J., Kramer, B., Xu, S., Chen, H., Bai, Y.: Predictive fault management in the dynamic environment of ip networks. In: 2004 IEEE International Workshop on IP Operations and Management, pp. 233–239 (2004)
22. Dusia, A., Sethi, A.S.: Recent advances in fault localization in computer networks. IEEE Commun. Surv. Tutor. **18**(4), 3030–3051 (2016)
23. Eugenio, M., Cayamcela, M., Lim, W.: Artificial intelligence in 5G technology: a survey. In: 2018 International Conference on Information and Communication Technology Convergence (ICTC) (2018)
24. Fabrègue, M., Braud, A., Bringay, S., Le Ber, F., Teisseire, M.: Mining closed partially ordered patterns, a new optimized algorithm. Knowl.-Based Syst. **79**, 68–79 (2015)
25. Feng, W., Teng, Y., Man, Y., Song, M.: Cell outage detection based on improved BP neural network in LTE system (2018)
26. Fournier-Viger, P., Cheng, C., Cheng, Z., Lin, J.C.W., Selmaoui-Folcher, N.: Mining significant trend sequences in dynamic attributed graphs. Knowl.-Based Syst. **182** (2019)
27. Fournier-Viger, P., Gomariz, A., Šebek, M., Hlosta, M.: VGEN: fast vertical mining of sequential generator patterns. In: International Conference on Data Warehousing and Knowledge Discovery, pp. 476–488. Springer (2014)
28. Fournier-Viger, P., Li, J., Lin, J.C.W., Chi, T.T., Kiran, R.U.: Mining cost-effective patterns in event logs. Knowl.-Based Syst. **191**, 105241 (2020)
29. Fournier-Viger, P., Lin, J.C.W., Truong-Chi, T., Nkambou, R.: A survey of high utility itemset mining. In: High-Utility Pattern Mining, pp. 1–45. Springer (2019)
30. Fournier-Viger, P., Wang, Y., Chun-Wei, J., Luna, J.M., Ventura, S.: Mining cross-level high utility itemsets. In: Proceedings of 33rd International Conference on Industrial, Engineering and Other Applications of Applied Intelligent Systems. Springer (2020)

31. Fournier-Viger, P., Wu, C.W., Gomariz, A., Tseng, V.S.: VMSP: Efficient vertical mining of maximal sequential patterns. In: Canadian Conference on Artificial Intelligence, pp. 83–94. Springer (2014)
32. Fournier-Viger, P., Yang, P., Lin, J.C.W., Yun, U.: Hue-span: fast high utility episode mining. In: Proceedings of 14th International Conference on Advanced Data Mining and Applications, pp. 169–184. Springer (2019)
33. Fournier-Viger, P., Yang, Y., Yang, P., Lin, J.C.W., Yun, U.: TKE: Mining top-k frequent episodes. In: Proceedings of 33rd International Conference on Industrial, Engineering and Other Applications of Applied Intelligent Systems. Springer (2020)
34. Fournier-Viger, P., Wu, C.W., Tseng, V.S., Cao, L., Nkambou, R.: Mining partially-ordered sequential rules common to multiple sequences. IEEE Trans. Knowl. Data Eng. **27**(8), 2203–2216 (2015)
35. Fournier-Viger, P., Lin, J.C.W., Kiran, R.U., Koh, Y.S., Thomas, R.: A survey of sequential pattern mining. Data Sci. Pattern Recogn. **1**(1), 54–77 (2017)
36. Fumarola, F., Lanotte, P.F., Ceci, M., Malerba, D.: Clofast: closed sequential pattern mining using sparse and vertical id-lists. Knowl. Inform. Syst. **48**(2), 429–463 (2016)
37. Gers, F.A., Schmidhuber, J., Cummins, F.: Learning to forget: continual prediction with LSTM. Neural Computation **12**(10) (2000)
38. Hashmi, U.S., Darbandi, A., Imran, A.: Enabling proactive self-healing by data mining network failure logs. In: 2017 International Conference on Computing, Networking and Communications (ICNC), pp. 511–517. IEEE (2017)
39. Hatonen, K., Klemettinen, M., Mannila, H., Ronkainen, P., Toivonen, H.: TASA: Telecommunication alarm sequence analyzer or how to enjoy faults in your network. In: Proceedings of NOMS'96-IEEE Network Operations and Management Symposium, vol. 2, pp. 520–529. IEEE (1996)
40. Hood, C.S., Ji, C.: Proactive network-fault detection (telecommunications). IEEE Trans. Reliab. **46**(3), 333–341 (1997)
41. Jain, A.K.: Data clustering: 50 years beyond k-means. Pattern Recogn. Lett. **31**(8), 651–666 (2010)
42. Jentner, W., Keim, D.A.: Visualization and visual analytic techniques for patterns. In: High-Utility Pattern Mining, pp. 303–337. Springer (2019)
43. Johnsson, A., Meirosu, C.: Towards automatic network fault localization in real time using probabilistic inference. In: 2013 IFIP/IEEE International Symposium on Integrated Network Management (IM 2013), pp. 1393–1398. IEEE (2013)
44. Katzela, I., Schwartz, M.: Schemes for fault identification in communication networks. IEEE/ACM Trans. Network. **3**(6), 753–764 (1995)
45. Kaytoue, M., Pitarch, Y., Plantevit, M., Robardet, C.: Triggering patterns of topology changes in dynamic graphs. In: Proceedings of the 2014 IEEE/ACM International Conference on Advances in Social Networks Analysis and Mining, pp. 158–165. IEEE (2014)
46. Khanafer, R.M., Solana, B., Triola, J., Barco, R., Moltsen, L., Altman, Z., Lazaro, P.: Automated diagnosis for UMTS networks using Bayesian network approach. IEEE Trans. vehic. Technol. **57**(4), 2451–2461 (2008)
47. Kiciman, E., Fox, A.: Detecting application-level failures in component-based internet services. IEEE Trans. Neural Netw. **16**(5), 1027–1041 (2005)
48. Klemettinen, M., Mannila, H., Toivonen, H.: Rule discovery in telecommunication alarm data. J. Netw. Syst. Manag. **7**(4), 395–423 (1999)
49. Kogeda, P., Agbinya, J.I.: Prediction of faults in cellular networks using Bayesian network model. In: International conference on Wireless Broadband and Ultra Wideband Communication. UTS ePress (2006)
50. Lamport, L.: Time, clocks, and the ordering of events in a distributed system. In: Concurrency: The Works of Leslie Lamport, pp. 179–196. ACM (2019)
51. Le, B., Duong, H., Truong, T., Fournier-Viger, P.: Fclosm, Fgensm: two efficient algorithms for mining frequent closed and generator sequences using the local pruning strategy. Knowl. Inform. Syst. **53**(1), 71–107 (2017)

52. LeCun, Y., Bengio, Y., Hinton, G.: Deep learning. Nature **521**(7553), 436–444 (2015)
53. Lee, G., Yun, U., Kim, D.: A weight-based approach: frequent graph pattern mining with length-decreasing support constraints using weighted smallest valid extension. Adv. Sci. Lett. **22**(9), 2480–2484 (2016)
54. łgorzata Steinder, M., Sethi, A.S.: A survey of fault localization techniques in computer networks. Sci. Comput. Program. **53**(2), 165–194 (2004)
55. Li, H., Wang, Y., Zhang, N., Zhang, Y.: Fuzzy maximal frequent itemset mining over quantitative databases. In: Asian Conference on Intelligent Information and Database Systems, pp. 476–486. Springer (2017)
56. Liu, W., Wang, Z., Liu, X., Zeng, N., Liu, Y., Alsaadi, F.E.: A survey of deep neural network architectures and their applications. Neurocomputing **234**, 11–26 (2017)
57. Lozonavu, M., Vlachou-Konchylaki, M., Huang, V.: Relation discovery of mobile network alarms with sequential pattern mining. In: 2017 International Conference on Computing, Networking and Communications (ICNC), pp. 363–367. IEEE (2017)
58. Luna, J.M., Fournier-Viger, P., Ventura, S.: Frequent itemset mining: a 25 years review. Wiley Interdisc. Rev.: Data Mining Knowl. Disc. **9**(6), e1329 (2019)
59. Luo, C., Chung, S.M.: Efficient mining of maximal sequential patterns using multiple samples. In: Proceedings of the 2005 SIAM International Conference on Data Mining, pp. 415–426. SIAM (2005)
60. Mannila, H., Toivonen, H., Verkamo, A.I.: Discovery of frequent episodes in event sequences. Data Mining Knowl. Disc. **1**(3), 259–289 (1997)
61. Marilly, E., Aghasaryan, A., Betge-Brezetz, S., Martinot, O., Delegue, G.: Alarm correlation for complex telecommunication networks using neural networks and signal processing. In: IEEE Workshop on IP Operations and Management, pp. 3–7. IEEE (2002)
62. Mendes, L.F., Ding, B., Han, J.: Stream sequential pattern mining with precise error bounds. In: 2008 Eighth IEEE International Conference on Data Mining, pp. 941–946. IEEE (2008)
63. Mismar, F.B., Evans, B.L.: Deep q-learning for self-organizing networks fault management and radio performance improvement. In: 2018 52nd Asilomar Conference on Signals, Systems, and Computers, pp. 1457–1461. IEEE (2018)
64. Nguyen, L.T., Vo, B., Nguyen, L.T., Fournier-Viger, P., Selamat, A.: Etarm: an efficient top-k association rule mining algorithm. App. Intell. **48**(5), 1148–1160 (2018)
65. Pei, J.: Mining sequential patterns efficiently by prefix-projected pattern growth. In: International Conference of Data Engineering (ICDE2001) (2001)
66. Pei, J., Wang, H., Liu, J., Wang, K., Wang, J., Yu, P.S.: Discovering frequent closed partial orders from strings. IEEE Trans. Knowl. Data Eng. **18**(11), 1467–1481 (2006)
67. Rashid, B., Rehmani, M.H.: Applications of wireless sensor networks for urban areas: a survey. J. Netw. Comput. Appl. **60**, 192–219 (2016)
68. Reidemeister, T., Munawar, M.A., Jiang, M., Ward, P.A.: Diagnosis of recurrent faults using log files. In: Proceedings of the 2009 Conference of the Center for Advanced Studies on Collaborative Research, pp. 12–23. IBM Corporation (2009)
69. Ruiz, M., Fresi, F., Vela, A.P., Meloni, G., Sambo, N., Cugini, F., Poti, L., Velasco, L., Castoldi, P.: Service-triggered failure identification/localization through monitoring of multiple parameters. In: Proceedings of 42nd European Conference on Optical Communication, pp. 1–3. VDE (2016)
70. Sozuer, S., Etemoglu, C., Zeydan, E.: A new approach for clustering alarm sequences in mobile operators. In: NOMS 2016-2016 IEEE/IFIP Network Operations and Management Symposium, pp. 1055–1060. IEEE (2016)
71. Srikant, R., Agrawal, R.: Mining sequential patterns: Generalizations and performance improvements. In: International Conference on Extending Database Technology, pp. 1–17. Springer (1996)
72. Truong, T., Duong, H., Le, B., Fournier-Viger, P.: Efficient vertical mining of high average-utility itemsets based on novel upper-bounds. IEEE Trans. Knowl. Data Eng. **31**(2), 301–314 (2018)

73. Truong-Chi, T., Fournier-Viger, P.: A survey of high utility sequential pattern mining. In: High-Utility Pattern Mining, pp. 97–129. Springer (2019)
74. Wang, J., He, C., Liu, Y., Tian, G., Peng, I., Xing, J., Ruan, X., Xie, H., Wang, F.L.: Efficient alarm behavior analytics for telecom networks. Inform. Sci. **402**, 1–14 (2017)
75. Wang, Z., Zhang, M., Wang, D., Song, C., Liu, M., Li, J., Lou, L., Liu, Z.: Failure prediction using machine learning and time series in optical network. Opt. Exp. **25**(16), 18553–18565 (2017)
76. Wietgrefe, H., Tuchs, K.D., Jobmann, K., Carls, G., Fröhlich, P., Nejdl, W., Steinfeld, S.: Using neural networks for alarm correlation in cellular phone networks. In: International Workshop on Applications of Neural Networks to Telecommunications (IWANNT), pp. 248–255. Citeseer (1997)
77. Wietgrefe, H.: Investigation and practical assessment of alarm correlation methods for the use in gsm access networks. In: NOMS 2002. IEEE/IFIP Network Operations and Management Symposium. Management Solutions for the New Communications World (Cat. No. 02CH37327), pp. 391–403. IEEE (2002)
78. Wong, W.E., Debroy, V.: A survey of software fault localization. Department of Computer Science, University of Texas at Dallas, Technical Report, UTDCS-45 **9** (2009)
79. Wu, X., Zhu, X., Wu, G., Ding, W.: Data mining with big data. IEEE Trans. Knowl. Data Eng. **26**, 97–107 (2014)
80. Xu, Y., Zeng, M., Liu, Q., Wang, X.: A genetic algorithm based multilevel association rules mining for big datasets. Math. Prob. Eng. (2014)
81. Xu, D., Tian, Y.: A comprehensive survey of clustering algorithms. Ann. Data Sci. **2**(2), 165–193 (2015)
82. Xu, L., He, W., Li, S.: Internet of things in industries: a survey. IEEE Trans. Indus. Inform. **10**, 2233–2243 (2014)
83. Yi, S., Zhao, T., Zhang, Y., Ma, S., Yin, J., Che, Z.: Seqgen: mining sequential generator patterns from sequence databases. Adv. Sci. Lett. **11**(1), 340–345 (2012)
84. Yu, C.B., Hu, J.J., Li, R., Deng, S.H., Yang, R.M.: Node fault diagnosis in WSN based on RS and SVM. In: Proceedings of 2014 International Conference on Wireless Communication and Sensor Network, pp. 153–156 (2014)
85. Yuan, C., Lao, N., Wen, J.R., Li, J., Zhang, Z., Wang, Y.M., Ma, W.Y.: Automated known problem diagnosis with event traces. ACM SIGOPS Oper. Syst. Rev. **40**(4), 375–388 (2006)
86. Zeadally, S., Hunt, R., Chen, Y.S., Irwin, A., Hassan, A.: Vehicular ad hoc networks (VANETS): status, results, and challenges. Telecommun. Syst. **50**, 217–241 (2012)
87. Zhang, Z., Mehmood, A., Shu, L., Huo, Z., Zhang, Y.L., Mukherjee, M.: A survey on fault diagnosis in wireless sensor networks. IEEE Access **6**, 11349–11364 (2018)
88. Zidi, S., Moulahi, T., Alaya, B.: Fault detection in wireless sensor networks through SVM classifier. IEEE Sens. J. **18**(1), 340–347 (2017)

Deep Bidirectional Gated Recurrent Unit for Botnet Detection in Smart Homes

Segun I. Popoola, Ruth Ande, Kassim B. Fatai, and Bamidele Adebisi

Abstract Bidirectional gated recurrent unit (BGRU) can learn hierarchical feature representations from both past and future information to perform multi-class classification. However, its classification performance largely depends on the choice of model hyperparameters. In this paper, we propose a methodology to select optimal BGRU hyperparameters for efficient botnet detection in smart homes. A deep BGRU multi-class classifier is developed based on the selected optimal hyperparameters, namely, rectified linear unit (ReLU) activation function, 20 epochs, 4 hidden layers, 200 hidden units, and Adam optimizer. The classifier is trained and validated with a batch size of 512 to achieve the right balance between performance and training time. Deep BGRU outperforms the state-of-the-art methods with true positive rate (TPR), false positive rate (FPR), and Matthews coefficient correlation (MCC) of $99.28 \pm 1.57\%$, $0.00 \pm 0.00\%$, and $99.82 \pm 0.40\%$. The results show that the proposed methodology will help to develop an efficient network intrusion detection system for IoT-enabled smart home networks with high botnet attack detection accuracy as well as a low false alarm rate.

Keywords Botnet detection · Internet of Things · Smart homes · Deep learning · Gated recurrent unit

S. I. Popoola (✉) · R. Ande · B. Adebisi
Department of Engineering, Manchester Metropolitan University, Manchester, UK
e-mail: segun.i.popoola@stu.mmu.ac.uk

R. Ande
e-mail: ruth.e.ande@stu.mmu.ac.uk

B. Adebisi
e-mail: b.adebisi@mmu.ac.uk

S. I. Popoola · R. Ande · K. B. Fatai
Artificial Intelligence for Cybersecurity Research, Cyraatek, Manchester, UK
e-mail: fbk6619@cyraatek.com

© The Author(s), under exclusive license to Springer Nature Switzerland AG 2021
H. Chiroma et al. (eds.), *Machine Learning and Data Mining for Emerging Trend in Cyber Dynamics*, https://doi.org/10.1007/978-3-030-66288-2_2

1 Introduction

Embedded devices are interconnected to each other and further connected to the Internet to form an Internet of Things (IoT) [1, 2]. Smart home paradigm exploits the enormous capabilities of IoT technologies to develop intelligent appliances and applications such as smart televisions, smart fridges, smart lighting, and smart security alarm systems [3, 4]. Primarily, IoT-enabled devices in homes autonomously communicate with residents and other IoT-enabled devices over the Internet. Unfortunately, invalidated assumptions and incompatibility in the integration of multiple IoT technologies, standards, proprietary communication protocols, and heterogeneous platforms have exposed smart homes to critical security vulnerabilities [5]. Most of the IoT devices and applications that are in use today were developed with little or no consideration for cybersecurity [6]. Hence, IoT devices tend to be easier to compromise than traditional computers [7].

Cyber attackers exploit the lack of basic security protocols in IoT devices to gain unauthorized remote access and control over insecure network nodes [8, 9]. Compromised IoT devices (i.e., bots) in smart homes can be connected to a master bot in a remote location. This kind of connection helps hackers to form a coordinated network of bots (botnets) [10, 11]. Botnets launch large-scale distributed denial of service (DDoS) attacks with massive traffic volume [10, 12]. Other botnet scenarios include port scanning, operating system (OS) fingerprinting, information theft, and keylogging [13]. Existing security solutions that are primarily designed for traditional computer networks may not be efficient for IoT botnet detection in smart home scenarios due to the unique characteristics of IoT devices and their systems [13].

Traffic patterns of different botnet attack scenarios can be detected in IoT network traffic data using machine learning (ML) approach. Various shallow learning techniques have been proposed to detect botnet attacks in IoT networks. These include support vector machine (SVM) [13–20], decision trees (DT) [14, 15, 19, 21–25], random forest (RF) [8, 15, 18, 21, 23, 26], bagging [15], k-nearest neighbor (k-NN) [15, 19, 21, 23, 24, 26], artificial neural network (ANN) [22, 25, 27], Naïve Bayes (NB) [22, 23, 25], isolation forest [16], feedforward neural network (FFNN) [18], k-means clustering [28, 29], and association rule mining (ARM) [25]. However, data generation in IoT networks is expected to be *big* in terms of volume, variety, and velocity [30]. Therefore, machine learning techniques with shallow network architecture may not be suitable for botnet detection in big IoT data applications.

Deep learning (DL) offers two main advantages over shallow machine learning, namely, hierarchical feature representation and automatic feature engineering, and improvement in classification performance owing to deeper network architecture. DL techniques have demonstrated good capability for botnet detection in big IoT data applications. State-of-the-art DL techniques for IoT botnet detection include deep neural network (DNN) [14], convolutional neural network (CNN) [8, 31–34], recurrent neural network (RNN) [13, 29], long short-term memory (LSTM) [13, 29, 35, 36], and bidirectional LSTM (BLSTM) [36, 37]. The classification performance of DL methods depends on the choice of optimal model hyperparameters. However, model hyperparameters are often selected by trial and error methods in previous studies.

Gated recurrent unit (GRU) is a variant of RNN, and it is suitable for large-scale sequential data processing [38]. Bidirectional GRU (BGRU) has a unique advantage of accessing both past and present information to make an accurate decision for efficient classification performance with lower computational demands [39]. To the best of our knowledge, previous researches have not investigated the capability of BGRU for botnet detection in a smart home scenario. In this paper, we aim to find the optimal hyperparameters for efficient deep BGRU-based botnet detection in IoT-enabled smart homes. The main contributions of this paper are summarized as follows:

a. A methodology is proposed to determine the most suitable hyperparameters (activation function, epoch, hidden layer, hidden unit, batch size, and optimizer) for optimal BGRU-based multi-class classification.
b. A deep BGRU model is designed based on the selected model hyperparameters to distinguish normal network traffic from IoT botnet attack scenarios.
c. The proposed methodology is implemented, and the deep BGRU model is developed with the bot-IoT dataset.
d. We evaluate the performance of the deep BGRU model based on training loss, validation loss, true positive rate (TPR), false positive rate (FPR), Matthews correlation coefficient (MCC), and training time.

The remaining part of this paper is organized as follows: Sect. 2 describes the proposed methodology for selection of optimal BGRU hyperparameters; the method is employed to develop a deep BGRU model for efficient IoT botnet detection; the results of extensive model simulations are presented in Sect. 3, and Sect. 4 concludes the paper.

2   Deep BGRU Method for Botnet Detection in IoT Networks

In this section, we describe the concept of BGRU, the proposed methodology for optimal BGRU hyperparameters, and the development of efficient deep BGRU classifiers for botnet detection in the context of a smart home. The overview of the framework is shown in Fig. 1.

2.1   Bidirectional Gated Recurrent Unit

GRU is a variant of RNN. This hidden unit achieves performance similar to LSTM with a simplified gated mechanism and lower computation requirements [40]. Unlike

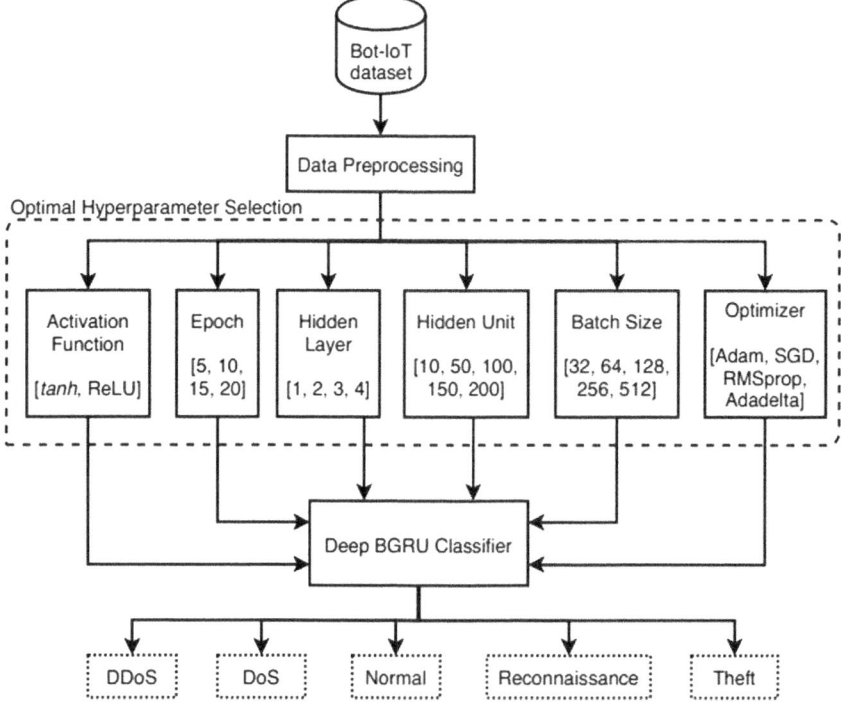

Fig. 1 Optimal hyperparameters of BGRU

LSTM, GRU discards the memory unit and replaces the input and forget gates with an update gate. Figure 2 shows the standard structure of a GRU. A GRU has two gates, namely, the reset gate (r_i) and the update gate (z_i). These gates depend on past hidden state $(\boldsymbol{h}_{t-1})$ and the present input (x_t). The reset gate determines whether the past hidden state should be ignored or not, while the update gate changes the

Fig. 2 The architecture of GRU

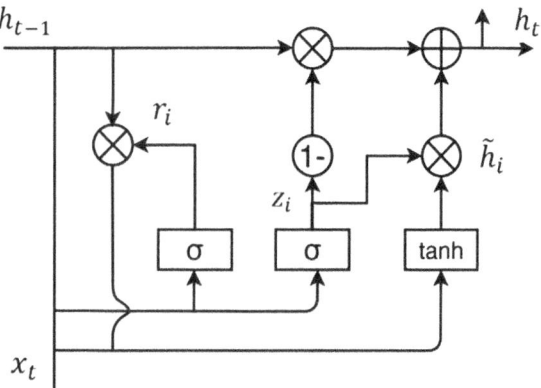

past hidden state to a new hidden state ($\tilde{h}_i$). The past hidden state is ignored when $r_i \simeq 0$ such that information that is not relevant to the future is dropped and a more compact representation is obtained. The update gate regulates the amount of data that is transmitted from the past hidden state to the present hidden state.

GRU is a unidirectional RNN, i.e., it employs a single hidden layer, and its recurrent connections are in the backward time direction only. GRU cannot update the present hidden state based on the information in the future hidden state. Interestingly, BGRU updates its current hidden state based on both the past and the future hidden state information [41]. A single BGRU has two hidden layers, which are both connected to the input and output. The first hidden layer establishes recurrent connections between the past hidden states and the present hidden state in the backward time direction. On the other hand, the second hidden layer establishes recurrent connections between the present hidden state and the future hidden states in the forward time direction. The computation of BGRU parameters is obtained by (1)–(9):

$$\overleftarrow{r_i} = \sigma\left(\left[\overleftarrow{\boldsymbol{W}}_r x\right]_i + \left[\overleftarrow{\boldsymbol{U}}_r \overleftarrow{\boldsymbol{h}}_{(t-1)}\right]_i + \overleftarrow{\boldsymbol{b}}_r\right), \tag{1}$$

$$\overleftarrow{z_i} = \sigma\left(\left[\overleftarrow{\boldsymbol{W}}_z x\right]_i + \left[\overleftarrow{\boldsymbol{U}}_z \overleftarrow{\boldsymbol{h}}_{(t-1)}\right]_i + \overleftarrow{\boldsymbol{b}}_z\right), \tag{2}$$

$$\overleftarrow{\tilde{h}}_i^{(t)} = \phi\left(\left[\overleftarrow{\boldsymbol{W}} x\right]_i + \left[\overleftarrow{\boldsymbol{U}}\left(\overleftarrow{r} \odot \overleftarrow{\boldsymbol{h}}_{(t-1)}\right)\right]_i\right), \tag{3}$$

$$\overleftarrow{h}_i^{(t)} = \overleftarrow{z}\overleftarrow{h}_i^{(t-1)} + \left(1 - \overleftarrow{z}_i\right)\overleftarrow{\tilde{h}}_i^{(t)}, \tag{4}$$

$$\overrightarrow{r_i} = \sigma\left(\left[\overrightarrow{\boldsymbol{W}}_r x\right]_i + \left[\overrightarrow{\boldsymbol{U}}_r \overrightarrow{\boldsymbol{h}}_{(t+1)}\right]_i + \overrightarrow{\boldsymbol{b}}_r\right), \tag{5}$$

$$\overrightarrow{z_i} = \sigma\left(\left[\overrightarrow{\boldsymbol{W}}_z x\right]_i + \left[\overrightarrow{\boldsymbol{U}}_z \overrightarrow{\boldsymbol{h}}_{(t+1)}\right]_i + \overrightarrow{\boldsymbol{b}}_z\right), \tag{6}$$

$$\overrightarrow{\tilde{h}}_i^{(t)} = \phi\left(\left[\overrightarrow{\boldsymbol{W}} x\right]_i + \left[\overrightarrow{\boldsymbol{U}}\left(\overrightarrow{r} \odot \overrightarrow{\boldsymbol{h}}_{(t+1)}\right)\right]_i\right), \tag{7}$$

$$\overrightarrow{h}_i^{(t)} = \overrightarrow{z_i} \overrightarrow{h}_i^{(t-1)} + \left(1 - \overrightarrow{z_i}\right)\overrightarrow{\tilde{h}}_i^{(t)}, \tag{8}$$

$$\tilde{y}_i = \vartheta\left(\boldsymbol{W}_y \overleftarrow{\boldsymbol{h}}_t^{(t)} + \boldsymbol{U}_y \overrightarrow{\boldsymbol{h}}_t^{(t)} + \boldsymbol{b}_y\right), \tag{9}$$

where $x, r, z, h, \tilde{y}$ and i are the input, reset gate, update gate, hidden state, output, and hidden unit index, respectively; $\overleftarrow{(\cdot)}$ and $\overrightarrow{(\cdot)}$ represent the parameters of the hidden layers in the backward and forward time directions, respectively; $\boldsymbol{W}_{(\cdot)}$ and $\boldsymbol{U}_{(\cdot)}$ are the weight matrices while $b_{(\cdot)}$ is the bias vector; $\sigma(\cdot)$ is a logistic sigmoid activation function; $\phi(\cdot)$ is either hyperbolic tangent (*tanh*) or rectified linear unit (ReLU) activation function; and $\vartheta(\cdot)$ is a softmax activation function.

2.2 The Proposed Method for Selection of Optimal BGRU Hyperparameters

The proposed method for optimal selection of BGRU hyperparameters is presented in Algorithm 1. Network traffic features are considered as sequential data given by (10):

$$X = \begin{bmatrix} x_{1,1} & x_{1,2} & \cdots & x_{1,\mu} \\ \vdots & \vdots & \ddots & \vdots \\ x_{\delta,1} & x_{\delta,2} & \cdots & x_{\delta,\mu} \end{bmatrix} \tag{10}$$

where μ is the number of network traffic features, and δ is the number of network traffic samples. The network traffic features in the training, validation, and testing sets are represented by X_{tr}, X_{va}, and X_{te}, respectively. The ground truth labels for training, validation, and testing are represented by y_{tr}, y_{va}, and y_{te}, respectively.

The selection of optimal BGRU hyperparameters will lead to efficient detection and classification of IoT botnet attacks. These hyperparameters include activation function (a_f), epoch (e_p), hidden layer (h_l), hidden unit (h_u), batch size (b_s), and optimizer (o_p). The optimal choice is made from a set of commonly used hyperparameters through extensive simulations. The collection of the hyperparameters is given by (11)–(16):

$$a_f = \begin{bmatrix} a_{f,1}, a_{f,2}, \ldots, a_{f,n} \end{bmatrix}, \tag{11}$$

$$e_p = \begin{bmatrix} e_{p,1}, e_{p,2}, \ldots, e_{p,m} \end{bmatrix}, \tag{12}$$

$$h_l = \begin{bmatrix} h_{l,1}, h_{l,2}, \ldots, h_{l,k} \end{bmatrix}, \tag{13}$$

$$h_u = \begin{bmatrix} h_{u,1}, h_{u,2}, \ldots, h_{u,q} \end{bmatrix}, \tag{14}$$

$$b_s = \begin{bmatrix} b_{s,1}, b_{s,2}, \ldots, b_{s,v} \end{bmatrix}, \tag{15}$$

$$o_p = \begin{bmatrix} o_{p,1}, o_{p,1}, \ldots, o_{p,g} \end{bmatrix}, \tag{16}$$

where n is the number of activation functions in a_f; m is the number of epochs in e_p; k is the number of hidden layers in h_l; q is the number of hidden units in h_u; v is the number of batch sizes in b_s; and g is the number of optimizers in o_p. The default hyperparameters are the first elements in each of the sets.

Algorithm 1. Optimal BGRU hyperparameter selection method

Input: $X_{tr}, y_{tr}, X_{va}, y_{va}, X_{te}, y_{te}$

Initialization: $c, \alpha, d, j, \beta, \gamma = 1$

Output: $\tilde{a}_f, \tilde{e}_p, \tilde{h}_l, \tilde{h}_u, \tilde{b}_s, \tilde{o}_p$

1. **def** $model()$
2. $N = BGRU(a_{f,c}, h_{l,d}, h_{u,j})$
3. $l_{tr}, l_{va}, \tilde{y} = BPTT(N, X_{tr}, y_{tr}, X_{va}, y_{va}, e_{p,\alpha}, b_{s,\beta}, o_{p,\gamma})$
4. $TPR, FPR, MCC = evaluate\_classifier(\tilde{y}, y_{te})$
5. $parameter = selector(l_{tr}, l_{va}, TPR, FPR, MCC)$
6.   **return** $parameter$
7. **for** $c = 1$ **to** n
8. $\tilde{a}_f = model()$
9. **for** $\alpha = 1$ **to** m
10. $\tilde{e}_p = model()$
11. **for** $d = 1$ **to** k
12. $\tilde{h}_l = model()$
13. **for** $j = 1$ **to** q
14. $\tilde{h}_u = model()$
15. **for** $\beta = 1$ **to** v
16. $\tilde{b}_s = model()$
17. **for** $\gamma = 1$ **to** g
18. $\tilde{o}_p = model()$

The development of the BGRU model for efficient IoT botnet detection in smart homes involves two main processes, namely, the choice of suitable deep network architecture and loss minimization through model training and validation. Deep network architecture (N) for BGRU is determined by $a_{f,c}$, $h_{l,d}$, and $h_u j$. The selected BGRU architecture is trained with X_{tr}, y_{tr}, X_{va}, y_{va}, $e_{p,\alpha}$, $b_{s,\beta}$, and $o_{p,\gamma}$ using back-propagation through time (BPTT) algorithm [42]. Loss minimization during training and validation is assessed based on the values of training loss (l_{tr}) and validation loss (l_{va}). A categorical cross-entropy loss function was used for loss minimization in a multi-class classification scenario.

The performance of BGRU classifier is based on TPR, FPR, and MCC when evaluated with highly imbalanced testing data (X_{te}, y_{te}). The definition of these performance metrics is given by (17)–(20) [43]:

$$\text{TPR} = \frac{\text{TP}}{\text{TP} + \text{FN}} \tag{17}$$

$$\text{FPR} = \frac{\text{FP}}{\text{FP} + \text{TN}}, \tag{18}$$

$$\lambda = 2\left[\frac{\text{TP} + \text{FN}}{\text{TP} + \text{FN} + \text{FP} + \text{TN}}\right] - 1 \tag{19}$$

$$\text{MCC} = \frac{1}{2}\left\{\left[\frac{\text{TPR} + \text{TNR} - 1}{\text{TPR} + (1 - \text{TNR})\left(\frac{1-\lambda}{1+\lambda}\right)}\right] + 1\right\}, \tag{20}$$

where true positive (TP) is the number of attack samples that are correctly classified; false positive (FP) is the number of normal network traffic samples that are misclassified as attacks; true negative (TN) is the number of normal network traffic samples that are correctly classified; false negative (FN) is the number of attack samples that are misclassified as normal network traffic; λ is the class imbalance coefficient. For each of the simulation scenarios, an optimal BGRU hyperparameter is expected to produce the lowest l_{tr}, l_{va}, and FPR as well as the highest TPR and MCC.

2.3   Deep BGRU Classifier for IoT Botnet Detection

Bot-IoT dataset [13] is made up of network traffic samples that were generated from real-life IoT testbed. The testbed is realistic because IoT devices were included. These IoT devices include a weather station, a smart fridge, motion-activated lights, a remote-controlled garage door, and an intelligent thermostat. This network of IoT devices is considered to be a good representation of an IoT-enabled smart home. The Bot-IoT dataset also contains recent and complex IoT botnet attack samples covering five common scenarios, namely, DDoS, DoS, reconnaissance, and information theft. Accurate ground truth labels are given to the IoT botnet attack samples. The samples of DDoS attack, DoS attack, normal traffic, reconnaissance attack, and information theft attack in the Bot-IoT dataset are 1,926,624, 1,650,260, 477, 91,082, and 79, respectively.

Network traffic samples, IoT botnet attack samples, and ground truth labels were pre-processed into appropriate formats suitable for deep learning. First, the complete dataset was randomly divided into a training set (70%), validation set (15%), and testing set (15%) as suggested in the literature [44, 45]. Non-numeric elements in feature matrices (X_{tr}, X_{va}, X_{te}) and label vectors (y_{tr}, y_{va}, y_{te}) were encoded using integer encoding and binary encoding methods, respectively. Furthermore, the elements of feature matrices were transformed using a min–max normalization method such that the value of each element falls between 0 and 1 [46].

The proposed method for the selection of optimal BGRU hyperparameters was implemented. All simulations were performed at a learning rate of 0.0001. The default hyperparameters for the simulations were the ReLU activation function, five epochs, a single hidden layer, 200 hidden units, a batch size of 128, and Adam optimizer. We investigated the suitability of the following: *tanh* and ReLU activation functions; epochs of 5, 10, 15, and 20; hidden layers of 1, 2, 3, and 4; hidden units of 10, 50, 100, 150, and 200; batch sizes of 32, 64, 128, 256, and 512; and finally, the optimizers (Adam, SGD, RMSprop, and Adadelta). Model training, validation, and testing were implemented using Keras library developed for Python programming running on Ubuntu 16.04 LTS workstation with the following specifications: RAM (32 GB), processor (Intel Core i7-9700 K CPU @ 3.60 GHz × 8), Graphics (GeForce RTX 2080 Ti/PXCIe/SSE2), and 64-bit operating system. The optimal BGRU hyper-parameters $\left(\tilde{a}_f, \tilde{e}_p, \tilde{h}_l, \tilde{h}_u, \tilde{b}_s, \tilde{o}_p\right)$ were used to develop a multi-class classifier for efficient IoT botnet detection in smart homes.

3 Results and Discussion

In this section, we evaluate the effectiveness of the method proposed for the selection of optimal BGRU hyperparameters in our attempt to develop an efficient IoT botnet detection system for smart home network security. Specifically, we examine the influence of different activation functions, number of epochs, number of hidden layers, number of hidden units, batch sizes, and optimizers on the performance of BGRU-based multi-class classifier.

3.1 Influence of Activation Functions on Classification Performance

To determine the right activation function for multi-class classification, ReLU and *tanh* activation functions were independently employed in two distinct BGRU neural networks, namely, BGRU-ReLU and BGRU-*tanh*. Apart from the activation function that differs, each of the BGRU neural networks is made up of a single hidden layer with 200 hidden units. BGRU-ReLU and BGRU-*tanh* were separately trained and validated with five epochs, 128 batch size, and Adam optimizer.

Training and validation losses in BGRU-ReLU and BGRU-*tanh* were analyzed to understand the extent of model underfitting and overfitting, respectively. Figure 3 shows that the ReLU activation function is more desirable than *tanh* activation function. Generally, training and validation losses reduced in both BGRU-ReLU and BGRU-*tanh* as the number of epochs increased from 1 to 5. However, training and validation losses were lower in BGRU-ReLU than in BGRU-*tanh* throughout the 5-epoch period. At the end of the experiment, we observed that training loss in

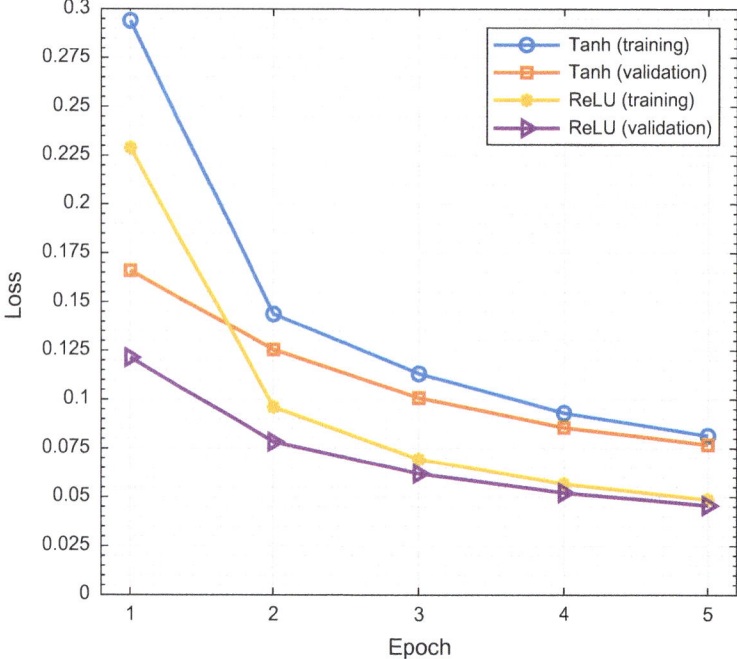

Fig. 3 Training and validation losses of two activation functions

BGRU-ReLU reduced to 0.1000, while validation loss reduced to 0.0721. Also, BGRU-ReLU reduced the average training and validation losses in BGRU-*tanh* by 31.16% and 35.07%, respectively. The reduction in training and validation losses implies that the likelihood of model underfitting and overfitting is minimal when the ReLU activation function was used in BGRU. Model underfitting will lead to poor classification accuracy while model overfitting will adversely affect the generalization ability of BGRU classifier when applied to previously unseen network traffic samples. Consequently, the adoption of the ReLU activation function in BGRU will help to achieve high classification accuracy and good generalization ability required for efficient IoT botnet detection in smart homes.

Multi-class classification performance of BGRU-ReLU and BGRU-*tanh* was evaluated with respect to the ground truth labels based on TPR, FPR, and MCC. TPR, also known as sensitivity or recall, is the percentage of samples that were correctly classified; FPR, also known as fall-out or false alarm ratio (FAR), is the percentage of samples that were wrongly classified; and MCC is a balanced measure that accounts for the impact of class imbalance on classification performance. Table 1 shows that BGRU-ReLU performed better than BGRU-*tanh*. BGRU-ReLU increased TPR and MCC in BGRU-*tanh* by 24.37% and 25.47%, respectively, while FPR was reduced by 40.86%. The lower the TPR and MCC values, the higher the chances that the BGRU classifier will fail to detect IoT botnet in smart homes. Also, the higher the

Table 1 Performance of BGRU with different activation functions

Metric	AF	DDoS	DoS	Normal	Reconn.	Theft	Mean ± Stdev
TPR	*tanh*	96.55	98.60	63.53	99.90	0.00	71.72 ± 42.85
	ReLU	98.01	98.86	74.12	99.99	75.00	89.20 ± 13.38
FPR	*tanh*	1.33	3.30	0.00	0.00	0.00	0.93 ± 1.45
	ReLU	1.08	1.90	0.00	0.00	0.00	0.60 ± 0.87
MCC	*tanh*	98.16	97.34	89.85	99.97	0.00	77.06 ± 43.25
	ReLU	98.72	98.38	93.05	99.99	93.30	96.69 ± 3.27

FPR value, the higher the probability that the BGRU classifier will produce a false alarm, i.e., it is more likely that the classifier will wrongly classify incoming network traffic or IoT botnet attack. Therefore, the choice of the ReLU activation function in BGRU will ensure a high detection rate and reduce false alarm in botnet detection system developed for smart homes.

3.2 Influence of the Number of Epochs on Classification Performance

In this subsection, we determine the optimal number of epochs required for efficient BGRU-based IoT botnet detection in smart homes. Four single layers BGRU neural networks were trained and validated with 5, 10, 15, and 20 epochs to produce BGRU-EP5, BGRU-EP10, BGRU-EP15, and BGRU-EP20 classifiers, respectively. Each of these classifiers utilized 200 hidden neurons, the ReLU activation function, a batch size of 128, and Adam optimizer.

Training and validation losses in BGRU-EP5, BGRU-EP10, BGRU-EP15, and BGRU-EP20 were analyzed to understand the extent of model underfitting and overfitting, respectively. Figure 4 shows that the lowest average training and validation losses were realized with 20 epochs. In general, training and validation losses reduced in all the classifiers throughout the epoch period. However, average training and validation losses were lower in BGRU-EP20 than in BGRU-EP5, BGRU-EP10, and BGRU-EP15. At the end of the experiment, we observed that training loss in BGRU-EP20 reduced to 0.0095, while validation loss reduced to 0.0094. BGRU-EP20 reduced the average training losses in BGRU-EP5, BGRU-EP10, BGRU-EP15 by 58.89%, 37.99%, and 17.87%, respectively, while the average validation losses were reduced by 53.80%, 35.22%, and 15.82%, respectively. The reduction in training and validation losses implies that the likelihood of model underfitting and overfitting is best minimized when the number of epochs in BGRU was 20. In other words, a sufficiently large number of epochs in BGRU will facilitate high classification accuracy and good generalization ability that are required for efficient IoT botnet detection in smart homes.

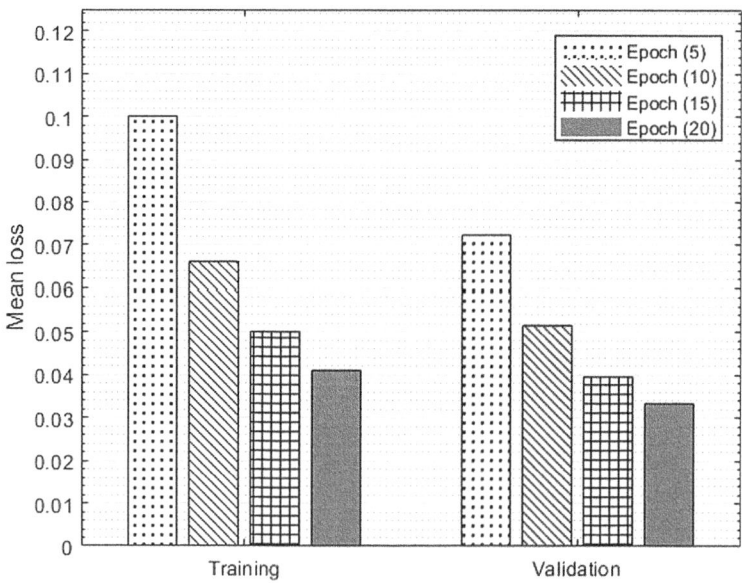

Fig. 4 Mean training and validation loss of different epochs

Multi-class classification performance of BGRU-EP5, BGRU-EP10, BGRU-EP15, and BGRU-EP20 was evaluated with respect to the ground truth labels based on TPR, FPR, and MCC. Table 2 shows that BGRU-EP20 performed better than BGRU-EP5, BGRU-EP10, and BGRU-EP15. BGRU-EP20 increased TPR by 7.97%, 3.65%, and 1.05% relative to BGRU-EP5, BGRU-EP10 and BGRU-EP15, respectively; FPR decreased by 87.27%, 73.08%, and 46.15%, respectively; and MCC increased by

Table 2 Performance of BGRU at different number of epochs

Metric	Epoch	DDoS	DoS	Normal	Reconn.	Theft	Mean ± Stdev
TPR	5	97.97	99.18	76.47	99.98	75.00	89.72 ± 12.80
	10	99.01	99.60	81.18	99.99	87.50	93.46 ± 8.63
	15	99.44	99.90	80.00	99.98	100.00	95.86 ± 8.87
	20	99.72	99.94	84.71	99.98	100.00	96.87 ± 6.80
FPR	5	0.78	1.94	0.00	0.00	0.00	0.55 ± 0.85
	10	0.38	0.95	0.00	0.00	0.00	0.26 ± 0.41
	15	0.10	0.54	0.00	0.00	0.00	0.13 ± 0.23
	20	0.06	0.27	0.00	0.00	0.00	0.07 ± 0.12
MCC	5	98.92	98.43	93.72	99.99	93.30	96.87 ± 3.12
	10	99.48	99.24	95.05	100.00	96.77	98.11 ± 2.11
	15	99.79	99.60	94.72	99.99	100.00	98.82 ± 2.30
	20	99.89	99.80	96.02	99.99	100.00	99.14 ± 1.75

2.34, 1.05, and 0.32%. Therefore, a sufficiently large number of epochs in BGRU will ensure a high detection rate and reduce false alarm in the botnet detection system developed for smart homes.

3.3   *Influence of the Number of Hidden Layers on Classification Performance*

In this subsection, we determine the optimal number of hidden layers required for efficient BGRU-based IoT botnet detection in smart homes. Four BGRU neural networks with 1, 2, 3, and 4 hidden layers formed BGRU-HL1, BGRU-HL2, BGRU-HL3, and BGRU-HL4 classifiers, respectively, when trained using 200 hidden neurons, ReLU activation function, five epochs, a batch size of 128, and Adam optimizer.

Training and validation losses in BGRU-HL1, BGRU-HL2, BGRU-HL3, and BGRU-HL4 were analyzed to understand the extent of model underfitting and over-fitting, respectively. Figures 5 and 6 show that the lowest training and validation losses were realized with four hidden layers. In general, training and validation losses reduced in all the classifiers throughout the five-epoch period. However, training and validation losses were lower in BGRU-HL4 than in BGRU-HL1, BGRU-HL2, and BGRU-HL3. At the end of the experiment, we observed that training loss in

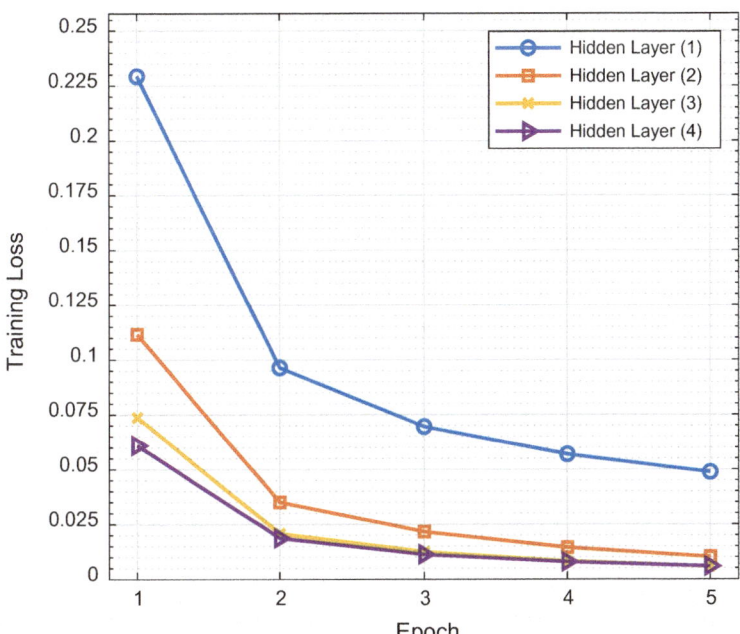

Fig. 5   Training loss of different number of hidden layers

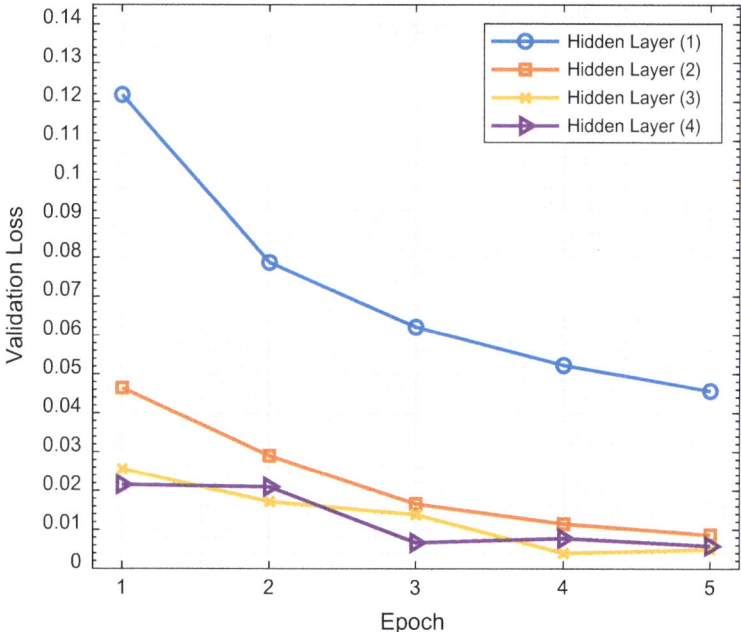

Fig. 6 Validation loss of different number of hidden layers

BGRU-HL4 reduced to 0.0057, while validation loss reduced to 0.0060. BGRU-HL4 reduced the average training losses in BGRU-HL1, BGRU-HL2, and BGRU-HL3 by 79.22%, 45.93%, and 13.43%, respectively, while the average validation losses were reduced by 82.44%, 43.88%, and 4.67%, respectively. The reduction in training and validation losses implies that the likelihood of model underfitting and overfitting is best minimized when the number of hidden layers in BGRU was 4. In other words, a sufficiently deep BGRU will facilitate high classification accuracy and good generalization ability that are required for efficient IoT botnet detection in smart homes.

Multi-class classification performance of BGRU-HL1, BGRU-HL2, BGRU-HL3, and BGRU-HL4 was evaluated with respect to the ground truth labels based on TPR, FPR, and MCC. Table 3 shows that BGRU-HL4 performed better than BGRU-HL1, BGRU-HL2, and BGRU-HL3. BGRU-HL4 increased TPR by 9.80, 1.71, and 1.69% relative to BGRU-HL1, BGRU-HL2, and BGRU-HL3, respectively; FPR decreased by 87.27%, 12.50%, and 0%, respectively; and MCC increased by 2.79%, 0.45%, and 0.44%. Therefore, a sufficiently deep BGRU will ensure a high detection rate and reduce false alarm in the botnet detection system developed for smart homes.

Table 3 Performance of BGRU at different number of hidden layers

Metric	Layer	DDoS	DoS	Normal	Reconn.	Theft	Mean ± Stdev
TPR	1	97.97	99.18	76.47	99.98	75.00	89.72 ± 12.80
	2	99.83	99.74	84.71	99.99	100.00	96.85 ± 6.79
	3	99.92	99.72	84.71	99.99	100.00	96.87 ± 6.80
	4	99.67	99.98	92.94	99.99	100.00	98.51 ± 3.12
FPR	1	0.78	1.94	0.00	0.00	0.00	0.55 ± 0.85
	2	0.24	0.16	0.00	0.00	0.00	0.08 ± 0.11
	3	0.27	0.08	0.00	0.00	0.00	0.07 ± 0.12
	4	0.02	0.32	0.00	0.00	0.00	0.07 ± 0.14
MCC	1	98.92	98.43	93.72	99.99	93.30	96.87 ± 3.12
	2	99.78	99.82	96.02	100.00	100.00	99.12 ± 1.74
	3	99.78	99.87	96.02	100.00	100.00	99.13 ± 1.74
	4	99.90	99.77	98.20	100.00	100.00	99.57 ± 0.77

3.4 Influence of Hidden Units on Classification Performance

In this subsection, we determine the optimal number of hidden units required for efficient BGRU-based IoT botnet detection in smart homes. Five single layers BGRU neural networks with 10, 50, 100, 150, and 200 hidden units formed BGRU-HU1, BGRU-HU2, BGRU-HU3, BGRU-HU4, and BGRU-HU5 classifiers respectively when trained using ReLU activation function, five epochs, a batch size of 128 and Adam optimizer.

Training and validation losses in BGRU-HU1, BGRU-HU2, BGRU-HU3, BGRU-HU4, and BGRU-HU5 were analyzed to understand the extent of model underfitting and overfitting, respectively. Figures 7 and 8 show that the lowest training and validation losses were realized with 200 hidden units. In general, training and validation losses reduced in all the classifiers throughout the five-epoch period. However, training and validation losses were lower in BGRU-HU4 than in BGRU-HU1, BGRU-HU2, and BGRU-HU3. At the end of the experiment, we observed that training loss in BGRU-HU4 reduced to 0.0486, while validation loss reduced to 0.0458. BGRU-HU5 reduced the average training losses in BGRU-HU1, BGRU-HU2, BGRU-HU3, and BGRU-HU4 by 53.54%, 31.72%, 21.14%, and 11.82%, respectively, while the average validation losses were reduced by 54.83%, 32.94%, 21.85%, and 12.07%, respectively. The reduction in training and validation losses implies that the likelihood of model underfitting and overfitting is best minimized when the number of hidden units in BGRU was 200. In other words, a sufficiently large number of hidden units in BGRU will facilitate high classification accuracy and good generalization ability that are required for efficient IoT botnet detection in smart homes.

Multi-class classification performance of BGRU-HU1, BGRU-HU2, BGRU-HU3, BGRU-HU4, and BGRU-HU5 was evaluated with respect to the ground truth labels based on TPR, FPR, and MCC. Table 4 shows that BGRU-HU5 performed

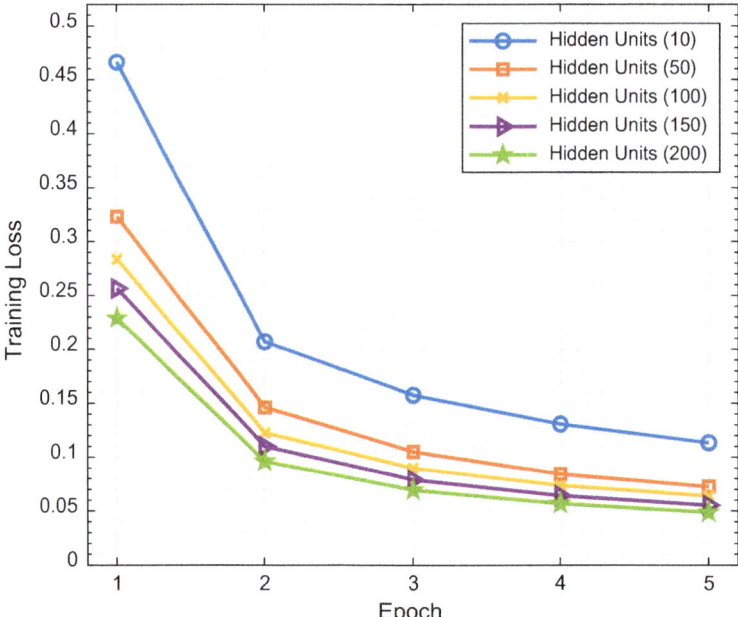

Fig. 7 Training loss of different number of hidden units

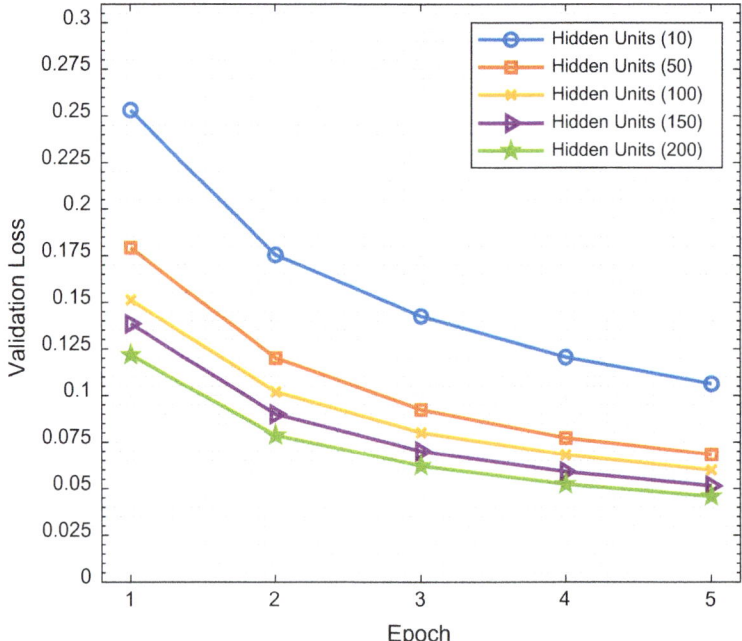

Fig. 8 Validation loss of different number of hidden units

Table 4 Performance of BGRU at different number of hidden units

Metric	Units	DDoS	DoS	Normal	Reconn.	Theft	Mean ± Stdev
TPR	10	95.41	97.98	0.00	99.78	0.00	58.64 ± 53.55
	50	97.84	98.30	61.18	99.83	0.00	71.43 ± 43.11
	100	97.91	98.57	69.41	99.98	75.00	88.17 ± 14.73
	150	98.01	98.86	74.12	99.99	75.00	89.20 ± 13.38
	200	97.97	99.18	76.47	99.98	75.00	89.72 ± 12.80
FPR	10	1.92	4.40	0.00	0.01	0.00	1.27 ± 1.94
	50	1.62	2.07	0.00	0.00	0.00	0.74 ± 1.02
	100	1.36	2.00	0.00	0.00	0.00	0.67 ± 0.95
	150	1.08	1.90	0.00	0.00	0.00	0.60 ± 0.87
	200	0.78	1.94	0.00	0.00	0.00	0.55 ± 0.85
MCC	10	97.43	96.42	0.00	99.94	0.00	58.76 ± 53.65
	50	98.28	98.12	89.11	99.96	0.00	77.09 ± 43.30
	100	98.49	98.23	91.66	99.99	93.30	96.33 ± 3.63
	150	98.72	98.38	93.05	99.99	93.30	96.69 ± 3.27
	200	98.92	98.43	93.72	99.99	93.30	96.87 ± 3.12

better than BGRU-HU1, BGRU-HU2, BGRU-HU3, and BGRU-HU4. BGRU-HU5 increased TPR by 53, 25.61, 1.75, and 0.58% relative to BGRU-HU1, BGRU-HU2, BGRU-HU3, and BGRU-HU4, respectively; FPR decreased by 56.69%, 25.68%, 17.91%, and 8.33%, respectively; and MCC increased by 64.86, 25.66, 0.56, and 0.19%. Therefore, a sufficiently large number of hidden units in BGRU will ensure a high detection rate and reduce false alarm in the IoT botnet detection system developed for smart homes.

3.5 Influence of Batch Size on Classification Performance

In this subsection, we determine the optimal batch size required for efficient BGRU-based IoT botnet detection in smart homes. Five single layers BGRU neural networks with batch sizes of 32, 64, 128, 256, and 512 formed BGRU-B32, BGRU-B64, BGRU-B128, BGRU-B256, and BGRU-B512 classifiers, respectively, when trained using 200 hidden units, ReLU activation function, five epochs, and Adam optimizer.

Training and validation losses in BGRU-B32, BGRU-B64, BGRU-B128, BGRU-B256, and BGRU-B512 were analyzed to understand the extent of model underfitting and overfitting, respectively. Figures 9 and 10 show that the lowest training and validation losses were realized with a batch size of 32. In general, training and validation losses reduced in all the classifiers throughout the five-epoch period. However, training and validation losses were lower in BGRU-B32 than in BGRU-B64, BGRU-B128, BGRU-B256, and BGRU-B512. At the end of the experiment, we observed

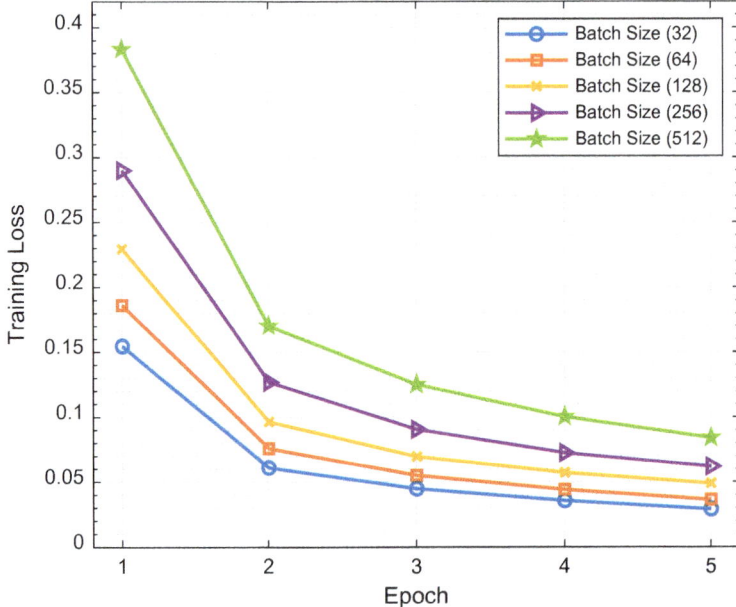

Fig. 9 Training loss of different batch sizes

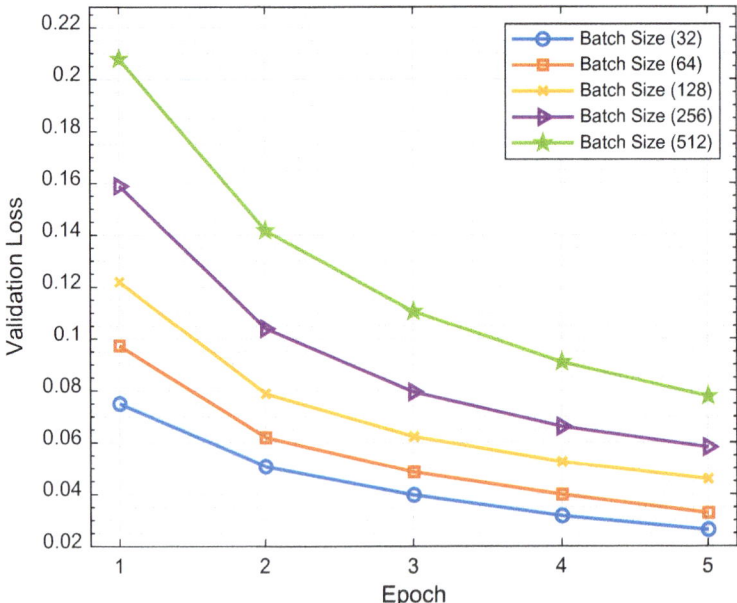

Fig. 10 Validation loss of different batch sizes

that training loss in BGRU-B32 reduced to 0.0287, while validation loss reduced to 0.0262. BGRU-B32 reduced the average training losses in BGRU-B64, BGRU-B128, BGRU-B256, and BGRU-B512 by 18.49%, 35.38%, 49.58%, and 62.48%, respectively, while the average validation losses were reduced by 20.52%, 38.3%, 52.24%, and 64.55%, respectively. The reduction in training and validation losses implies that the likelihood of model underfitting and overfitting is best minimized when the batch size in BGRU was 32. In other words, a sufficiently small batch size in BGRU will facilitate high classification accuracy and good generalization ability that are required for efficient IoT botnet detection in smart homes.

Multi-class classification performance of BGRU-B32, BGRU-B64, BGRU-B128, BGRU-B256, and BGRU-B512 was evaluated with respect to the ground truth labels based on TPR, FPR, and MCC. Table 5 shows that BGRU-B32 performed better than BGRU-B64, BGRU-B128, BGRU-B256, and BGRU-B512. BGRU-B32 increased TPR by 0.42%, 0.86%, 12.80%, and 26.49% relative to BGRU-B64, BGRU-B128, BGRU-B256, and BGRU-B512, respectively; FPR decreased by 35%, 52.73%, 60%, and 69.05%, respectively; and MCC increased by 0.21%, 0.42%, 3.84%, and 26.24%. Therefore, a sufficiently small batch size in BGRU will ensure a high detection rate and reduce false alarm in the botnet detection system developed for smart homes. Figure 11 shows that training time decreased as the batch size increased. BGRU-B32 took the longest time (101.35 min) to train, while the shortest training time of 6.70 min was achieved in BGRU-B512.

Table 5 Performance of BGRU at different batch sizes

Metric	Units	DDoS	DoS	Normal	Reconn.	Theft	Mean ± Stdev
TPR	32	99.12	99.51	78.82	99.99	75.00	90.49 ± 12.47
	64	98.60	99.29	77.65	99.99	75.00	90.11 ± 12.63
	128	97.97	99.18	76.47	99.98	75.00	89.72 ± 12.80
	256	97.70	98.93	67.06	99.90	37.50	80.22 ± 27.57
	512	97.60	98.01	62.35	99.76	0.00	71.54 ± 42.95
FPR	32	0.47	0.85	0.00	0.00	0.00	0.26 ± 0.38
	64	0.68	1.34	0.00	0.00	0.00	0.40 ± 0.60
	128	0.78	1.94	0.00	0.00	0.00	0.55 ± 0.85
	256	1.02	2.21	0.00	0.00	0.00	0.65 ± 0.98
	512	1.89	2.31	0.00	0.00	0.00	0.84 ± 1.16
MCC	32	99.44	99.28	94.39	100.00	93.30	97.28 ± 3.18
	64	99.16	98.88	94.06	100.00	93.30	97.08 ± 3.14
	128	98.92	98.43	93.72	99.99	93.30	96.87 ± 3.12
	256	98.68	98.18	90.94	99.97	80.62	93.68 ± 8.11
	512	98.02	97.88	89.48	99.94	0.00	77.06 ± 43.27

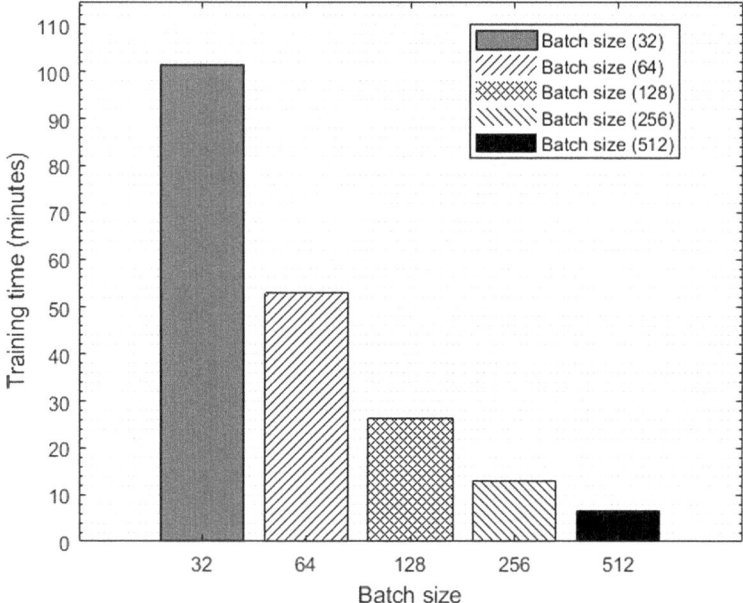

Fig. 11 Training time of different batch sizes

3.6 Influence of Optimizers on Classification Performance

In this subsection, we determine the most suitable optimizer required for efficient BGRU-based IoT botnet detection in smart homes. Four single layers BGRU neural networks with Adam, SGD, RMSprop, and Adadelta optimizers formed BGRU-OP1, BGRU-OP2, BGRU-OP3, and BGRU-OP4 classifiers, respectively, when trained using ReLU activation function, five epochs, and a batch size of 128.

Training and validation losses in BGRU-OP1, BGRU-OP2, BGRU-OP3, and BGRU-OP4 were analyzed to understand the extent of model underfitting and overfitting, respectively. Figures 12 and 13 show that the lowest training and validation losses were realized with Adam optimizer. In general, training and validation losses reduced in all the classifiers throughout the five-epoch period. However, training and validation losses were lower in BGRU-OP1 than in BGRU-OP2, BGRU-OP3, and BGRU-OP4. At the end of the experiment, we observed that training loss in BGRU-OP1 reduced to 0.0486, while validation loss reduced to 0.0458. BGRU-OP1 reduced the average training losses in BGRU-OP2, BGRU-OP3, and BGRU-OP4 by 86.54%, 4.22%, and 90.40%, respectively, while the average validation losses were reduced by 89.70%, 5.49%, and 92.31%, respectively. The reduction in training and validation losses implies that the likelihood of model underfitting and overfitting is best minimized when Adam optimizer was employed in BGRU. In other words, the use of Adam optimizer in BGRU will facilitate high classification accuracy and good generalization ability that are required for efficient botnet detection in smart homes.

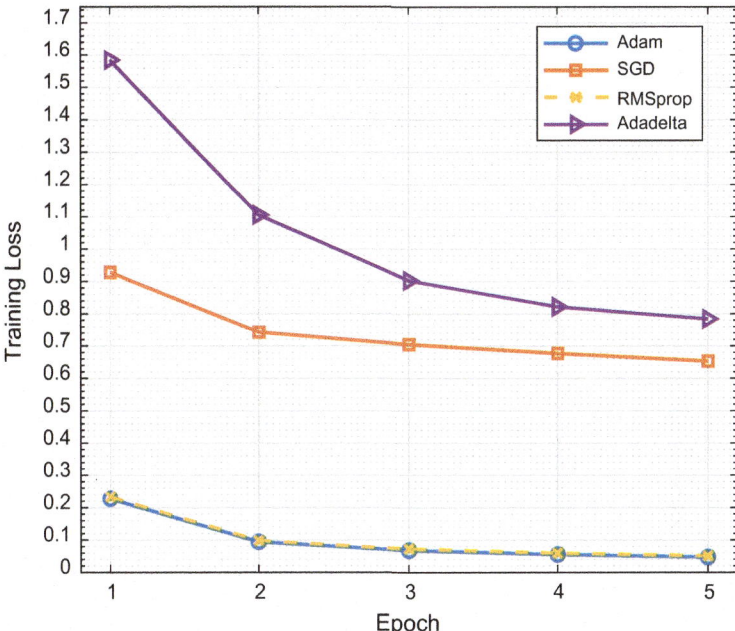

Fig. 12 Training loss of different optimizers

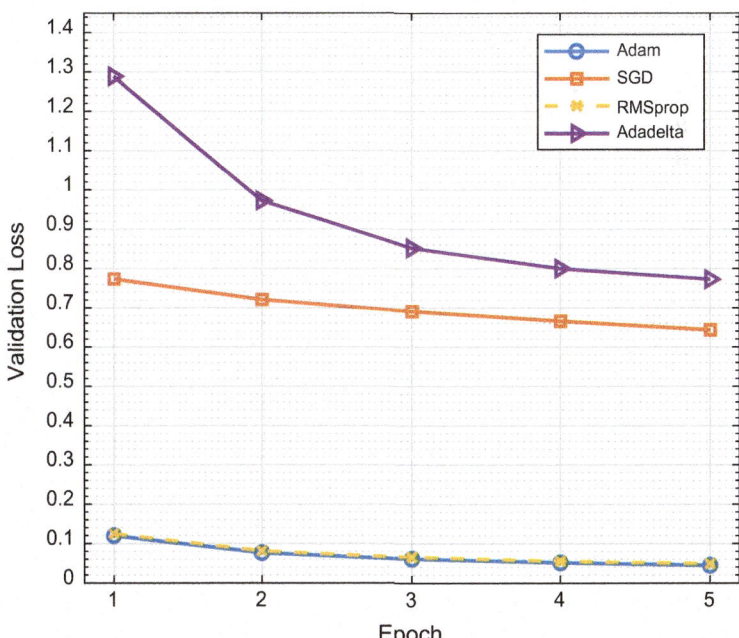

Fig. 13 Validation loss of different optimizers

Table 6 Performance of BGRU for different optimizers

Metric	Optimizer	DDoS	DoS	Normal	Reconn.	Theft	Mean ± Stdev
TPR	Adam	97.97	99.18	76.47	99.98	75.00	89.72 ± 12.80
	SGD	82.42	66.80	0.00	0.00	0.00	29.84 ± 41.24
	RMSprop	98.35	98.21	22.35	99.82	0.00	63.75 ± 48.64
	Adadelta	99.89	3.63	0.00	0.00	0.00	20.70 ± 44.29
FPR	Adam	0.78	1.94	0.00	0.00	0.00	0.55 ± 0.85
	SGD	35.06	18.19	0.00	0.00	0.00	10.65 ± 15.75
	RMSprop	1.70	1.59	0.00	0.01	0.00	0.66 ± 0.90
	Adadelta	92.35	3.74	0.00	0.00	0.00	19.22 ± 40.92
MCC	Adam	98.92	98.43	93.72	99.99	93.30	96.87 ± 3.12
	SGD	72.17	76.90	0.00	0.00	0.00	29.81 ± 40.86
	RMSprop	98.35	98.43	73.64	99.95	0.00	74.07 ± 42.83
	Adadelta	52.78	49.79	0.00	0.00	0.00	20.51 ± 28.11

Multi-class classification performance of BGRU-OP1, BGRU-OP2, BGRU-OP3, and BGRU-OP4 was evaluated with respect to the ground truth labels based on TPR, FPR, and MCC. Table 6 shows that BGRU-OP1 performed better than BGRU-OP2, BGRU-OP3, and BGRU-OP4. BGRU-OP1 increased TPR by 200.67%, 40.74%, and 333.43% relative to BGRU-OP2, BGRU-OP3, and BGRU-OP4, respectively; FPR decreased by 94.84%, 16.67%, and 97.14%, respectively; and MCC increased by 224.96%, 30.78%, and 372.31%. Therefore, the adoption of Adam optimizer in BGRU will ensure a high detection rate and reduce false alarm in the botnet detection system developed for smart homes.

3.7 Performance of Deep BGRU-Based Multi-class Classifier

In this subsection, we evaluate the suitability of deep BGRU for IoT botnet detection in smart homes. A deep BGRU multi-class classifier was developed with the optimal hyperparameters in Sects. 3.1–3.6, namely, ReLU activation function, 20 epochs, 4 hidden layers, 200 hidden neurons, a batch size of 512, and Adam optimizer.

Training and validation losses in deep BGRU multi-class classifiers were analyzed to understand the extent of model underfitting and overfitting, respectively. Figure 14 shows that the training and validation losses were shallow when the optimal BGRU hyperparameters were used. Training and validation losses reduced throughout the five-epoch period. At the end of the experiment, we observed that training loss in deep BGRU multi-class classifiers reduced to 0.0018, while validation loss reduced to 0.0006. The reduction in training and validation losses implies that the likelihood of model underfitting and overfitting is minimized when the optimal hyperparameters were employed in BGRU. In other words, the choice of the optimal BGRU

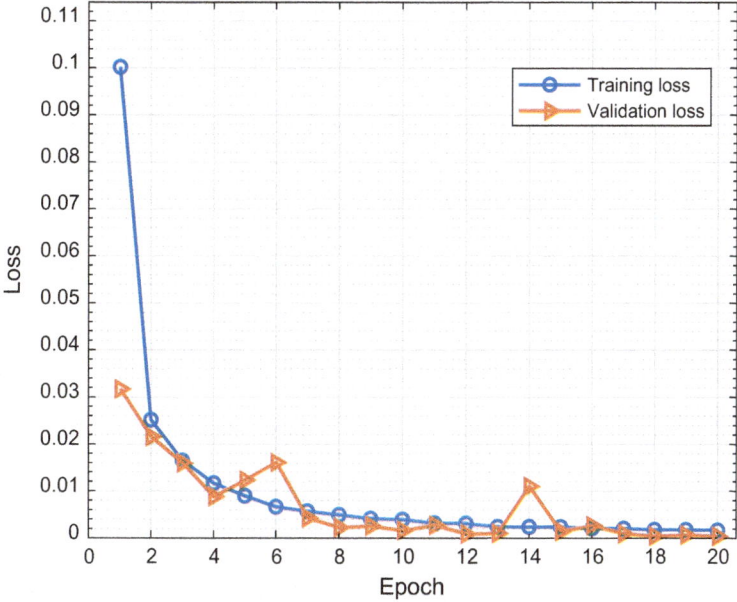

Fig. 14 Training and validation losses of deep BGRU classifier

hyperparameters will facilitate high classification accuracy and good generalization ability that are required for efficient IoT botnet detection in smart homes. The time needed to train the optimal deep BGRU classifier was 33.95 min.

Multi-class classification performance of deep BGRU multi-class classifier was compared with the state-of-the-art methods based on TPR, FPR, and MCC. Tables 7, 8, and 9 show that deep BGRU multi-class classifier outperforms mixture localization-based outliers (MLO) [47], SVM [48], RF [49], artificial immune system (AIS) [50], and feedforward neural network (FFNN) [51]. Deep BGRU multi-class classifier achieved high detection accuracy and low false alarm with true positive rate (TPR), false positive rate (FPR), and Matthews coefficient correlation (MCC) of 99.28 ± 1.57%, 0.00 ± 0.00%, and 99.82 ± 0.40%.

Table 7 TPR of multi-class classifiers for IoT botnet detection in smart homes

Ref	DDoS	DoS	Normal	Reconn.	Theft	Mean ± Stdev
MLO [47]	98.24	98.20	0.00	95.67	95.86	77.59 ± 43.39
SVM [48]	98.95	99.87	87.97	99.33	100.00	97.22 ± 5.19
RF [49]	99.40	0.00	0.00	99.37	99.38	59.63 ± 54.43
AIS [50]	100.00	98.53	0.00	98.22	98.90	79.13 ± 44.24
FFNN [51]	99.414	0.00	98.071	98.38	88.92	76.96 ± 43.23
Deep BGRU	**100.00**	**99.99**	**96.47**	**99.96**	**100.00**	**99.28 ± 1.57**

Table 8 FPR of multi-class classifiers for IoT botnet detection in smart homes

Ref	DDoS	DoS	Normal	Reconn.	Theft	Mean ± Stdev
SVM [48]	0.00	0.26	0.00	5.96	0.01	1.56 ± 2.36
Deep BGRU	**0.01**	**0.00**	**0.00**	**0.00**	**0.00**	**0.00 ± 0.00**

Table 9 MCC of multi-class classifiers for IoT botnet detection in smart homes

Ref	DDoS	DoS	Normal	Reconn.	Theft	Mean ± Stdev
SVM [48]	99.74	99.81	96.90	96.82	100.00	98.65 ± 1.64
Deep BGRU	**99.99**	**99.99**	**99.11**	**99.99**	**100.00**	**99.82 ± 0.40**

4 Conclusion

In this paper, an optimal model was developed for efficient botnet detection in IoT-enabled smart homes using deep BGRU. A methodology was proposed to determine the optimal BGRU hyperparameters (activation function, epoch, hidden layer, hidden unit, batch size, and optimizer) for multi-class classification. The proposed methodology was implemented, and the classification performance was jointly assessed based on training loss, validation loss, accuracy, TPR, FPR, MCC, and training time. Extensive simulation results showed that: (a) ReLU performed better than *tanh* activation functions; (b) classification performance improved with an increase in the numbers of epochs, hidden layers, and hidden units; (c) the performance of BGRU improved as the batch size becomes smaller, but this comes with a significant increase in training time; (d) Adam outperformed SGD, RMSprop, and Adadelta optimizers. Finally, the combination of ReLU activation function, 20 epochs, 4 hidden layers, 20 hidden units, a batch size of 512, and Adam optimizer achieved the best multi-class classification performance as follows: low training loss (0.0107 ± 0.0219), validation loss (0.0072 ± 0.0086), FPR ($0.00 \pm 0.00\%$); high accuracy (99.99%), TPR (99.28 ± 1.57), and MCC (99.82 ± 0.40).

Acknowledgments The authors would like to thank Cyraatek UK and the Faculty of Science & Engineering, Manchester Metropolitan University, for the research grant provided.

References

1. Stoyanova, M., Nikoloudakis, Y., Panagiotakis, S., Pallis, E., Markakis, E.K.: A survey on the Internet of Things (IoT) forensics: challenges, approaches and open issues. IEEE Commun. Surv. Tutori. (2020). https://doi.org/10.1109/COMST.2019.2962586
2. Alam, S., Siddiqui, S.T., Ahmad, A., Ahmad, R., Shuaib, M.: Internet of Things (IoT) enabling technologies, requirements, and security challenges. In: Kolhe, M., Tiwari, S., Trivedi, M., Mishra, K. (eds.) Advances in Data and Information Sciences, vol. 94. pp. 119–126. Springer (2020)

3. Zaidan, A., Zaidan, B.: A review on intelligent process for smart home applications based on IoT: coherent taxonomy, motivation, open challenges, and recommendations. Artif. Intell. Rev. **53**(1), 141–165 (2020)
4. Bhattacharyya, R., Das, A., Majumdar, A., Ghosh, P.: Real-time scheduling approach for IoT-based home automation system. In: Data Management, Analytics and Innovation, pp. 103–113. Springer (2020)
5. Mahadewa, K., Wang, K., Bai, G., Shi, L., Liu, Y., Dong, J.S., Liang, Z.: Scrutinizing implementations of smart home integrations. IEEE Trans. Software Eng. (2019). https://doi.org/10.1109/TSE.2019.2960690
6. Singh, J., Pasquier, T., Bacon, J., Ko, H., Eyers, D.: Twenty security considerations for cloud-supported Internet of Things. IEEE Internet Things J. **3**(3), 269–284 (2015)
7. Yin, L., Luo, X., Zhu, C., Wang, L., Xu, Z., Lu, H.: ConnSpoiler: disrupting C&C communication of IoT-based Botnet through fast detection of anomalous domain queries. IEEE Trans. Indus. Inf. **16**(2), 1373–1384 (2020). https://doi.org/10.1109/TII.2019.2940742
8. Pour, M.S., Mangino, A., Friday, K., Rathbun, M., Bou-Harb, E., Iqbal, F., Samtani, S., Crichigno, J., Ghani, N.: On data-driven curation, learning, and analysis for inferring evolving Internet-of-Things (IoT) botnets in the wild. Comput. Secur. **91**, 101707 (2020)
9. Russell, B.: IoT cyber security. In: Intelligent Internet of Things, pp. 473–512. Springer (2020)
10. Alieyan, K., Almomani, A., Abdullah, R., Almutairi, B., Alauthman, M.: Botnet and Internet of Things (IoTs): a definition, taxonomy, challenges, and future directions. In: Security, Privacy, and Forensics Issues in Big Data, pp. 304–316. IGI Global (2020)
11. Al-Duwairi, B., Al-Kahla, W., AlRefai, M.A., Abdelqader, Y., Rawash, A., Fahmawi, R.: SIEM-based detection and mitigation of IoT-botnet DDoS attacks. Int. J. Electr. Comput. Eng. **10**, 2088–8708 (2020)
12. Gupta, B.B., Dahiya, A., Upneja, C., Garg, A., Choudhary, R.: A comprehensive survey on DDoS attacks and recent defense mechanisms. In: Handbook of Research on Intrusion Detection Systems, pp. 186–218. IGI Global (2020)
13. Koroniotis, N., Moustafa, N., Sitnikova, E., Turnbull, B.: Towards the development of realistic botnet dataset in the Internet of things for network forensic analytics: Bot-iot dataset. Fut. Gen. Comput. Syst. **100**, 779–796 (2019)
14. Asadi, M., Jamali, M.A.J., Parsa, S., Majidnezhad, V.: Detecting Botnet by using particle swarm optimization algorithm based on voting system. Fut. Gen. Comput. Syst. **107**, 95–111 (2020)
15. Nguyen, H.-T., Ngo, Q.-D., Nguyen, D.-H., Le, V.-H.: PSI-rooted Subgraph: A Novel Feature for IoT Botnet Detection Using Classifier Algorithms. ICT Express (2020)
16. Nõmm, S., Bahşi, H.: Unsupervised anomaly based Botnet detection in IoT networks. In: 2018 17th IEEE International Conference on Machine Learning and Applications (ICMLA) 2018, pp. 1048–1053. IEEE (2018)
17. Al Shorman, A., Faris, H., Aljarah, I.: Unsupervised intelligent system based on one class support vector machine and Grey Wolf optimization for IoT Botnet detection. J. Amb. Intell. Human. Comput. 1–17 (2019)
18. Yang, Y., Wang, J., Zhai, B., Liu, J.: IoT-based DDoS attack detection and mitigation using the edge of SDN. In: International Symposium on Cyberspace Safety and Security 2019, pp. 3–17. Springer (2019)
19. D'hooge, L., Wauters, T., Volckaert, B., De Turck, F.: In-depth comparative evaluation of supervised machine learning approaches for detection of cybersecurity threats. In: Proceedings of the 4th International Conference on Internet of Things, Big Data and Security 2019 (2019)
20. Gurulakshmi, K., Nesarani, A.: Analysis of IoT Bots against DDoS attack using machine learning algorithm. In: 2018 2nd International Conference on Trends in Electronics and Informatics (ICOEI) 2018, pp. 1052–1057. IEEE (2018)
21. Nomm, S., Guerra-Manzanares, A., Bahsi, H.: Towards the Integration of a post-Hoc interpretation step into the machine learning workflow for IoT Botnet detection. In: 2019 18th IEEE International Conference on Machine Learning And Applications (ICMLA) 2019, pp. 1162–1169. IEEE (2019)

22. Moustafa, N., Turnbull, B., Choo, K.-K.R.: An ensemble intrusion detection technique based on proposed statistical flow features for protecting network traffic of Internet of Things. IEEE Internet Things J. **6**(3), 4815–4830 (2018)

23. Wildani, I., Yulita, I.: Classifying Botnet attack on Internet of Things device using random forest. In: IOP Conference Series: Earth and Environmental Science 2019, vol. 1, p. 012002. IOP Publishing (2019)

24. Bahşi, H., Nõmm, S., La Torre, F.B.: Dimensionality reduction for machine learning based IoT botnet detection. In: 2018 15th International Conference on Control, Automation, Robotics and Vision (ICARCV) 2018, pp. 1857–1862. IEEE (2018)

25. Koroniotis, N., Moustafa, N., Sitnikova, E., Slay, J.: Towards developing network forensic mechanism for Botnet activities in the IoT based on machine learning techniques. In: International Conference on Mobile Networks and Management 2017, pp. 30–44. Springer (2017)

26. Guerra-Manzanares, A., Bahsi, H., Nõmm, S.: Hybrid feature selection models for machine learning based Botnet detection in IoT networks. In: 2019 International Conference on Cyberworlds (CW) 2019, pp. 324–327. IEEE (2019)

27. Soe, Y.N., Santosa, P.I., Hartanto, R.: DDoS Attack Detection Based on Simple ANN with SMOTE for IoT Environment. In: 2019 Fourth International Conference on Informatics and Computing (ICIC) 2019, pp. 1–5. IEEE (2019)

28. Haq, S., Singh, Y.: Botnet detection using machine learning. In: 2018 Fifth International Conference on Parallel, Distributed and Grid Computing (PDGC) 2018, pp. 240–245. IEEE (2018)

29. Bansal, A., Mahapatra, S.: A comparative analysis of machine learning techniques for Botnet detection. In: Proceedings of the 10th International Conference on Security of Information and Networks 2017, pp. 91–98 (2017)

30. Amanullah, M.A., Habeeb, R.A.A., Nasaruddin, F.H., Gani, A., Ahmed, E., Nainar, A.S.M., Akim, N.M., Imran, M.: Deep learning and big data technologies for IoT security. Comput. Commun. (2020)

31. Jung, W., Zhao, H., Sun, M., Zhou, G.: IoT Botnet detection via power consumption modeling. Smart Health **15**, 100103 (2020)

32. Le, H.-V., Ngo, Q.-D., Le, V.-H.: Iot Botnet detection using system call graphs and one-class CNN classification. Int. J. Innov. Technol. Explor. Eng. **8**(10), 937–942

33. Liu, J., Liu, S., Zhang, S.: Detection of IoT Botnet based on deep learning. In: 2019 Chinese Control Conference (CCC) 2019, pp. 8381–8385. IEEE (2019)

34. Nguyen, H.-T., Ngo, Q.-D., Le, V.-H.: IoT Botnet detection approach based on PSI graph and DGCNN classifier. In: 2018 IEEE International Conference on Information Communication and Signal Processing (ICICSP) 2018, pp. 118–122. IEEE (2018)

35. Hwang, R.-H., Peng, M.-C., Nguyen, V.-L., Chang, Y.-L.: An LSTM-based deep learning approach for classifying malicious traffic at the packet level. Appl. Sci. **9**(16), 3414 (2019)

36. McDermott, C.D., Majdani, F., Petrovski, A.V.: Botnet detection in the Internet of things using deep learning approaches. In: 2018 International Joint Conference on Neural Networks (IJCNN), pp. 1–8. IEEE (2018)

37. McDermott, C.D., Petrovski, A.V., Majdani, F.: Towards situational awareness of botnet activity in the Internet of things. In: 2018 International Conference on Cyber Situational Awareness, Data Analytics And Assessment (Cyber SA), Glasgow, UK, pp. 1–8. IEEE (2018)

38. Sachin, S., Tripathi, A., Mahajan, N., Aggarwal, S., Nagrath, P.: Sentiment analysis using gated recurrent neural networks. SN Comput. Sci. **1**(2), 1–13 (2020)

39. Liu, C., Liu, Y., Yan, Y., Wang, J.: An intrusion detection model with hierarchical attention mechanism. IEEE Access (2020)

40. Cho, K., Van Merriënboer, B., Gulcehre, C., Bahdanau, D., Bougares, F., Schwenk, H., Bengio, Y.: Learning phrase representations using RNN encoder-decoder for statistical machine translation. arXiv preprint arXiv:1406.1078 (2014)

41. Schuster, M., Paliwal, K.K.: Bidirectional recurrent neural networks. IEEE Trans. Signal Process. **45**(11), 2673–2681 (1997)

42. Werbos, P.J.: Backpropagation through time: what it does and how to do it. Proc. IEEE **78**(10), 1550–1560 (1990)
43. Luque, A., Carrasco, A., Martín, A., de las Heras, A.: The impact of class imbalance in classification performance metrics based on the binary confusion matrix. Pattern Recogn. **91**, 216–231 (2019)
44. Baloglu, U.B., Talo, M., Yildirim, O., San Tan, R., Acharya, U.R.: Classification of myocardial infarction with multi-lead ECG signals and deep CNN. Pattern Recogn. Lett. **122**, 23–30 (2019)
45. Hartmann, C., Opritescu, D., Volk, W.: An artificial neural network approach for tool path generation in incremental sheet metal free-forming. J. Intell. Manuf. **30**(2), 757–770 (2019)
46. Patro, S., Sahu, K.K.: Normalization: a pre-processing stage. arXiv preprint arXiv:1503.06462 (2015)
47. AlKadi, O., Moustafa, N., Turnbull, B., Choo, K.-K.R.: Mixture localization-based outliers models for securing data migration in cloud centers. IEEE Access **7**, 114607–114618 (2019)
48. Khraisat, A., Gondal, I., Vamplew, P., Kamruzzaman, J., Alazab, A.: A novel ensemble of hybrid intrusion detection system for detecting internet of things attacks. Electronics **8**(11), 1210 (2019)
49. Soe, Y.N., Feng, Y., Santosa, P.I., Hartanto, R., Sakurai, K.: Towards a lightweight detection system for cyber attacks in the IoT environment using corresponding features. Electronics **9**(1), 144 (2020)
50. Aldhaheri, S., Alghazzawi, D., Cheng, L., Alzahrani, B., Al-Barakati, A.: DeepDCA: novel network-based detection of IoT attacks using artificial immune system. Appl. Sci. **10**(6), 1909 (2020)
51. Ge, M., Fu, X., Syed, N., Baig, Z., Teo, G., Robles-Kelly, A.: Deep learning-based intrusion detection for IoT networks. In: 2019 IEEE 24th Pacific Rim International Symposium on Dependable Computing (PRDC) 2019, pp. 256–25609. IEEE (2019)

Big Data Clustering Techniques: Recent Advances and Survey

Ibrahim Hayatu Hassan, Mohammed Abdullahi,
and Barroon Isma'eel Ahmad

Abstract Clustering as an unsupervised machine learning technique has appeared as a great learning method to examine correctly the huge volume of dataset produced by today's applications. There is a huge amount of information in the field of clustering and a lot of attempts have already been made to identify and evaluate it for a wide number of applications, but the major problems with the application of classical clustering algorithm for big data analysis are its high complicity, massive volume, variety, and generation rate. So, the classical clustering methods are becoming increasingly inept in processing such data. This poses exciting challenges for researchers to develop modern scalable and efficient methods of clustering that can extract useful information from these vast amounts of data produced in different areas of life. Here in this paper we present a classification for the review of big data clustering algorithms by identifying major research subjects. We then present an up-to-date review of the study works within each of the subject.

Keywords Big data · Clustering · MapReduce · Spark · MPI · GPU

1 Introduction

Clustering is an unsupervised machine learning method used to divide unlabeled dataset into groups called clusters, which comprise data elements that are dissimilar from those in other groups and similar to each other in the same cluster [1]. This has been considered as an important technique for extracting knowledge from databases and has been widely applied to a variety of scientific and business areas [2]. In the

H. Ibrahim Hayatu
Institute for Agricultural Research, Ahmadu Bello University, Zaria, Nigeria
e-mail: ibrogo@gmail.com

H. Ibrahim Hayatu · A. Mohammed (✉) · A. Barroon Isma'eel
Department of Computer Science, Ahmadu Bello University, Zaria, Nigeria
e-mail: abdullahilwafu@abu.edu.ng

A. Barroon Isma'eel
e-mail: barroonia@gmail.com

© The Author(s), under exclusive license to Springer Nature Switzerland AG 2021
H. Chiroma et al. (eds.), *Machine Learning and Data Mining for Emerging Trend in Cyber Dynamics*, https://doi.org/10.1007/978-3-030-66288-2_3

present digital age, based on the enormous advancement and growth of the internet and web technologies, the data produced by machines and devices has reached a huge volume and is expected to increase in the coming years [3]. Data has become the most integral part of the knowledge-driven society and the economy of today. The goal of big data clustering is to summarize, segment, and group the enormous volumes and data diversities generated in groups of similar items at an accelerated rate. In exploratory data analysis, this has turned out to be one of the most important methods. Sadly, the classical clustering methods are becoming increasingly ineffective in the processing of these data due to their high complicity, large volume, variety, and generation rate. This poses exciting opportunities for researchers to develop scalable and efficient methods of clustering that can extract useful information from these vast amounts of data produced in different areas of life [4, 5]. Therefore, the proposed paper presents an up-to-date review of the research works presented to handle big data clustering. The remainder of this paper is structured as follows: Sect. 2 includes a summary of the different approaches to clustering. Section 3 provides an overview of different big data clustering approaches. Section 4 describes numerous applications in the real world that used specific types of the big data clustering. Finally, the paper is summarized in Sect. 5.

2   Clustering Techniques

A number of clustering algorithms have been proposed in the literature. These algorithms are typically grouped according to the basic principle that the clustering algorithm is based on [6]. That leads to the following groups being identified as shown in Fig. 1.

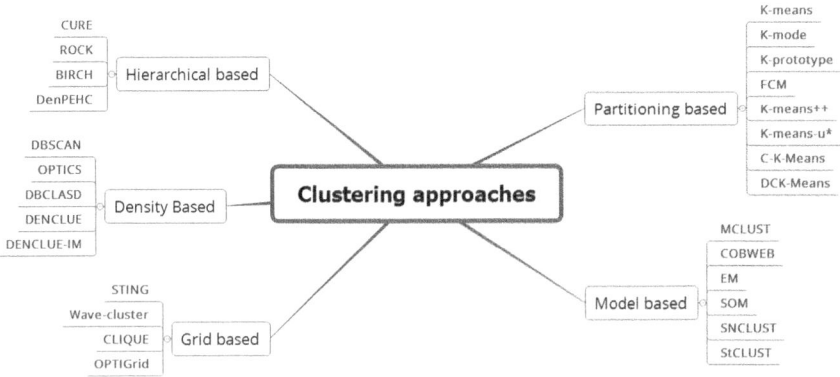

Fig. 1 Grouping of clustering approaches

2.1 Partitioning Methods

Clustering partitioning algorithms organize items into a given number of groups by maximizing a specific objection function capturing the grouping structure. Most partitioning algorithms start via an initial assignment and then use an iterative method which transfers items from cluster to cluster to maximize the objective function. Convergence is therefore local and the best global solution that cannot be guaranteed. These methods required that the number of clusters should be known at the initial stage, and this is normally set by the user [6]. Most of the research in clustering comes from the family of partitioning algorithms. K-means [7], k-mode [8], k-prototype [9], and fuzzy c-means [10] are the examples of the partitioning methods.

2.2 Hierarchical Clustering Methods

Hierarchical clustering techniques build a clusters hierarchy in a top-down approach (Divisive) or using bottom-up approach (Agglomerative) [11]. All these approaches rely on building a similarity matrix between all of the data items; the choice of the metric for the construction of the matrix may affect the shape of the clusters. Examples of common similarity metric that can be employed are: cosine and Jaccard distance metric. Linkage criteria that calculate the distance between groups of objects is a function of the similar distance between objects. But the problem with hierarchical clustering algorithm is that they have large time complexity "$O(n^3)$" and needs "$O(n^2)$" space, for which n represents the number of data items. It can be never undone once the merge or the split step is completed. CURE [12], ROCK [13], and BIRCH [14] are some of the well-known hierarchical clustering algorithms.

2.3 Density-Based Methods

Data items are clustered in density-based method on the basis of their boundary, connectivity, and regions of density. A cluster described as a dense component that is connected grows in any direction to which density leads. Such methods are related to neighbours closest to the point. Density-based algorithms are capable of identifying clusters of arbitrary types and providing natural immunity to outliers [15]. Therefore, the total density of a particular point is examined to evaluate the features of the dataset that influence a particular data point. DBSCAN [16], OPTICS [17], DBCLASD [18], and DENCLUE [19] are the most widely used density-based clustering algorithms.

2.4   Grid-Based Algorithms

The data item space in grid-based technique is split into grids. The key benefit for this method was its great ability to process, and this is because it runs once over the dataset to measure the statistical grid values. Grid-based clustering is performed on the grid, instead of clustering directly from the database. Quality is heavily dependent on the grid size, which is usually smaller than that of the database. However, the use of a single uniform grid might be insufficient for highly irregular data distributions to find a reliable clustering performance or to meet the time limitation. STING [20], CLIQUE [21], OptiGrid [22], and Wave-Cluster [23] are the common examples of this approach.

2.5   Model-Based Methods

Model-based techniques attempt to adjust the fit between some predetermined mathematical formula and data based on the idea that data is created by a combination of simple distributions of probability. This results in the automated detection of number of clusters based on standard statistics, bringing outliers into perception, and thereby building a reliable clustering strategy. Statistical and neural networks are the two major approaches on the model-based techniques. MCLUST by Fraley and Raftery [24] is regarded as the most common model-based algorithm but there are other algorithms like COBWEB [25] (conceptual clustering), EM [26], and neural network methods like self-organizing feature maps [27]. Neural network system uses a combination of linked input/output units with each connection's weight associated. Neural network uses several attributes for clustering which made them famous. The statistical methods employ probability procedures when evaluating clusters. Probabilistic definitions are usually used to describe any idea derived from them [15].

3   Big Data Clustering Approaches

Based on the combination of conventional clustering and acceleration techniques, many approaches have been proposed in the literature to manage huge data. This approach focuses on increasing the speed of the clustering process by decreasing the complicacy of the computation. On the basis of the acceleration method that is used to enhance the scalability, big data clustering methods can be classified into parallel techniques, data reduction-based techniques, and centre reduction-based techniques as shown in Fig. 2.

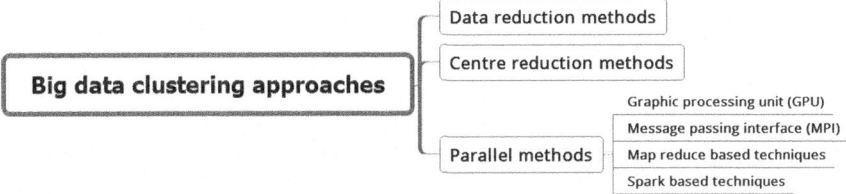

Fig. 2 An overview of big data clustering

3.1 Data Reduction-Based Methods

This technique attempts to reduce the number of data items during cluster development in order to hasten the clustering technique. The MinBatch k-means method (MBKM) was introduced by Sculley [28]. Its basic idea is the use of random small split of fixed size data items which can be placed in memory. This technique is based on the fact that random lots appear to be noisier than single data points. Alternative to k-means for massive-scale data processing named recursive partition k-means (RPKM) has been introduced by Capó et al. [29]. The idea of this technique is to estimate the k-means for the entire dataset by attempting the iterative application of a weighted version of k-means over a small number of data subsets. RPKM's key steps are defined as follows: first, the data are divided into a number of sets of data in which each set is defined by a member and their corresponding weight. Secondly, over the collection of members, weighted variant of k-means is used.

3.2 Centre-Based Reduction Methods

This technique tries to decrease the amount of evaluations when searching for the nearby cluster centres, which are the most time-consuming step. Kanungo et al. [30] proposed the use of kd-tree structure to speed up k-means (KdtKM). kd-Tree is defined as a binary tree which divides the data space constructed by piercing hyperplanes. The k in kd-tree signify data dimension. Chen et al. [31] proposed a fast density peak clustering for large-scale data based on kNN. Density peak (DPeak) clustering algorithm is not valid for large-scale data, due to two quantities, i.e., both of which are acquired with complexity O(n2) by brute force algorithm. Hence, a simple but fast DPeak was proposed, namely FastDPeak, which runs in projected time in the intrinsic dimensionality around O(nlog(n)). This replaces density with kNN-density, which is determined by fast kNN algorithms such as cover tree, resulting in enormous density calculation improvements. Based on kNN-density, local density peaks and non-local density peaks are defined, and a fast algorithm is also proposed with complexity O(n), which uses two different strategies to calculate for them. Experimental tests show FastDPeak to be successful and to outperform other DPeak variants.

3.3 Parallel Techniques

Parallelization technique is among the most frequently used method to reduce the time complexity of traditional clustering techniques. Parallelization is described as a method that splits the task into chunks and executes them in parallel. In the literature, many parallel clustering techniques were proposed. Various frameworks like graphics processing unit [32], message passing interface [33], MapReduce [34], or Spark [35] may be used for parallelization. In this part, we review various research works on parallel clustering techniques with focus on such frameworks.

Graphical Processing Unit-Based Techniques (GPU). GPU is a customized technology built to speed up graphical activities like photo and video formatting. A parallel programming paradigm called compute unified device architecture (CUDA) was used to shorten the development implementations of parallel GPU without going into the details of the technology. Particularly in comparison to a central processing unit (CPU), GPU consists of a huge number of processing cores. It also offers two parallelization stages. GPU will have many multiprocessors on the initial stage; then at the next stage every multiprocessor will also have several streaming processors. Having followed this layout, the GPU application is split into chunks running on streaming processors, and then these chunks are gathered to form chunks blocks running on a multiprocessor [32]. Compared with the few terabytes of RAM now allowed in servers, the comparatively small amount of video random access memory (VRAM) on a GPU card has prompted some to assume that GPU acceleration is restricted to applications with "small data." But, the assumption lacks two common practices in applications of "big data." One is that processing a whole dataset at once is rarely needed to achieve the desired effect. GPU VRAM, system RAM, and storage [direct attached storage (DAS), storage area networks (SAN), network-attached storage (NAS), etc.] can provide virtually unlimited large data workload volume. For machine learning, for instance, the training data can be streamed from memory or storage as required. Live streams of data coming from the internet of things (IoT) or other applications such as the Kafka or Spark may also be consumed in a common "incremental" method. The second technique is being able to scale up and out GPU-accelerated configurations. Numerous GPU cards can be cited in a single server, and several servers can be organized in a cluster. This scaling resulted in more cores and more memory, all functioning concurrently and enormously in parallel to process data at unparalleled speed. Therefore, the only real limit to GPU acceleration's potential computing power is the budget. However, whatever the budget available, it will still be possible for a GPU to speed up configuration and produce more flops per dollar because CPUs are and will remain much more costly than GPU. The GPU database offers a simple and potentially significant price/performance advantage in a single server or cluster [36]. To quicken data clustering via GPU, several works have been proposed, such as the work of Che et al. [37] that introduced a k-means algorithm based on GPU (GPUKM). First, the initial centroid is posted into the GPU's shared memory and the input data is split and send to each of the multiprocessor. And from there, the distance from each corresponding data item is computed by each multiprocessor and allocated to the nearest cluster. Depending on the subset of the data items, a local cluster centroid is computed. The central processing unit (CPU) updates the new cluster centroids after each data items is allocated to cluster centroids and submits them to the multiprocessors once more. The above steps will be called multiple times using the same technique until convergence is achieved. The parallel k-means reduces the time complexity of the algorithm. Andrade et al. [38] proposed a GPU-enhanced algorithm for density-based clustering (G-DBSCAN). This is split into two phases: the first phase involves building graph. Each objects represent a node and an edge is created between two objects if the distance is lower than or equal to a predefined thresholds. Once the graph is set, the second phase is the identification of clusters that use breath first search algorithm (BFS) to traverse the graph produced from the first phase. Both phases of the algorithm have been parallelized to achieve better performance. Result indicates that in comparison to the serial implementation, G-DBSCAN is 112 times faster.

A GPU fuzzy c-means technique (GPUFCM) was proposed by Al-Ayyoub et al. [39]. The starting cluster locations are first stored in the shared memory. It then produces from the dataset the starting participation matrix and the starting cluster centroids. After that, by calculating distances, each multiprocessor calculates partial participation. It then calculates the participation values by adding partial participation. The added participation values are transferred from the GPU to the CPU to calculate new cluster centroids from which the algorithm will move to the next iteration. Cuomo et al. [40] suggest an enhanced GPU-based k-means algorithm. The proposed work adopted a parallel processing structure of the graphical processing unit (GPU). The design proposed was optimized to handle CPU space inefficiency and data transmission time for the host device. A parallelize kernel k-means using CPU and GPU was proposed by Baydoun et al. [41]. Kernel k-means involves several computational phases and has additional computational criteria. As a consequence, kernel k-means has not seen the same value and much can still be achieved with respect to its parallelization and stable implementations. Several databases are used, with varying number of features and patterns. The findings show that CUDA typically has the best runtimes with speed-ups ranging from two to more than 200 times over a single-core CPU implementation depending on the dataset used. Shahrezaei and Tavoli [42] studied parallelization of k-means ++ with CUDA. k-means ++ is an algorithm developed to boost the method of locating initial seeds in the algorithm k-means. In this algorithm, the initial seeds are selected consecutively through a probability proportional to the distance from the nearest node. The most critical problem of this algorithm is that it reduces the speed of clustering while running in serial way. This work parallelized the k-means ++ algorithm's most time-consuming steps to increase the runtime of the k-means algorithm while using the k-means ++ seeding technique. In the k-means ++ algorithm, the distance of all points to the selected centre should be determined and the seeds are selected according to a probability formula. Those are the most difficult things for the k-means ++ algorithm being parallelized. This work's parallelization process starts with the partitioning of the points into l different points. Then the measurement of the distance of points to the centres remains parallel until the end of the algorithm. The results of this work show that the algorithm performs more efficiently over GPU than CPU. It has also been shown that constant memory and texture memory have greater effectiveness in this research than global memory. Even with GPU's certified success in managing massive-scale dataset, it tends to suffer from memory threshold. For instance, it may not handle terabyte data with a bound of 12 GB of memory for every GPU. The efficiency of the GPU-based technique goes way down once the size of the data surpasses the GPU memory capacity. Therefore, developers should first configure the memory before implementing programs via GPU. The drawback to this is the small number to usable GPU tools and algorithms.

Message Passing Interface Techniques. MPI is a language-independent parallel computing communication protocol that supports point-to-point and collective communication [43]. High performance, scalability, and portability are MPI targets. MPI is typically the prevalent model used in high-performance computing [44] and provides portability among parallel programs running on distributed memory systems. But the standard presently does not really support fault tolerance [44] because it primarily addresses problems with high performance computing (HPC). Another MPI downside is that leveraging the parallelism of multicore architectures for shared memory multiprocessing, for example, is not ideal for small grain degree of parallelism. Most researchers seek to combine the MPI with other application programming interface (API), such as the OpenMP, to overcome this issue. OpenMP is an API that enables multiplatform multiprocessing programming for shared memory [45] on most processor architecture and operating systems. OpenMP for its high performance is becoming the standard for parallel shared memory computing but it is not suitable for distributed memory system. The combination of these methods therefore allows for two levels of granularity: a small grain parallel with OpenMP and a large grain parallel with MPI. MPI is implemented in the design of master/slave whereby the master transmits tasks to slaves and collects computed outcomes. However, the job of the slave is to accept tasks, process, and then send the output to the master. Many techniques of clustering that use the MPI paradigm were proposed. For example, Kwok et al. [46] proposed a parallel fuzzy c-means algorithm based on the MPI paradigm (MPIFCM). The master first divides the input data into chunks and sends it to slaves. Each slave gets the associated chunk; it updates the membership matrix based on the distance measurement. The master then receives all the information needed for the slaves to measure the new cluster. These steps are repeated until convergence. A comparative study was made between the MPIFCM and the parallel k-means, which shows a similar structure of parallelism between the two algorithms. A particular implementation of the proposed algorithm is used and tested to cluster a large dataset. It is shown that the MPIFCM algorithm has nearly ideal speed-ups for large dataset and performs equally well when demanding more clusters. Experimentally, the scale-up performance regarding the size of the dataset is also shown to be excellent. Zhang et al. [47] developed a k-means (MPIKM) based on MPI. In this technique, the input data is shared among the slaves. After that, the master chooses k data items as the starting centroid and sends them to the slaves. Each one of the slaves then determines the distance between each data object and the cluster centroids and assigns it to their nearest cluster. The master then receives all the necessary information from the slaves. The entire process is reiterated until convergence. The experimental results based on the seven datasets used have shown that by increasing the dataset the running time of the proposed MPIKM algorithm is slightly lower than that of the standard k-means. The result also shows that for a small-scale dataset, it is not a good choice to use the proposed algorithm because the time of dividing the dataset and assigning the task into each process occupies a certain proportion. Savvas and Sofianidou [48] proposed a new semi-parallel variant of the k-means algorithm for n-dimensional data objects via MPI. The first step of the proposed strategy is for the master node to discover the resources available and to divide the data into equal or nearly equal parts. The master node passes the data subsets to the workers who in turn start applying the original sequential k-means algorithm on it after obtaining the data. The worker nodes move the local centroids and the number of data points allocated to each centroid to the master node after its termination, and their job ends here. The master node calculates the global centroids which apply the weighted arithmetic mean after collecting this information. The master node at first sorts out the centroids and then divides them into k subgroups and determine the global centroids (by the use of weighted arithmetic mean). The centroids generated from the sequential algorithm by the proposed technique use one-dimensional data points. On increasing the size of n-dimensional data points, the technique generates centroids very similar to the original algorithm with similarities varying from 91 to 100%. Shan et al. [49] proposed a new parallel k-means algorithm for high-dimensional text data using both GPU and MPI. The GPU part design of this work focuses on the measure of similarity between text objects, that is, the computation of the matrix. The original complexity of computation is $O(nd)$. In parallel design, it is designed to start n threads, which runs in parallel. Complexity becomes $O(d)$ which greatly reduces runtime of the algorithm. The portion of the MPI divides nodes into one control node and multiple compute nodes. In order to test the runtime of the proposed technique, the algorithm was tested on CPU, GPU, and GPU + MPI platforms. The results show that the use of MPI and GPU can not only increase the performance of the algorithm, but can also be easily potted to the GPU cluster. However, even with the effectiveness of the MPI paradigm for massive-scale data handling, it is affected by fault intolerance limit. MPI seems to have no fault management tool. A machine downfall in the system can make the entire system to be closed down. In MPI-based techniques developers, therefore, must introduce a fault tolerance method in the design of master/slave to handle machine failures.

MapReduce-Based Techniques. MapReduce has recently become one of the most widely used parallel programming paradigms for handling massive-scale data along clusters systems. It is defined for its strong openness, which enables designers to easily and accurately parallelize algorithms. The MapReduce paradigm consists of two parts, which are Map and Reduce. The map component runs the map functions to create a set of intermediate <key/value> pairs that process each <key/value> in parallel. All intermediate values connected to the same intermediate key are grouped together, while the reduce part executes the reduce function to combine all intermediate values linked to the same intermediate key. The key factors that make MapReduce a good paradigm for handling massive data are the fault tolerance, linear scalability, and simple programming paradigm [34]. A large number of research works have suggested in the literature that clustering method is suitable via MapReduce. For instance, Zhao et al. [50] implemented k-means based on MapReduce (MRKM) in Hadoop framework. This approach divides the input dataset into blocks, where each block is sent to the mapper and each data element is assigned to its closest cluster based on the distance computation by the map function. By computing the mean data item of each cluster, the reduce function updates the new cluster centroids at the reduce phase. Then the new cluster centroids are taken to the Hadoop distributed file system (HDFS) for the next iteration which is to be used by the map function. In conclusion, until convergence, this whole process will continue. This way the time complexity of the k-means algorithm is reduced. Kim et al. [51] proposed an effective clustering algorithm based on density for massive data using MapReduce (DBCURE-MR). Although the conventional density-based methods discover every cluster one after the other, the proposed algorithm discovers multiple clusters together in parallel. The DBCURE parallelization using MapReduce was performed in four stages. First, the matrix of neighbourhood covariances was calculated in parallel for every point. Similarity joins were performed in the second stage to explore all pairs of points, each of which is inside each of their neighbourhood. In the third stage, core clusters were identified. The core clusters were continuously fusing through the last stage to create the final clusters. Experimental results indicate that DBCURE-MR proficiently discovers specific groups and scales well with MapReduce paradigm. Ludwig [52] proposed a MapReduce fuzzy c-means clustering (MRFCM). Two MapReduce jobs were used in this work. First MapReduce job computes the cluster centre matrix and the second MapReduce calculates the distances that will be used to update the matrix. Using this method the time complexity of the fuzzy c-means was reduced in comparison with the serial fuzzy c-means algorithm. Another MapReduce-based k-prototype (MRKP) for clustering massive-scale mixed data was also proposed by Ben HajKacem et al. [53]. Shahrivari and Jalili [54] implemented a single-pass model based on MapReduce termed as (MRK-means) using the re-clustering strategy. Unlike other MapReduce-based k-means implementations, MRK-means needs only to read the dataset once and is therefore multiple times faster than the original k-means. MRK-means time complicacy is linear which is smaller than that of iterative k-means. Because of the use of the seeding method, MRK-means also lead to clusters of greater quality. Dongbo et al. [55] also proposed a canopy k-means (CK-means) clustering algorithm based on MapReduce. Two MapReduce are being used in this work. The first MapReduce estimates the number of clusters and centroids to be used by the second MapReduce using a pre-clustering algorithm. The second MapReduce performed the clustering of k-means by dividing the dataset in which each partition is associated with a map function. The map function assigns every data point of the associated segment to the nearest cluster by measuring the distance between the objects and the cluster centres. In the reduction function the cluster centres are then updated by computing the cumulative average of all data points in each cluster. The new centres are then moved to the HDFS, which the map function can use for the next iteration. The whole process is repeated until convergence, which reduces the time complexity of the algorithm compared to the original k-means algorithm. Nonetheless, the algorithm's immunity to noise is low and the threshold values of the canopy algorithm are very difficult to determine, which has some effect on the result. In the work of Valcarce et al. [56], an implementation of posterior likelihood clustering and recommendation-relevance models based on MapReduce was proposed. In the current environment where the number of recommendations hitting the big data scale is rising day after day, high efficiency estimates are not sufficient. This research addresses one urgent and important critical need for recommendation systems, which is scalability of algorithms. The research adapted these highly effective algorithms to the functional MapReduce paradigm, which was earlier proved as an appropriate tool to allow scalability of the recommender. A good scalability behaviour was achieved with respect to the number of nodes in the MapReduce cluster. To evaluate large datasets, the impulse for today's scenario is to boost the conventional methods. Tripathi et al. [57] suggest an efficient clustering tool called an enhanced grey wolf optimizer (MR-EGWO) based on MapReduce for clustering large-scale datasets. The method implemented a new type of grey wolf optimizer, called enhanced grey wolf optimizer (EGWO), in which grey wolf's hunting strategy is hybridized with binomial crossover and levy flight steps are induced to boost the searching capability for pray. The suggested variant is further used to improve the clustering method. The EGWO's clustering performance is evaluated on seven UCI benchmark datasets and compared to the five current clustering techniques, namely k-means, particle swarm optimization (PSO), gravitational search algorithm (GSA), bat algorithm (BA), and grey wolf optimizer (GWO). The EGWO's convergence behaviour and consistency were validated through the convergence graph and boxplots. Therefore, in the Hadoop context the proposed EGWO is parallelized on the MapReduce model and called MR-EGWO to manage the large-scale datasets. Furthermore, the MR-EGWO's clustering accuracy is also validated in terms of F-measure and compared to four state-of-the-art MapReduce-based algorithms, namely, parallel k-means, parallel K-PSO, MapReduce-based artificial bee colony optimization (MR-ABC), and dynamic frequency-based parallel k-bat algorithm (DFBPKBA). Experimental findings confirm that the proposed technique is promising and effective substitute for the efficient and large-scale data clustering. Classical clustering methods generally represent each cluster in all dimensions without any difference and depend only on an individual dimension of projection as the weight of an attribute ignores relevance among attributes. Pang et al. [58] address these two problems with the use of multiattribute weights using a MapReduce-based subspace clustering algorithm called PUMA. The subspaces attribute is developed into the PUMA by calculating a value weight attribute dependent on the probability of co-occurrence of attribute values between different dimensions. PUMA obtains parallel sub-clusters of each computing node corresponding to the respective attribute subspaces. Finally, by applying the hierarchical clustering method, PUMA tests different scale clusters to iteratively combine sub-clusters. PUMA was deployed on an Hadoop cluster of 24-nodes. Findings show that using multiattribute weights with subspace clustering can achieve better clustering accuracy in both the artificial and the real-world large datasets. The results also show that in terms of number of nodes PUMA achieves great success in terms of extensibility, scalability, and the almost linear speed-up. In addition, experimental findings show that PUMA is realistic, efficient, and functional for expert systems such as automatic abstracting, acquiring information, recommending systems, and disambiguating word sense. While MapReduce tends to be ideal for clustering large-scale data, it tends to suffer from the inefficiency of performing iterative operations [59]. The whole data must be read and written back to the HDFS at each iteration of the technique. Consequently, multiple I/O memory operations occur during every iteration, and this significantly reduces the effectiveness of MapReduce-based technique.

Spark-Based Techniques. Spark is a distributed model of data processing developed to resolve the MapReduce limitations. It has been introduced as part of the Hadoop, and is designed to run with Hadoop, specifically by accessing HDFS data. This paradigm is based on resilient distributed datasets (RDDs) which is a particular form of data structure used to coordinate computations transparently. Spark offers a collection of in-memory operators in addition to the standard MapReduce to analyse data faster in distributed settings [35]. A number of research works on big data clustering with Spark paradigm were proposed. For instance, Zayani et al. [60] recommended the parallelization of overlapping k-means clustering technique based on the Spark paradigm (SOKM). The proposed technique can carry out parallel clustering processes that result to pre-disjoint data segmentation. Current DBSCAN parallel algorithms produce data partitions in which the initial data are usually split into many disjoint partitions. But by the increase in data dimensions, dividing and reorganizing huge dimensional space can take much time. In order to solve this problem, Luo et al. [61] propose a Spark-based parallel DBSCAN method (S-DBSCAN) that can rapidly split the data and combine the results of the clustering. The suggested work is broken down into three phases. The first phase is partitioning the set of data based on the strategy of random samples. The aim is to calculate the number of partitions based on the actual computing nodes, and on this basis, the original random data will operate through a custom random function. The original random data would be exporting to each slice. Slice of data points has approximately the same data volume and is similar to a simple random sampling. When the number of samples each slice extracts is large enough, it has a similar distribution with the original raw data. After the data is split, the local DBSCAN is calculated in parallel, so partial clustering results are produced. At the map task, partial cluster is created. The partial cluster is saved to HDFS as a new RDD, and at the reduced task each partial cluster centroid is computed. At the last point, the results of global clustering were created by fusing the partial cluster. They suggested a merger strategy based on the centroids. The idea is to measure the distance in the same partition between each of the two partial clusters, and use a sorting technique to determine the minimum distance. Moreover, it determines the minimum value by sorting the minimum distance. Thirdly, threshold is set to merge partial clusters. Finally, construct a matrix with a centroid distance and traverse every element in the matrix. If the distance is below the threshold then add it to the merge queue until each element is visited. The experimental results indicate a large speed-up of the S-DBSCAN. For the same number of nodes, the larger the dataset, the greater the speed-up. This completely illustrates the superior efficacy of the S-DBSCAN algorithm compared to other parallel DBSCAN algorithms when dealing with huge data. Ben HajKacem et al. [62] presented a k-prototypes clustering technique for massive-scale data with mixed attributes (KP-S). An RDD object with the input dataset from the m partitions of the dataset was generated to parallelize the k-prototype algorithm using Spark. To accomplish this, a Spark framework operation testfile() was used. The map task selects a partition of the dataset, performs the k-prototypes algorithm on that partition, and releases the intermediate centres extracted and their weights as the output. MapPartition() transformation was used in this work which runs the k-prototype algorithm separately on each RDD block. Upon completion of the map phase, a set of intermediate weighted centres is obtained as the map phase output and this collection of centres is released to a single reduce step. The reduction step takes the set of intermediate centres and their weights, performs the k-prototype method on them again, and releases the final centres as the output. ReduceByKey() transformation from the Spark framework was used to simplify implementation. When the final cluster centres are established, the closest cluster centre is assigned to each data point. The obtained results show that KP-S always completes many times faster than existing methods. For instance, with 100 clusters on the Poker dataset, the KP-S algorithm can decrease runtime by 94.47 and 85.72% in comparison to k-prototypes and k-prototype using MapReduce (KP-MR), respectively. In all of the analysis, over 95% of the running time is spent in the map process, which indicates that KP-S is actually performed in memory. An effective parallel density peak clustering technique using GraphX was implemented based on Spark technology to decrease high time complexity of density peak clustering method by Liu et al. [63]. This implementation is based on construction of graph, computing the truncated distance, computing the local density and computing the distance from higher density points. Construction of graph consists of three phases. First, importing vertex and edge data stored on HDFS or other file systems separately to vertex RDD and edge RDD, and setting each edge's initial value to a constant. Secondly, measuring the distance of each edge based on a distance measure formula updating the value of each edge by the distance. Thirdly, the combination of vertex RDD and edge RDD in GrapgX to form a graph. When the vertex set and edge set are updated, the initial value for each edge is set to 1. When measuring the distance of each edge, the distance is changed to the value of each edge. To reduce the computation load, the truncated distance is measured before calculating the local density. After the truncated distance is measured, the local density and the distance from points of higher density is determined. Experimental result shows that the Spark implementation can improve significantly (10x) compared to MapReduce implementation. K-prototype is an iterative algorithm that requires some iteration for generating best results. Conversely, MapReduce has a significant issue with iterative algorithms. As a result, the entire dataset at each iteration must be loaded into the main memory from the file system. Then, the data must be written to the file system again after it is processed. Consequently, several I/O disk operations occur during each iteration, and this reduces runtime. To overcome this problem, Ben HajKace et al. [64] proposed a scalable random dampling k-prototypes implemented on the Spark paradigm. A reservoir random sampling technique was used for selecting random data sample from the original dataset. The data sampling MapReduce job first generates an RDD object with input data X form by m partitions. In the map phase, each partition is then processed to generate the intermediate data samples. The reduce phase then receives the set of intermediate data samples to produce the final data sample. The map function takes a partition during the map process and applies the random reservoir sampling algorithm to create an intermediate data sample of size r. The map function then releases the intermediate data samples as a single-phase reduce output. The reduce phase gathers the intermediate data samples generated in the map phase to create the final data sample. Upon generating the data sample, the second MapReduce job checks for cluster centres. First, it generates an RDD object with the data sample S form by m partition. Then, each partition is analysed in map phase to generate the intermediate centres. The reduction step afterwards processed the collection of intermediate centres to produce the final clusters. The map function selects a partition during the map process and executes the k-prototypes on that partition to obtain the cluster centres. To achieve good quality, for each obtained centre a weight that signifies the number of assign data points was recorded. The number of allocated data points per cluster centre reflects the centre's importance. At last, the map function releases the intermediate centres extracted and their weights as the output to a single reduce step. In this implementation, mapPartition(func) transformation was used separately on every partition of the RDD. The reduce step takes the intermediate centres and their weights, runs the k-prototypes algorithm on them again, and returns the final centres as the output. ReduceByKey(func) transformation from the Spark framework was used to simplify the implementation. The experimental results obtained show that the method proposed often finishes many times faster than the classical k-prototypes and the k-prototype based on MapReduce. The method proposed was found to be 14 times faster than the classical k-prototypes and four times faster than the MapReduce-based method. However, it was found that most of the running time was spent on the step of the map which indicates that the proposed method is actually being performed in memory. In the field of agricultural image segmentation, the fuzzy c-means (FCM) algorithm has been commonly used because it provides easy computation and high quality segmentation. However, the sequential FCM is too slow due to its large amount of computation to complete the segmentation task in an appropriate time. Liu et al. [65] suggest a parallel FCM segmentation algorithm based on the Apache Spark for clustering agricultural images. The input image is first converted to the Lab colour space from the RGB colour space and generates point cloud data. Point cloud data is then split and stored in separate computer nodes where the membership degrees of pixel points to separate cluster centres are determined and the cluster centres are modified iteratively in parallel form until the stop condition is met. In the RDD, point cloud data is finally restored after clustering to reconstruct the segmented image. The FCM output is measured on the Spark platform and achieves an average speed of 12.54

on 10 computing nodes. Experimental results show that the Spark-based parallel fuzzy c-means algorithm can achieve substantial speed-up improvements, and the agricultural image test set provides a 128% better performance improvement than the Hadoop-based solution. This work shows that the Spark-based parallel FCM algorithm provides faster segmentation speed for big data on agricultural images and has better scale-up and size-up. Yu et al. [66] presented an effective three-way cluster ensemble approach based on Spark to solve the difficulty of clustering on large-scale data; this methodology has the potential to deal with both soft and hard clustering. A three-way cluster ensemble based on Spark was suggested, inspired by the principle of three-way decisions, and a distributed three-way clustering algorithm for k-means was developed. Also introduced was the concept of cluster unit, which represents the minimum structure of distribution of granularity decided upon by all members of the ensemble. This research also incorporates quantitative tests to determine the relationship between units and between clusters. Eventually, a consensus-based clustering algorithm was suggested, and several three-way decision strategies were developed to assign small cluster units and non-unit objects. The experimental tests using 19 real-world datasets were used with various metrics such as ARI, ACC, NMI and F1-Measure to verify the feasibility of the proposed method. The experimental result shows that the proposed technique can manage large-scale data effectively, and the proposed consensus clustering algorithm has a lower time cost without loss of the consistency of the clusters.

4 Big Data Clustering Applications

Most applications in the real world generate vast amounts of data. Some of these are (but not restricted to) healthcare applications, IoT applications, detection of anomalies. Here we present a list of applications for big data clustering as shown in Fig. 3.

Fig. 3 Taxonomy of big data clustering applications

4.1 Healthcare

Several researchers have used different forms of clustering strategies for big data in occupational medicine [67], acute inflammation [68], and grouping of cancer samples [69]. McParland and Gormley [70] and Mcparland et al. [71] develop a clustering algorithm for studying high-dimensional categorical genotypic and numerical phenotypic data. The research contributes to a greater understanding of metabolic syndrome. Narmadha et al. [72] use a hybrid fuzzy k harmonic means (HFKHM) for the categorization of tumours as benign or malignant. Su et al. [73] also employs a hierarchical clustering to continually group respondents based on similarity structures of 27 different factors of asthma symptoms and 14 different product applications.

4.2 Internet of Things (IoT)

Amini et al. [74] suggested a hybrid clustering algorithm built on density-based clustering method for the IoT streaming data. The proposed method consists of three phases under which the new data point is connected to a grid or embedded into a current mini cluster, the outliers are eliminated and arbitrary shape clusters are formed through the use of an enhanced DBSCAN. A new data clustering framework is used for big sensory data generated by IoT applications by Karyotis et al. [75]. This framework can also be used for community detection and performing more energy-efficient smart-city/building sensing. Clustering algorithms have been widely used in IoT applications such as looking for similar sensing patterns, identifying outliers, and dividing massive real-time behavioural groups [76].

4.3 Anomaly Detection

Fanaee-T and Gama [77] uses an expectation maximization to model an anomaly detection using structural time series for industrial Ethernet congestion into four sections cantered on a method with a specific significance for detection. This model helps to improve the efficacy of identifying irregular and low false alarm levels. A new developed clustering algorithm was also used in the work of Alguliyev et al. [78] for detection of anomalies. Yin et al. [79] used a self-organizing maps to detect anomalous risk in mobile apps

4.4 Social Media

Alsayat and El-Sayed [80] use an improved k-means algorithm to identify communities by clustering posts from huge streams of social data. In the work of Gurusamy et al. [81], k-means algorithm is also used to exploit social network users' attitudes. On the basis of textual similarity, spectral k-means clustering is used to group people in social network by Singh et al. [82]. A varied density-based spatial clustering for twitter data algorithm that extracts clusters from geo tagged twitter posts using heterogeneity in space. The algorithm uses incremental spline interpolation to ascertain various cluster detection search radii [83]. They were successfully tested for event detection in social media using geo tagged twitter post received in the course of a storm in the United States.

5 Discussion

Parallel methods for big data clustering have been categorized in this work into four GPU-based, MPI-based, MapReduce-based, and Spark-based methods groups. But before applying clustering to any of the parallel frameworks, consideration of certain challenges for each of the framework is necessary. GPU suffers from limited memory. If the data size exceeds the GPU memory size, the performance reduces the GPU-based approach significantly. For example, it is not acceptable to handle terabyte data with a maximum of 12 GB of memory per GPU. Moreover, MPI has no fault control function at all. One system failure in the network will cause the entire network to shut down. Practitioners also need to incorporate some kind of fault tolerance system inside the software to resolve mistakes. The MapReduce architecture looks better than MPI as it is defined by a straightforward structure for programming, linear scalability, and tolerance of faults. It is, however, inappropriate to run iterative algorithms, because the entire dataset must be read and written to disks at each iteration and this results in high (I/O) operations. This significantly degrades the performance of MapReduce-based method. Finally, Spark framework is an alternative to MapReduce which is designed to overcome the disk I/O limitations and improve the performance of MapReduce framework. A recent survey outlined the various frameworks for big data analytics and presents the advantages and drawbacks of each of these frameworks based on different metrics such as scalability, data I/O rate, fault tolerance, and iterative task support [84, 85].

While most of the big data partition clustering methods described in this work provide an efficient analysis for large-scale data for users, some parameters have to be calculated before the learning is finished. The methods include configuring the number of clusters in advance, which is not a trivial task in real-life applications where the number of clusters predicted is typically unknown. One could use various model heuristics as a solution to determine the optimal number [86, 87]. For instance, with an increased number of clusters, the user may check various clusters and then take the clustering that has the best balance between minimizing the objective function and number of clusters. In addition, the partitioning methods must initialize the centres of the clusters. However, initialized centres of high quality are critical, both for the accuracy and performance of traditional clustering methods. In order to overcome this problem, users may use random sampling methods to obtain cluster centres or initialization techniques that exploit the fairly widespread use of good clustering [88–90]. The outcome of the described methods, using centre initialization techniques, often converges to a local optimum of the objective criterion, rather than the global optimum. To resolve this problem, users should combine traditional methods with heuristic techniques to avoid optimum local clustering results [3, 21, 29, 91–93].

6 Challenges and Future Research Work

Despite the growth in these technologies and algorithms to handle big data, there are still challenges such as:

- Scalability and storage issues: Data rises are much higher than existing computing systems. Storage devices are not adequately capable of processing these data [94–97]. A processing system needs to be built which not only addresses the needs of today but also future needs.
- Analytical timeliness: Data value decreases over time. Many applications, such as telecom, insurance, and banking fraud detection, require real-time or near-real-time transactional data processing [94, 95].
- Heterogeneous data representation: Data derived from different sources are of heterogeneous type. It is difficult to store and process unstructured data such as images, videos, and social media data using conventional methods such as SQL. Smartphones are now capturing and exchanging images, audios, and videos at an exponentially higher pace, pushing our brains to further process. However, there is a lack of effective storage and processing of the medium for representing pictures, audios, and videos [94, 95, 98].
- Privacy and security: New devices and technologies like cloud computing provide a gateway to access and to store information for analysis. This integration of IT architectures will pose greater risks to data security and intellectual property. Access to personal information like buying preferences and call detail records will lead to increase in privacy concerns [96, 99]. Researchers have technological infrastructure to access data from any source of data, including social networking sites, for potential use, although users are unaware of the benefits that can be made from the information they post [100]. The disparity between privacy and convenience is unknown to big data researchers.

Based on this review, the following future works have been formulated.

- Hybrid approaches that combine multiple acceleration techniques showed great promise for big data clustering. For the development of a reliable hybrid approach for big data clustering, further work is needed, however.
- Due to the effectiveness of the big data parallel methods, a method for the parallelization of the hybrid methods is highly needed.

7 Conclusion

This paper offers an in-depth analysis of the big data clustering algorithms to guide the selection of the big data algorithms. We also include categorization of various big data clustering research works. At the end, we describe various applications where big data clustering was used. In sum up, though conventional sampling and dimension reduction algorithms are still useful, but they don't have enough power in managing massive quantities of data because even after sampling a petabyte of data, it is still very large and clustering algorithms cannot cluster it, so the future of clustering is related to distributed computing. Although parallel clustering is potentially very useful for clustering, but the complexity of implementing such algorithms is still a challenge.

References

1. Bakr, A.M., Ghanem, N.M., Ismail, M.A.: Efficient incremental density-based algorithm for clustering large datasets. Alexandria Eng. J. 1147–1152 (2015)
2. Jia, H., Cheung, Y.-M.: Subspace clustering of categorical and numerical data with an unknown number of clusters. IEEE Trans. Neural Netw. Learn. Syst. 1–17 (2017)
3. Benabdellah, A.C., Benghabrit, A., Bouhaddou, I.: A survey of clustering algorithms for an industrial contex. In: Second International Conference on Intelligent Computing in Data Sciences (ICDS 2018), pp. 291–302 (2019)
4. Oussous, A., Benjelloun, F.-Z., Lahcen, A. A., Belfkih, S.: Big data technologies: a survey. J. King Saud Univ. Comput. Inform. Sci. 432–448 (2018)
5. Nasraoui, O., Ben N'Cir, C.-E.: Clustering Methods for Big Data Analytics Techniques, Toolboxes and Applications. Springer, Switzerland (2019)
6. Sanse, K., Sharma, M.: Clustering methods for big data analysis. Int. J. Adv. Res. Comput. Eng. Technol. (IJARCET) 642–648 (2015)
7. MacQueen, J.: Some methods for classification and analysis of mult-variate observations. In: Proceedings of the Fifth Berkeley Symposium on Mathematical Statistics and Probability, pp. 281–297. University Califonia Press, USA (1967)
8. Khan, S.S., Kant, D.: Computation of initial modes for k-modes clustering algorithm using evidence accumulation. In: Proceedings 20th International Joint Conference, Artificial Intelligent, pp. 2784–2789. Kaufmann, USA (2007)
9. Zhexue, H.: Clustering large data sets with mixed numeric and categorical values. In: Proceedings of the First Pacific-Asia Conference on Knowledge Discovery and Data Mining, pp. 21–34 (1997)
10. Bezdek, J.C., Ehrlich, R., Full, W.: FCM: the fuzzy c-means clustering algorithm. J. Comput. Geosci. 191–203 (1984)

11. Ahmad, A., Khan, S.S.: Survey of state-of-the-art mixed data clustering algorithms. Inst. Electr. Electron. Eng. Access 31883–31902 (2019)
12. Guha, S., Rastogi, R., Shim, K.: CURE: An efficient clustering algorithm for large databases. In: Proceedings of the ACM SIGMOID Conference on Management of Data, pp. 73–84. ACM Press (1998)
13. Guha, S., Rastogi, R., Shim, K.: Rock: a robust clustering algorithm for categorical attributes. Inform. Syst. **25**(5), 345–366 (2000)
14. Zhang, T., Ramakrishnan, R., Livny, M.: BIRCH: A New Data Clustering Algorithm and Its Applications, pp. 141–182. Kluwer Academic Publishers. Manufactured in The Netherlands
15. Fahad, A., Alshatri, N., Tari, Z., Alamri, A., Kkalil, I., Zomaya, Y.A.,… Bouras, A. (2014). A survey of clustering algorithms for big data: taxonomy and empirical analysis. IEEE Trans. Emerg. Top. Comput. 267–279
16. Ester, M., Kriegel, H.-P., Sander, J., Xu, X.: A density-based algorithm for discovering clusters a density-based algorithm for discovering clusters in large spatial databases with noise. In: KDD' 96 Proceedings of the Second International Conference on Knowledge Discovery and Data Mining, pp. 226–231. AAAI Press, Portland (1996)
17. Ankerst, M., Breunig, M.M., Krie, H.-P., Sander, J.: OPTICS: ordering points to identify the clustering structure. In: ACM SIGMOID International Conference on Management of Data, pp. 49–60. ACM press
18. Xu, X., Ester, M., Kriegel, H.-P., Sander, J.: A distribution-based clustering algorithm for mining in large spatial databases. In: Published in the Proceedings of 14th International Conference on Data Engineering (ICDE'98). IEEE, Orlando, USA (1998)
19. Hinneburg, A., Keim, D.A.: An efficient approach to clustering in large multimedia databases with noise. J. Am. Assoc. Artif. Intell. 58–65 (1998)
20. Wang, W., Yang, J., Muntz, R.R.: STING: a statistical information grid approach to spatial data mining. In: Proceedings of the 23rd International Conference on Very Large Databases, pp. 186–195. Morgan Kaufmann Publishers Inc., San Francisco, CA, USA (1997)
21. Agrawal, R., Gehrke, J., Gunopulos, D., Raghavan, R.: Automatic subspace clustering of high dimensional data for data mining applications. In: ACM SIGMOID International Conference on Management of Data, pp. 94–105. ACM Press, Sanfrancisco (1998)
22. Hinneburg, A., Keim, D.A.: Optimal grid-clustering: towards breaking the curse of dimensionality in high-dimensional clustering. In: Proceeding of the 25th International Conference on Very Large Databases, pp. 506–517. Morgan Kaufman, San Francisco, CA, USA (1999)
23. Sheikholeslami, G., Chatterjee, S., Zhang, A.: WaveCluster: a wavelet-based clustering approach for spatial data in very large databases. VLDB J. 289–304 (2000)
24. Fraley, C., Raftery, E.: MCLUST: software for model-based cluster and discriminant analysis. Department of Statistics University of Washington, Seattle, WA, USA (1998)
25. Fisher, D.H.: Knowledge Acquisition Via Incremental Conceptual Clustering. Machine Learning, pp. 139–172. Kluwer Academic Publishers, Netherlands (1987)
26. Moon, T.K.: The expectation-maximization algorithm. IEEE Signal Process. Mag. 47–60 (1996)
27. Kohonen, T.: The self-organizing map. In: Proceedings of the IEEE, pp. 1464–1480, Finland (1990)
28. Sculley, D.: Web-scale k-means clustering. In: Proceedings of the 19th International Conference on World Wide Web, pp. 1177–1178. ACM, Raleigh, NC, USA (2010)
29. Capó, M., Pérez, A., Lozano, J.A.: An efficient approximation to the K-means clustering for massive data. Knowl.-Based Syst. 56–69 (2017)
30. Kanungo, T., Mount, D.M., Netanyahu, N.S., Piatko, C.D., Silverman, R., Wu, A.Y.: An efficient k-means clustering algorithm: analysis and implementation. IEEE Trans. Pattern Anal. Mach. Intell. 881–892 (2002)
31. Chen, Y., Hu, X., Fan, W., Shen, L., Zhang, Z., Liu, X.,… Li, H.: Fast density peak clustering for large scale data based on kNN. Knowl.-Based Syst. 104824 (2020)
32. Owens, J.D., Houston, M., Luebke, D., Green, S., Stone, J.E., Phillips, J.C.: GPU computing. Proc. IEEE 879–889 (2008)

33. Snir, M.: MPI—The Complete Reference: The MPI Core. MIT Press, Cambridge (1998)
34. Dean, J., Ghemawat, S.: MapReduce: simplified data processing on large clusters. In: 6th Symposium on Operating Systems Design and Implementation, pp. 107–113. ACM Press (2008)
35. Zaharia, M., Chowdhury, M., Franklin, M., Shenker, S., Stoica, I.: Spark: cluster computing with working set. HotCloud 10 (2010)
36. Eric, M., Roger, B.: Introduction to GPUs for Data Analytics Advances and Applications for Accelerated Computing. O'Relly, USA (2017)
37. Che, S., Boyer, M., Meng, J., Tarjan, D., Sheaffer, J.W., Skadron, K.: A performance study of general-purpose applications on graphics processors using CUDA. J. Parallel Distrib. Comput. 1370–1380 (2008)
38. Andrade, G., Ramos, G., Madeira, D., Sachetto, R., Ferreira, R., Rocha, L.: G-DBSCAN: a GPU accelerated algorithm for density-based clustering. Procedia Comput. Sci. 369–378 (2013)
39. Al-Ayyoub, M., Abu-Dalo, A.M., Jararweh, Y., Jarrah, M., Al Sa'd, M.: A GPU based implementations of the fuzzy C-means algorithms for medical image segmentation. J. Supercond. 3149–3162 (2015)
40. Cuomo, S., De Angelis, V., Farina, G., Marcellino, L., Toraldo, G.: A GPU-accelerated parallel K-means algorithm. Comput. Electr. Eng. 1–13 (2017)
41. Baydoun, M., Ghaziri, H., Al-Husseini, M.: CPU and GPU parallelized kernel K-means. J Supercomput. 3975–3998 (2018)
42. Shahrezaei, M.H., Tavoli, R.: Parallelization of K-means ++ using CUDA. arXiv (2019)
43. Sato, K., Moody, A., Mohror, K., Gamblin, T., de Supinski, B.R., Maruyama, N., Matsuoka, S.: Fmi: fault tolerant messaging interface for fast and transparent recovery. In: IEEE International Parallel and Distributed Processing Symposium, pp. 1225–1234 (2014)
44. Sur, S., Koop, M.J., Panda, D.K.: High-performance and scalable MPI over infiniband with reduced memory usage: an in-depth performance analysis. In: CM/IEEE Conference on Super Computing, p. 105. ACM (2006)
45. D'Urso, P.: Exponential distance-based fuzzy clustering for interval valued data. Fuzzy Optim. Decis. Making 51–70 (2017)
46. Kwok, T., Smith, K., Lozano, S., Taniar, D.: Parallel fuzzy c-means clustering for large data sets. In: Euro-Par 2002 Parallel Processing, pp. 27–58. Springer, Berlin, Heidelberg (2002)
47. Zhang, J., Wu, G., Hu, X., Li, S., Hao, S.: A parallel K-means Clustering Algorithm with MPI. In: Fourth International Symposium on Parallel Architectures, Algorithms and Programming, pp. 60–64. IEEE (2011). https://doi.org/10.1109/paap.2011.17
48. Savvas, I.K., Sofianidou, G.N.: A novel near-parallel version of k-means algorithm for n-dimensional data objects using MPI. Int. J. Grid Utility Comput. 80–91 (2016)
49. Shan, X., Shen, Y., Wang, Y.: A parallel K-means algorithm f or high dimensional text Data. In: IEEE International Conference on Consumer Electronics, Taiwan (2018)
50. Zhao, W., Ma, H., He, Q.: Parallel K-means clustering based on MapReduce. In: Proceedings of Cloud Computing, pp. 674–679. Springer, Berlin, Heidelberg (2009)
51. Kim, Y., Shim, K., Kim, M.-S., Lee, J.S.: DBCURE-MR: an efficient density-based clustering algorithm for large data using MapReduce. Inform. Syst. 15–35 (2014)
52. Ludwig, S.A.: MapReduce-based fuzzy c-means clustering algorithm: implementation and scalability. Int. J. Mach. Learn. Cybern. 923–934 (2015). https://doi.org/10.1007/s13042-015-0367-0
53. Ben HajKacem, M.A., Ben N'cir, C.-E., Essouss, N.: Parallel K-prototypes for clustering big data. In: Proceedings of Data Science and Advanced Analytics, pp. 628–637. Springer (2015)
54. Shahrivari, S., Jalili, S.: Single-pass and linear-time k-means clustering based on MapReduce. Inform. Syst. 1–12 (2016)
55. Dongbo, Z., Shou, Y., Xu, J.: An improved parallel K-means algorithm based on MapReduce. Int. J. Embedded Syst. 275–282 (2017)
56. Valcarce, D., Parapar, J., Barreiro, Á.: A MapReduce implementation of posterior probability clustering and relevance models for recommendation. Eng. Appl. Artif. Intell. 114–124 (2018)

57. Tripathi, A.K., Sharma, K., Bala, M.: A novel clustering method using enhanced grey wolf optimizer and MapReduce. Big Data Res. 93–100 (2018)
58. Pang, N., Zhang, J., Zhang, C., Qin, X., Cai, J.: PUMA: parallel subspace clustering of categorical data using multi-attribute weights. Exp. Syst. Appl. 233–245 (2019)
59. Ekanayake, J., Li, H., Zhang, B., Gunarathne, T., Bae, S.H., Qiu, J., Fox, G.: Twister: a runtime for iterative MapReduce. In: Proceedings of the 19th ACM International Symposium on High Performance Distributed Computing, pp. 810–818. ACM, New York (2010)
60. Zayani, A., Ben N'Cir, C.E., Essoussi, N.: Parallel clustering method for non-disjoint partitioning of large-scale data based on spark framework. In: Proceedings of IEEE International Conference on Big Data, pp. 1064–1069. IEEE, Piscataway (2016)
61. Luo, G., Luo, G., Gooch, T.F., Tian, L., Qin, K.: A parallel DBSCAN algorithm based on spark. In: IEEE International Conferences on Big Data and Cloud Computing (BDCloud), Social Computing and Networking (SocialCom), Sustainable Computing and Communications (SustainCom), pp. 548–553. IEEE, Atlanta, GA, USA (2016). https://doi.org/10.1109/bdcloud-socialcom-sustaincom.2016.85
62. Ben HajKacem, M.A., Ben N'cir, C.E., Essoussi, N.: KP-S: a spark-based design of the K-prototypes clustering for big data. In: Proceedings of ACS/IEEE International Conference on Computer Systems and Applications, pp. 1–7. IEEE, Hammamet, Tunisia (2017)
63. Liu, R., Li, X., Du, L., Zhi, S., Wei, M.: Parallel implementation of density peaks clustering algorithm based on spark. Procedia Comput. Sci. 442–447 (2017)
64. Ben HajKacem, M.A., Ben N'cir, C.-E., Essoussi, N.: Scalable random sampling K-prototypes using spark. In: International Conference, DaWaK 2018, pp. 317–326. Springer, Regensburg, Germany (2018)
65. Liu, B., He, S., He, D., Zhang, Y., Guizani, M.: A spark-based parallel fuzzy c-means segmentation algorithm for agricultural image big data. IEEE Access 42169–42180 (2019)
66. Yu, H., Chen, Y., Lingras, P., Wang, G.: A three-way cluster ensemble approach for large-scale data. Int. J. Approx. Reason. 32–49 (2019)
67. Saadaoui, F., Bertrand, P. R., Boudet, G., Rouffiac, K., Dutheil, F., Chamoux, A.: A dimensionality reduce clustering methodology for hetrogeneous occupational medicine data mining. IEEE Trans. Nanobiosci. 707–715 (2015)
68. Pathak, A., Pal, N.R.: Clustering of mixed data by integrating fuzzy, probabilistic and collaborative clustering framework. Int. J. Fuzzy Syst. 339–348 (2016)
69. Zainul Abidin, F.N., Westhead, D.R.: Flexible model-based clustering of mixed binary and continuous data: application to genetic regulation and cancer. Nucl. Acids Res. 1–11 (2016)
70. McParland, D., Gormley, I.C.: Model based clustering for mixed data: clustMD. Adv. Data Anal. Classif. 155–169 (2016). https://doi.org/10.1007/s11634-016-0238-x
71. Mcparland, D., Philips, C.M., Brennan, L., Roche, H.M., Gormley, I.C.: Clustering high-dimensional mixed data to uncover sub-phenotypes: joint analysis of phenotypic and genotypic data. J. Stat. Med. 4548–4569 (2017)
72. Narmadha, D., Balamurugan, A.A., Sundar, G.N., Priya, S.J.: Survey of clustering algorithms for categorization of patient records in healthcare. Indian J. Sci. Technol. 1–5 (2016)
73. Su, F.-C., Friesen, M.C., Humann, M., Stefaniak, A.B., Stanton, M.L., Liang, X.,… Virji, M.A.: Clustering asthma symptoms and cleaning and disinfecting activities and evaluating their associations among healthcare workers. Int. J. Hygiene Environ. Health, 1–11 (2019)
74. Amini, A., Saboohi, H., YingWah, T., Herawan, T.: A fast density-based clustering algorithm for real-time Internet of Things stream. Sci. World J. 1–11 (2014)
75. Karyotis, V., Tsitseklis, K., Sotiropoulos, K.: Big data clustering via community detection and hyperbolic network embedding in IoT applications. MDPI/Sens. 1–21 (2018)
76. Rui, T., Fong, S.: Clustering big IoT data by metaheuristic optimized mini-batch and parallel partition-based DGC in Hadoop. Fut. Gen. Comput. Syst. 1–33 (2018)
77. Fanaee-T, H., Gama, J.: Tensor-based anomaly detection: an interdisciplinary survey. Knowl.-Based Syst. 1–28 (2016)

78. Alguliyev, R., Aliguliyev, R., Sukhostat, L.: Anomaly detection in big data based on clustering. Stat. Optim. Inform. Comput. 325–340 (2017)
79. Yin, C., Zhang, S., Kim, K.-J.: Mobile anomaly detection based on improved self-organizing maps. Mob. Inform. Syst. 1–9 (2017)
80. Alsayat, A., El-Sayed, H.: Social media analysis using optimized K-means clustering. In: 2016 IEEE 14th International Conference on Software Engineering Research, Management and Applications (SERA), pp. 1–10. IEEE, Towson, MD, USA (2016)
81. Gurusamy, V., Kannan, S., Prabhu, J.R.: Mining the attitude of social network users using K-means. Int. J. Adv. Res. Comput. Sci. Softw. Eng. 226–230 (2017)
82. Singh, K., Kumar Shakya, H., Biswas, B.: Clustering of people in social network based on textual similarity. Perspect. Sci. 570–573 (2016)
83. Ghaemi, Z., Farnaghi, M.: A varied density-based clustering approach for event detection from heterogeneous Twitter data. Int. J. Geo-Inform. 1–18 (2019)
84. Mohebi, A., Aghabozorgi, S., Wah, T.Y., Herawan, T., Yahyapour, R.: Iterative big data clustering algorithms: a review. Softw. Pract. Exp. 107–129 (2016)
85. Singh, D., Reddy, C.K.: A survey on platforms for big data analytics. J. Big Data 8 (2015)
86. Likas, A., Vlassis, N., Verbeek, J.J.: The global k-means clustering algorithm. Pattern Recogn. 451–461 (2003)
87. Pelleg, D., Moore, A.: X-means: extending k-means with efficient estimation of the number of clusters. In: Proceedings of the 17th International Conference on Machine Learning, pp. 27–734 (2000)
88. Bradley, P.S., Fayyad, U.M.: Refining initial points for K-means clustering. In: Proceedings of the Fifteenth International Conference on Machine Learning, pp. 1–99. ICML (1998)
89. Emre Celebi, M., Kingravi,, H.A., Vela, P.A.: A comparative study of efficient initialization methods for the k-means clustering algorithm. Expert Syst. Appl. 200–210 (2013)
90. Li, Q., Wang, P., Wang, W., Hu, H., Li, Z., Li, J.: An efficient k-means clustering algorithm on MapReduce. In: Proceedings of Database Systems for Advanced Applications, pp. 357–371 (2014)
91. Bandyopadhyay, S., Maulik, U.: An evolutionary technique based on K-means algorithm for optimal clustering in RN. Inform. Sci. 221–237 (2002)
92. Esmin, A., Coelho, R.A., Matwin, S.: A review on particle swarm optimization algorithm and its variants to clustering high-dimensional data. Artif. Intell. Rev. 23–45 (2015)
93. Krishna, K., Narasimha Murty, M.: Genetic K-means algorithm. IEEE Trans. Syst. Man Cybern. 433–439 (1999)
94. Chen, M., Mao, S., Liu, Y.: Big data: a survey. Mob. Netw. Appl. 19(2), 171–209 (2014)
95. Li, H., Lu, X.: Challenges and trends of big data analytics. In: Ninth International Conference on P2P, Parallel, Grid, Cloud and Internet Computing (3PGCIC), Guangzhou, China. pp. 566–567 (2014)
96. Kaisler, S., Armour, F., Espinosa, J., Money, W.: Big data: issues and challenges moving forward. In: 46th Hawaii International Conference on System Sciences (HICSS), Hawaii, pp. 995–1004 (2013)
97. Assunção, M.D., Calheiros, R.N., Bianchi, S., Netto, M.A.S., Buyya, R.: Big data computing and clouds: trends and future directions. J. Parallel Distrib. Comput. 79–80, 3–15 (2015)
98. Cuzzocrea, A., Song, I.-Y., Davis, K.C.: Analytics over large scale multidimensional data: the big data revolution! In: Proceedings of the ACM 14th International workshop on Data Warehousing and OLAP (DOLAP'11), pp. 101–104. ACM, New York, NY, USA (2011)
99. Benjamins, V.R.: Big data: from hype to reality? In: Proceedings of the 4th International Conference on Web Intelligence, Mining and Semantics (WIMS14), pp. 2:1–2:2. ACM, New York, NY, USA (2014)
100. Boyd, D., Crawford, K.: Critical questions for big dat Information. Commun. Soc. 15(5), 662–679 (2012)

A Survey of Network Intrusion Detection Using Machine Learning Techniques

Thomas Rincy N and Roopam Gupta

Abstract Nowadays, a huge amount of information flows daily on public and private computer networks. Since sensitive information has a high probability of being transmitted, there is an important need to protect networks from intrusions. Hence, adopting an intrusion detection system is imperative. As the frequency of sophisticated attacks has been increasing tremendously over the past years, machine learning approaches were introduced to identify intrusion patterns and prevent sophisticated attacks. This survey provides an up-to-date review of leading-edge techniques used by intrusion detection systems that rely on machine learning techniques. Moreover, it introduces important key machine learning concepts such as ensemble learning and feature selection that are applied to protect networks from unauthorized access and make networks and computers safer. The article then reviews signature, anomaly, and hybrid intrusion detection systems that apply machine learning techniques. It is observed that hybrid network intrusion detection system may be the most effective. Then, the article examines the characteristics of popular benchmark datasets for evaluating intrusion detection systems such as NSL-KDD, Kyoto 2006 +, and KDD Cup-'99 and performance metrics to appraise intrusion detection results. Finally, the article discusses research opportunities in the field of intrusion detection.

Keywords Intrusion detection system · Machine learning · Ensemble learning · Feature selection · Datasets · Performance metrics

Thomas Rincy N (✉)
Department of Computer Science and Engineering, University Institute of Technology, Rajiv Gandhi Proudyogiki Vishwavidyalaya, Bhopal, MP, India
e-mail: rinc_thomas@rediffmail.com

Roopam Gupta
Department of Information Technology, University Institute of Technology, Rajiv Gandhi Proudyogiki Vishwavidyalaya, Bhopal, MP, India
e-mail: roopamgupta@rgtu.net

© The Author(s), under exclusive license to Springer Nature Switzerland AG 2021
H. Chiroma et al. (eds.), *Machine Learning and Data Mining for Emerging Trend in Cyber Dynamics*, https://doi.org/10.1007/978-3-030-66288-2_4

1   Introduction

In recent decades, the internet has greatly facilitated communication between individuals and businesses. Many services are now offered and a lot of important and confidential information is transmitted over the internet and private computer networks. In this context, an important issue is to protect the privacy and integrity of computer networks from intruders and to ensure that networks can continue to operate normally when they are under attack. A recent report by Cybersecurity Ventures estimated that by 2021, cybercrime damage will rise to $6 trillion annually from $3 trillion in 2015 [1]. Moreover, it is predicted that there will be 6 billion internet users worldwide by 2022, and global spending on cyber security will cumulatively exceed $1 trillion over the next five years.

Due to the increasing frequency of sophisticated cyber-attacks, network security has become an emerging area of research for computer networking. To prevent intrusions and protect computers, programs, data, and unauthorized access to systems, intrusion detection systems (IDS) have been designed. An IDS can identify an external or internal intrusion in an organization's computer network and raise an alarm if there exists a security breach in the network [2]. To allow IDS to operate automatically or semi-automatically, machine learning techniques have been employed. Machine learning [3] can detect the correlation between features and classes found in training data, and identify relevant subsets of attributes by feature selection and dimensionality reduction, then use the data to create a model for classifying data to perform predictions.

The rest of this article is organized as follows. Section 2 succinctly introduces machine learning concepts and techniques. Section 3 reviews machine learning based intrusion detection system and in particular discusses the importance of ensemble learning and feature selection techniques. Section 4 concentrates on hybrid intrusion detection systems. Section 5 describes the NSL-KDD, Kyoto 2006 + , and KDD Cup-'99 benchmark datasets and the various performance metrics to evaluate the IDS. Section 6 discusses the research opportunities. Finally, Sect. 7 draws a conclusion.

2   Machine Learning

Machine learning is prominently applied in the area of intrusion detection systems. It is the field through which the various computer algorithms are studied, that improves incrementally through the experience and has been an area of computational learning theory and pattern recognition in artificial intelligence research since the field of its inception [4]. It helps in predictions and decisions on data, by studying algorithms and building a model from the input data. Machine learning is classified into supervised learning, semi-supervised learning, un-supervised learning, and reinforcement learning [5]. Figure 1 shows the classification of machine learning types with their corresponding methods.

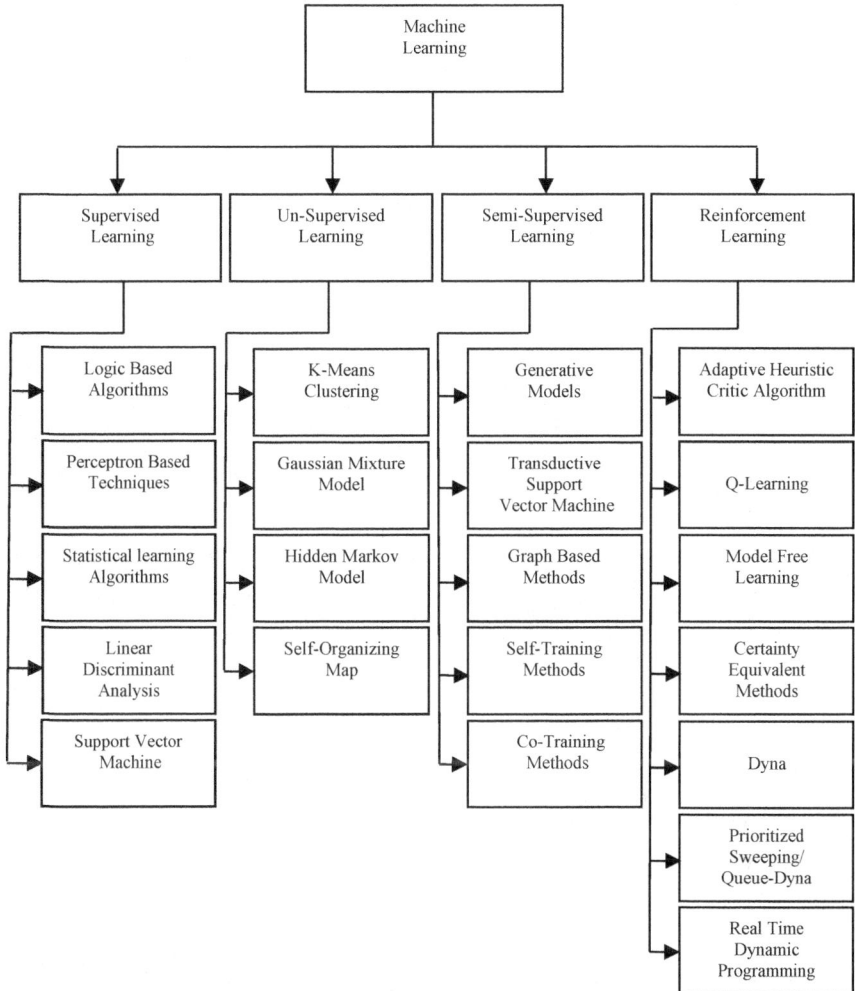

Fig. 1 Classification of machine learning system

2.1 Supervised Learning

Supervised learning is a machine learning task that assumes a function from the labeled training data. In supervised learning, there is an input variable (P) and an output variable (Q). From the input variable, the function of the algorithm is to study the mapping function to the output variable Q = f(P). Spervised learning aims to analyze the training data then produces a complete function that can be utilized to map the new instances. The learning algorithm should be able to analyze and generalize those class labels correctly from the unobserved instances. This section introduces the various algorithms used in supervised learning.

2.1.1 Logic Based Algorithms

This subsection describes the decision trees and rule-based classifiers.

Decision Trees

A decision tree [6] is a directed tree with a root node having no incoming edges. The remaining node has an incoming edge. Each of the leaf nodes is available with a label, non leaf nodes have a feature which is called a feature set. According to the distinct values in the feature set, the decision tree splits the data that falls inside the non leaf node. The testing of the feature is operated from the leaf node and its outcome is achieved until the leaf node has arrived.

Rule-Based Classifiers

Quinlan et al. [7] stated that, by transforming the decision tree into a distinct set of rules, a different path can be created for the set of rules. From the root of the tree to the leaf of the tree, a distinct rule for each path is created the decision tree is altered in to set of rules. Directly, from the training data, the rules can be inducted by different algorithms and apply these rules. The main goal is to construct the smallest set of rules that is similar to the training data.

2.1.2 Perceptron Based Techniques

Perceptron algorithm is applied to the batch of training instances for learning. The main objective is to run the algorithm frequently among the training data, till it discovers a prediction vector that is appropriate on all of the training data. For predicting the labels, this prediction rule is then applied to the test set [8]. The following subsection introduces perceptron based techniques.

Neural Network

The neural network conceptual model was developed in 1943 by McCulloch et al. [9]. It consists of different cells. The cell receives data from other cells, processes the inputs, and passes the outputs to other cells. Since then, there was intensive research to develop the ANNs. A perceptron [10] is a neural network that contains a single neuron that can receive more than one input to produce a single output.

A linear classifier contains a weight vector w and a bias b. From the given instance x, the predicted class label y is obtained according to

$$y = \text{sign}(w^{\text{T}}x + b)$$

With the help of weight vector w, the instance space is mapped into a one-dimensional space, afterwards to isolate the positive instances from negative instances.

2.1.3 Statistical Learning Algorithms

In statistical learning, there is a likelihood of a probability where each instance belongs to its class. The following subsection describes statistical learning algorithms.

Bayesian Network

Bayesian Network is the directed acyclic graph (DAG) [11], in which each edge correlates to the conditional dependency and each node correlates to a distinctive random variable. The Bayesian network performs the following task for learning. Firstly it learns the directed acyclic graph of the network. Secondly, the aim is to find the network parameters. Once the parameters or probabilistic parameters are found, then the parameters are induced in the set of tables, and joint distribution is remodeled by increasing exponentially the tables.

Naive-Bayesian Classifier

Naive Bayesian classifiers [12] are probabilistic classifiers with their relation related to Bayes theorem having a strong assumption of naive independence among its features. Bayes theorem can be stated in mathematical terms:

$$P(A/B) = P(A)(B/A)/P(B)$$

where A and B are events.

P(A) and P(B) are events.

P(A) and P(B) are the prior probabilities of A and B. P(A/B), is a posterior probability, of the probability of observing an event A, provided that B is true. P (B/A) is known as likelihood, the probability of observing an event B provided that A is true. Naive Bayes classifier has the advantage of requiring the least execution time required for training the data.

K-Nearest Neighbor Classifiers

K-NN [13] is a nonparametric technique applied for regression and classification. In the feature space, the input of k-NN contains the k closest training examples. Then the output will depend on whether k-NN is applied for regression or classification purposes. The output is the property value for the object in k-NN regression.

2.1.4   Linear Discriminant Analysis

Linear discriminant analysis is a method to find the linear combination of the features that separate the two or more classes of objects. The resulting combination can be used as a linear classifier. A linear classifier [14] contains a weight vector w and a bias b. From the given instance x, the predicted class label y is obtained according to

$$y = \text{sign}(w^{\mathrm{T}}x + b)$$

With the help of weight vector w, the instance space is mapped into a one-dimensional space, afterward to isolate the positive instances from negative instances.

2.1.5   Support Vector Machine

SVM [15] revolves around the margin on either side of a hyperplane that separates two data classes. To reduce an upper bound on the generalization error, the main idea is to generate the largest available distance between its instance on either side and separating hyperplane. The data points that lie on the margins of the optimum hyperplane are termed as Support Vector Points, and it is characterized as the linear combination of these points. An alternative data point is neglected. The different features available on the training data do not affect the complexity of SVM. This is the main reason the SVM is employed with learning tasks having a large number of features with respect to the number of training data.

Large problems of the SVM cannot be solved as it not only contains the large quadratic operations but also there exist numerical computation which makes the algorithm slow in terms of processing time. There is a variation of SVM called Sequential Minimal Optimization (SMO). SMO can solve the SVM quadratic problem without employing any extra matrix storage and without using any numerical quadratic programming optimization steps [16].

2.2 Un-Supervised Learning

In Un-Supervised learning the data is not labeled, more precisely, there are unlabeled data. Un-Supervised learning consists of an input variable (P), but there is no output variable. The representation is seen as a model of data. Un-Supervised learning aims to obtain the hidden structures from unlabeled data or to infer a model having the probability density of input data. This section reviews some of the basic algorithms used in Un-Supervised learning.

2.2.1 K- Means Clustering

K-means clustering [17] partitions the N observation in space into K clusters. The information and nearest mean belongs to this cluster and works as a model of the cluster. As a result, the data space splits into Voronoi cells. K-means algorithm is an iterative method, which starts with a random selection of the k-means v_1, v_2... v_k. With each number of iteration the data points are grouped by k-clusters, keeping with the closest mean to each of the points, mean is then updated according to the points within the cluster. The variant of K-means is termed as K-medoids. In K-medoids, instead of taking the mean, the larger part of the cluster having the centrally located data point is investigated as a reference point of the corresponding cluster [18].

2.2.2 Gaussian Mixture Model

The Gaussian mixtures were popularized by Duda and Hart in their seminal 1973 text, Pattern Classification and Scene Analysis [19]. A Gaussian Mixture is a function that consists of several Gaussians, each identified by $k \in \{1,...,K\}$, where K represents the number of clusters of a dataset. Each Gaussian is represented by the combination of mean and variance, consisting of a mixture of M Gaussian distributions. The weight of each Gaussian will be a third parameter related to each Gaussian distribution in a Gaussian mixture model (GMM).

When clustering is performed applying Gaussian Mixture Model, the goal is to obtain a criterion such as mean and covariance of each distribution and the weights, so that the resulting model fits optimally in the data. To optimally fit the data, one should enlarge the likelihood of the data given in the Gaussian Mixture Model. This can be achieved by using the iterative expectation maximization (EM) algorithm [20].

2.2.3 Hidden Markov Model

HMM [21] is a parameterized distribution for sequences of observations. A hidden Markov model is a Markov process that is divided into two components called observable components and unobservable or hidden components. That is, a hidden Markov model is the Markov process (Y_k, Z_k) k ≥ 0 on the state space $C \times D$, where we presume that we have a means of observing Y_k, but not Z_k as the signal process and C as the signal state space, while the observed component Y_k is called the observation process and D is the observation state space.

2.2.4   Self-organizing Map

A self-organizing map (SOM) [22] is a form of an artificial neural network, trained by applying unsupervised learning to yield a low dimensional, discretized representation of input space of the training samples called maps. SOM applies the competing learning while ANN applies the error-correction learning (backpropagation with gradient descent). To retain the topological properties of the input space the SOM applies the neighborhood function. SOM begins with initializing the weight vectors. Sample vectors are chosen randomly and the map of the weight vectors is inspected to find the best weight that represents that sample. The weight vector has neighboring weights that are close to each other. The chosen best weight sample and neighbors of that weight are rewarded and is more likely to become the randomly selected sample vector. This strategy allows the maps to grow and take the different shapes such as rectangular, square, and hexagonal in two-dimensional feature spaces.

2.3   Semi-supervised Learning

Semi-Supervised learning is the sequence of labeled and unlabeled data. The labeled data is very sparse while there is an enormous amount of unlabeled data. The data is used to create an appropriate model of the data classification. Semi-Supervised learning aims to classify the unlabeled data from the labeled data. This section explores some of the most familiar algorithms used in Semi-Supervised learning.

2.3.1   Generative Model

Generative model [23] considers a model $p(x, y) = p(y)p(x/y)$ where $p(x/y)$ is known as mixture distribution. The mixture components can be analyzed when there are large numbers of unlabeled data is available. Let $\{p_\theta\}$ be the distribution family and is denoted by parameter vector θ. θ can be identified only if $\theta_1 \neq \theta_2 \Rightarrow p_{\theta_1} \neq p_{\theta_2}$ to a mixture components transformation. The expectation–maximization (EM) algorithm is applied on the mixture of multinomial for a task of text classification [24].

2.3.2   Transductive Support Vector Machine

TSVM [25] extends the Support Vector Machine (SVM) having the unlabeled data. The idea is to have the maximal margin among the labeled and unlabeled data on its linear boundary by labeling the unlabeled data. Unlabeled data has the least generalization error on a decision boundary. The linear boundary is put away from the dense region by the unlabeled data. With all the available estimation solutions to TSVM, it is curious to understand just how valuable TSVM will be as a global optimum solution. Global optimal solution on small datasets is found in [26]. Overall an excellent accuracy is obtained on a small dataset.

2.3.3   Graph Based Methods

In Graph based methods [27] labeling information of each sample is proliferated to its neighboring sample until a global optimum state is reached. A Graph is constructed with nodes and edges where nodes are defined with labeled and unlabeled samples, while the edges define the resemblance among labeled and unlabeled data. The label of the data sample is advanced to its neighboring points. The graph based techniques are the focus of interest among researchers due to their better performance. Kamal and Andrew [28] propose semi-supervised learning as the graph mincut (known as st-cut) problem. Szummer et al. [29] performed a step Markov random walk on the graph.

2.3.4   Self Training Methods

Self-training is a methodology applied in semi-supervised learning. On a small amount of data, the classifier is trained and then applied to classify the unlabeled data. The most optimistic unlabeled points, together with their labels predicted, are then added to the training set. The classifier is again trained with the training dataset. This procedure goes on repeating itself. For teaching itself, the classifier had its own predictions. This methodology is called bootstrapping or self-teaching [30]. Various natural language processing tasks apply the methodology of self teaching.

2.3.5   Co-Training

Co-Training [31] is applied where there are a large amount of unlabeled data and a small amount of labeled data. In Co-Training, there is a general assumption that the features can be divided into feature subsets. Every feature subset is adequate for training the classifier, and for the given class the feature sets are relatively independent.

There are two classifiers primarily that are trained on a labeled data on the pair of feature subsets. With labeled and unlabeled data, each classifier trains the other classifier to retain the supplementary training sample from another classifier and this process gets repeated often until optimal features are found.

2.4 Reinforcement Learning

In reinforcement learning, the software agent gathers the information from the inter-action, with the environment to take actions that would maximize the reward. The environment is formulated as a markov decision process. In reinforcement learning, there is no availability of input/output variables. The software agent receives the input i, the present state of environment s, then the software agent determines an action a, to achieve the output. The software agent is not told which action would be best in terms of long term interest. The software agent needs to gather informa-tion about the states, actions, transition, rewards for optimal working. This section reviews algorithms used in reinforcement learning.

2.4.1 Adaptive Heuristic Critic

The adaptive heuristic critic algorithm is a policy iteration, adaptive version [32] is implemented by Sutton's $TD(0)$ algorithm [33]. Let $((a, w, t, a'))$ be the tuple representing a single transformation in an environment. Let a, be the state of an agent before its transition, w is the action of its choice, t is the award it receives instantaneously, and a' is culminating state. The policy value is studied by Sutton's $TD(0)$ algorithm which utilizes the update rule given by:

$$V(a) := V(a) + w(t + tV(a') - V(a)).$$

When the state a is visited, its estimated value is amended to be adjacent to $(t + tV(a')$ since t is the spontaneous reward received and $V(a')$ is the approximate rate of the next appearing state.

2.4.2 Q-Learning

Q-learning [34] is a type of model free reinforcement learning. It may also be known as an approach of asynchronous dynamic programming (DP). Q-learning allows the agents to have the ability to learn and to perform exemplary in the markovian field by recognizing the effects of its actions, which is no longer required by them to build domain maps. Q-learning finds an optimal policy and it boosts the predicted value of the total reward, from the beginning of the current state to any and all successive steps, for a finite markov decision process, given the infinite search time and a partially random policy. An optimal action-selection policy can be associated with Q-learning.

2.4.3 Certainty Equivalent Methods

The certainty equivalent method [35] is studied continuously throughout the agent lifespan. The present model is applied to calculate the optimum policy and functional value. The certainty model effectively uses the available data on rather side, it is computationally costly even for small state areas.

2.4.4 Dyna

Suttons Dyna Architecture [36] applies a yielding approach that tends to be more successful and is more computationally effective than certainty-equivalent methods. Markov decision process is defined by a tuple (a, w, t, a') and it acts as follows: The model gets transited state from a to a' upon on action and getting the reward t by taking action w in the state a. Updated models are then M and P. Dyna updates the state a by applying the rules that are an adaptation of the value iteration updating for the Q rules.

2.4.5 Prioritized Sweeping

The limitations with Dyna are proportionately undirected. When the target has just been reached it goes on to update its random state action pairs, instead of concentrating on the curious part of the state space. Prioritized sweeping addresses this problem by choosing the updates at random, the values are associated with space rather than state-action pairs. Additional information has to be stored in the model for having the fitting choices. Every state remembers its previous state, having the non zero transition probability. The priority of each state is set initially to zero. Prioritized sweeping updates the k states having the highest priority, rather than updating the k random state action pairs [37].

2.4.6 Real Time Dynamic Programming

The RTDP [38] applies Q learning. The computing is done on the areas, where the agent is coherent to occupy the state space. RTDP is specific to problems in which the agent tries to attain the particular goal state and the reward everywhere else is 0. RTDP finds the shortest path from the start state to the target state without visiting the unnecessary state space.

2.5   Ensemble Learning

Ensemble learning [39] is an imperative study in the domain of Machine Learning, Data mining, Pattern Recognition, Neural networks, and Artificial Intelligence. The goal of ensemble learning methods is to build the set of learners or classifiers then merge them so that it can make an accurate prediction, in contrast to single learner the accurate prediction may not be good as a set of learners. There are several learners in an ensemble learning called base learners. The goodness of ensemble learning is that they are capable of boosting the base learners. The base learners are sometimes called the weak learners. Therefore the ensemble learning is also called as learning of multiple classifier systems. In this section, few important concepts of ensemble learning techniques are discussed.

2.5.1   Boosting

Boosting [40] is an ensemble learning paradigm, which builds a strong classifier from the number of base learners. Boosting works sequentially by training the set of base learners to combine it for the prediction. The boosting algorithm takes the base learning algorithms repeatedly, having the different distributions or weighting of training data on the base learning algorithms. On running the boosting algorithm, the base learning algorithms will generate a weak prediction rule, until many rounds of steps the boosting algorithm combines the weak prediction rule into a single prediction rule that will be more accurate than the weak prediction rule.

2.5.2   Bagging

Bootstrap AGGregatING [41]. Bagging algorithms combine the bootstrap and aggregation and it represents parallel ensemble methods. Bootstrap sampling [42] is applied by bagging to acquire the subsets of data for training the base learning algorithms. Bagging applies the aggregating techniques such as voting for classification and averaging for regression. Bagging uses a precedent to its base classifiers to obtain its outputs, votes its labels, and then obtains the label as a winner for the prediction. Bagging can be applied with binary classification and multi-class classification.

2.5.3   Stacking

In Stacking [43] the independent learners are combined by the learner. The independent learners can be called cardinal learner, while the combined learner is called Meta learner. Stacking aims to train the cardinal learner with initial datasets to generate the contemporary dataset to be applied to the Meta learner. The output generated by the cardinal learner is regarded as input features, while the original labels are still

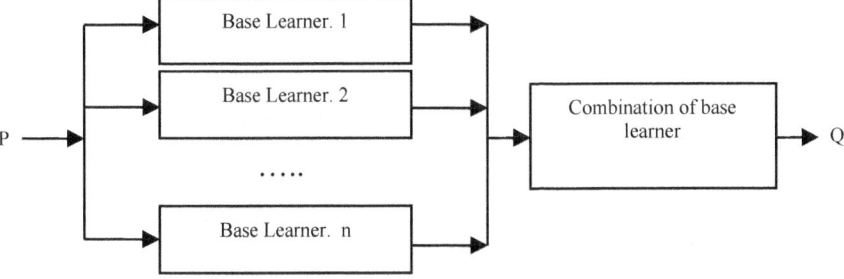

Fig. 2 A basic architecture of ensemble learning

regarded as labels of the contemporary dataset. The cardinal learner is obtained by applying different learning algorithms, often generating composite stacked ensembles, although we can construct uniformly stacked ensembles. The contemporary dataset has to be obtained by the cardinal learner otherwise, the dataset is the same for the cardinal and Meta learner there is speculation of over fitting. Wolpert et al. [44] emphasized various features for the contemporary dataset and categories of learning algorithms for the Meta Learner. Figure 2 shows the basic architecture of ensemble learning.

2.6 Feature Selection

Feature selection is a well-studied research topic in the field of artificial intelligence, machine learning, and pattern recognition. Feature selection removes the redundant, irrelevant, and noisy features from the original features of datasets by choosing the relevant features having the smaller subdivision of the dataset. By applying the various techniques of feature selection to the datasets, results in lower computational costs, higher classifier accuracy, reduced dimensionality, and a predictable model. Existing feature selection methods for machine learning fall into two broad categories. The learning algorithms that classify relevant and useful features applied to the data are termed as the wrapper method and those evaluate the merit of features by using heuristics based on general characteristics of data are termed as filter method [45]. This section focuses on the characteristics of feature selection algorithms, introduces the filter and wrapper approaches.

2.6.1 Characteristics of Feature Selection Algorithms

The feature selection algorithm does its task by performing the exploration in the space having feature subsets, and as a result, should define the four fundamental problems influencing the variety of the search [46].

- *Starting point.* The selection of points in a feature subset space affects the direction of the search. There are different choices to begin the search. The search space can start with no features to successfully adding the attributes. Within the search space, the search moves forward, the subsequent possibility is that the search space begins with all features and adequately removes them. Another alternating technique is to search the space in the middle and then moves outwardly from this point.
- *Search organisation.* Whenever there is data having the features containing the larger number, the searching of the feature subset within the space and time frame is decisive. There exists a 2^N possible subsets for the N number of features. As compared to exhaustive search techniques the heuristic search techniques are more practicable. Greedy hill climbing is a strategy that considers the single feature that can consecutively be added or deleted from the current feature subset. Whenever the algorithm favors the inclusion and elimination of the feature subset then it is termed as forward selection and backward elimination [47].
- *Evaluation strategy.* The biggest challenge among the feature selection algorithms for machine learning is the classification of feature subsets [48]. In this strategy, the algorithm eliminates the undesirable features before the learning starts. The algorithm applies the data having a heuristic based on general characteristics to calculate the goodness of the feature subset. When selecting the features, the bias of the induction algorithm should be considered. This method is called Wrapper, which applies an induction algorithm, with re-sampling techniques to obtain the accuracy of the feature subsets.
- *Stopping criterion.* An important decision of feature selection is to determine when to terminate the searching space of feature subsets. A feature selector should terminate the searching space of feature subsets by summing or eliminating the features when the feature selector finds no one of its substitutes. It enhances the quality of a prevailing feature subset. Concurrently, the algorithm may consider carrying on, altering the feature subset until the quality of the feature subset does not degrade.

2.6.2   Filter Methods

The earlier approaches that were introduced to feature selection are filter methods. Filter methods apply the data that has an examining property in general to calculate the goodness of feature subset, except a learning algorithm that evaluates the quality of the feature subsets. Filter algorithm executes many times faster than wrapper so there is a high probability of scaling the databases having a large number of features. Filter does not require the re-execution for the different learning algorithms. This strategy makes filter approaches quicker than wrapper methods. There are major drawbacks of the filter methods such as some filter algorithm is unable to handle the noisy data, some algorithms need the user to specify the level of noise, and in some

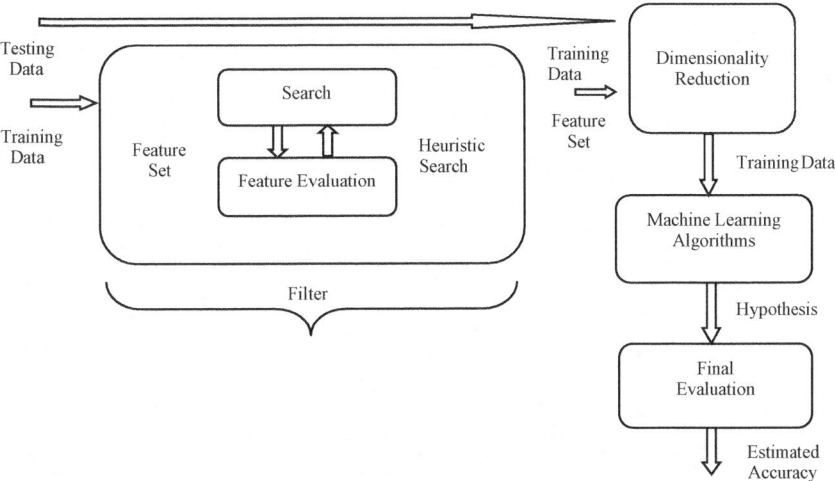

Fig. 3 A filter feature selector

cases, the features are ranked and the final choice is left to the user. In other cases, the users need to specify the required number of features. Filter methods are applied mainly where the data is having highly dimensionality rate. Figure 3 shows the filter feature selector.

2.6.3 Wrapper Methods

For the selection of best feature subsets, a wrapper method applies an induction algorithm for the feature selection.

The idea of wrapper approaches is that since induction methods are using the feature subset it should enhance the accuracy than a separate measure that applies distinct inductive bias [49]. Wrapper approaches are known to obtain better results than filter methods because they are specifically adapted to the interaction between the training data and induction algorithm. Whereas the wrapper methods are much slower than filter methods as the wrapper methods, frequently call the induction algorithm and reruns when a different induction algorithm is applied. Figure 4 shows the wrapper method.

2.6.4 Principal Component Analysis

Principal component analysis (PCA) [50] is an analytical procedure that converts the correlated variables into linearly uncorrelated variables, with the help of an orthogonal transformation. These are named as principal components. The PCA is a multivariate dimensionality reduction tool that extracts the features representing much of

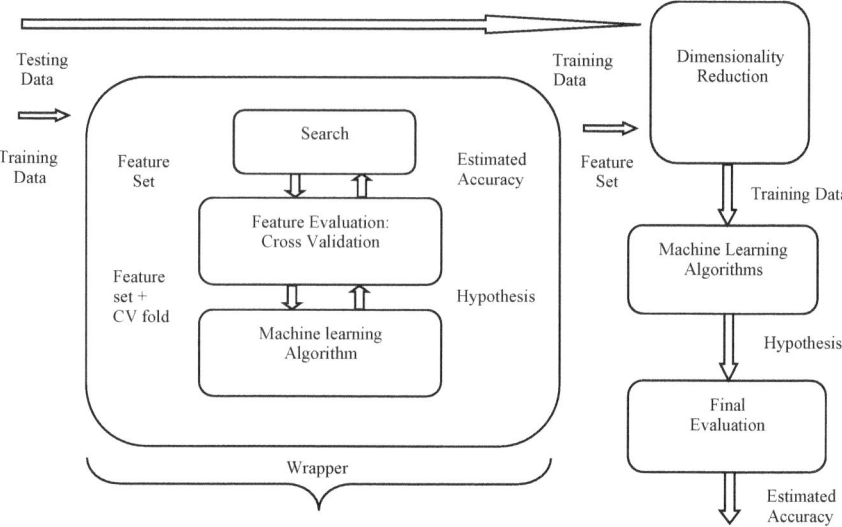

Fig. 4 A wrapper feature selector

the features from the given data and thus removing the unimportant features having less information without losing the crucial information in data. When real data is collected, the random variables describing the data attributes are presumed to be highly correlated. The correlation between random variables can be found in the covariance matrix. The aggregate of the variances will give the overall variability.

2.6.5   Classification and Regression Trees

CART [51] is a binary decision tree that is constructed by splitting a node into two child nodes repeatedly.

The whole learning samples contains at the beginning with the root node. The basic idea behind the tree growth is to choose a split among all the possible splits at each node so that the resulting child nodes obtained are the purest. CART algorithm considers the univariate splits i.e. each split depends on the value of one predictor variable. Classification analysis of the tree is performed when the predicted outcome is the class to which the data belongs, while the regression analysis of the tree is done when the predicted outcome can be considered as a real number.

3 Machine Learning Based Intrusion Detection System

3.1 Intrusion Detection System (IDS)

The IDS literature history starts with a paper by James Anderson [52]. The main challenges of the organization are to protect the information from network threats. A string of actions through which intruder gains control of the system [53]. The main aim of intrusion detection is to detect previously known and unknown attacks, to learn and comply with new attacks and detect intrusions in an appropriate period. IDS are classified according to three main approaches: Implementation, Architecture, and Detection Methods [54]. Figure 5 shows the classification of the Intrusion Detection System.

3.1.1 Implementation Classification

The Intrusion Detection System is mainly classified according to Host-Based Intrusion detection System (HIDS) and Network-Based Intrusion Detection Systems (NIDS).

Host-Based Intrusion Detection Systems

HIDS positions the sensors at an individual host of the network sensor. It collects the relevant information about data, based on various events. The sensor also records the data related to system logs, registry keys, and other logs generated by the operating system process called inspection logs or audit trail and any other unauthorized action. The HIDS rely heavily on inspection logs, degrading the performance of hosts related to bandwidth which can lead to high computational costs. The HIDS is better known

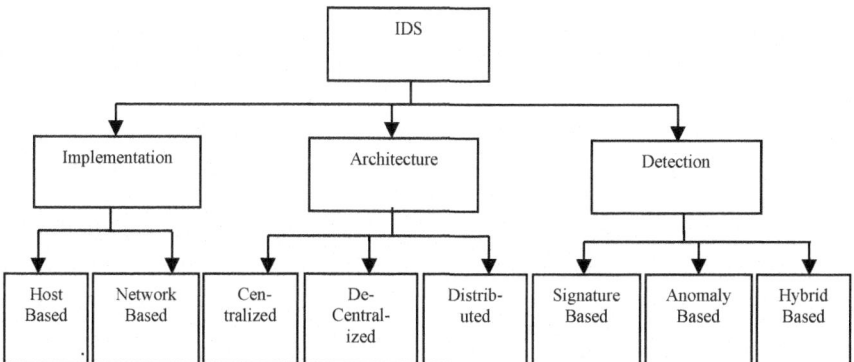

Fig. 5 Classification of intrusion detection system

to trace malicious activities regarding specific users and also keeping a track record of the behavior of individual users within an organization.

Network Based Intrusion Detection Systems

It collects the information from networks and checks for attacks or irregular behavior by examining the contents and header information and flow information of all packets across all networks. NIDS collects the information from networks examines the contents and the header information regarding all packets in various networks and then validates the irregular behavior of the network by verifying the various contents and header information of various packets whether there is an attack or not. The network sensor is having various sensors, a with built-in mechanism with attack signatures that define rules. These sensors also allow the user to define their signature for verifying the new attacks in the networks.

NIDS is computationally costly and also time-consuming, as it has to inspect each flow and header level of packets from various networks, on the other side the NIDS is operating system and platform-independent that does not require any modification when NIDS operates. This makes NIDS more scalable and robust compared with HIDS.

3.1.2   Architectural Classification

The architectural classification of IDS is classified into three categories:

Centralized IDS

The analysis and detection of monitored data are analyzed inside the CPU. There are various networks and different sensors present inside the centralized IDS. Inside the centralized IDS, different sensors monitor the data that are sent from various networks. Whenever there is a need for expansion of the network, the central processing unit is overloaded that is the major disadvantage of centralized IDS.

Decentralized IDS

The CPU overloading problem is solved in decentralized IDS architecture. The multiple sensors and multiple processing units are scattered across the network. The gathered data are send to the nearest processing unit for getting processed, before they are sent to the main processing unit.

Distributed IDS

In distributed IDS the work load of the CPU is distributed among all its peers. The multiple IDS are distributed over the large network and its fundamental principle is that the multiple intrusion detection systems will communicate through peer to peer architecture.

3.1.3   Detection Methods

The detection methods of intrusion detection systems are classified into three major types: Signature-based, Anomaly-based and Hybrid-based. This section discuss various machine learning approaches used in signature, anomaly and hybrid IDS.

Signature-Based IDS

The signature-based intrusion detection system stores the previously known or existing signature in a database and it looks for malignant activities and malignant packets in the network. It measures the similarity and patterns of the attacks within the existing signature that is saved in the database. If the similarity or pattern is found then an alarm is generated. Signature-based IDS are accurate in detecting the previously known attacks, but they cannot recognize the new attacks.

Some of the machine learning algorithms that were used in signature-based IDS are Bayesian network [55], Decision Trees [56], SVM [57], ANN [58], K-NN [59], and Graph-based approaches [60].

Anomaly-Based IDS

In anomaly-based IDS new action profiles are created and are applied to analyze the outliers that diverge from the profiles, Chandola et al. [61]. Anomaly-based intrusion detection system depends on analytical techniques to construct the various attack forecast models. It has the leverage of detecting anonymous attacks that do not have the extant signature in the database, but the major disadvantage is creating new action profiles. Sometimes the outliers that diverge from new action profiles are not always an attack, this creates incorrectly classified as an attack or False Positive. Some of the anomaly-based IDS that uses various machine learning techniques are Probability-based technique [62], Fuzzy rule classifiers using NN technique [63], Neural Network using back propagation technique [64], One class support vector machine [65], and K- NN [66].

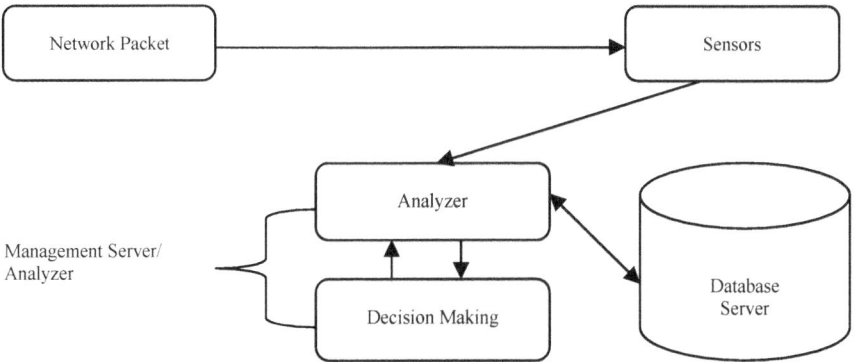

Fig. 6 A basic architecture of intrusion detection system

Hybrid-Based IDS

The hybrid-based IDS combines both signature and anomaly-based detection approaches to detect attacks. The main advantage of a hybrid intrusion detection system is that it combines the existing features from the anomaly and signature-based IDS. The downside of hybrid-based techniques is its increased execution time of applying both matchings of signatures and anomaly detection to evaluate the incoming of the network connections. The techniques that use the hybrid-based IDS using machine learning techniques are combined NN and SOM [67], combining ANN, SVM, and MARS [68], etc. Figure 6 shows the basic architecture of an intrusion detection system.

4   Hybrid Intrusion Detection Systems

Hybrid-based intrusion detection systems were enormously applied as they produce a better result. The authors concluded that the hybrid-based approaches can overcome the one approach over the other [69]. Neha and Shailendra [70] conclude that hybrid-based IDS are relevant methods to get good accuracy and proposed, intelligent water drops (IWD) algorithm described by Shah–Hosseini [71] based feature selection approach. This strategy applies the IWD algorithm, an optimization algorithm that is inspired by the nature of the selection of subset features, combined with SVM as a classifier to evaluate the chosen features. The search of the subset is performed by the IWD algorithm while the evaluation task of the subset is carried by the classifier. Overall this approach reduces the total of 45 features to 10 features from the input dataset. The proposed approach achieved a detection rate, precision, and accuracy of 99% on the KDD Cup-'99 dataset. Arif et al. [72] introduced a Hybrid approach for intrusion detection systems where PSO is applied to prune the node while pruned decision trees are applied as a classification technique for NIDS. Particle Swarm

Optimization is a population-based stochastic optimization technique based on bird flocking and fish schooling. The PSO algorithm works by having the population called the swarm and the potential solutions called particles. PSO performs the searching through a swarm of particles that updates through repetitions. Each particle advances in the direction to its previous best position and globally best position for finding the most promising solution. The above approach applies the single and multi-purpose particle swarm optimization algorithm. The evaluation is done on the KDD Cup-'99 dataset. Thirty arbitrary samples of the datasets are preferred from 10% of KDD Cup-'99 training and testing datasets. The estimate of records in individual training, as well as testing datasets, is 24,000 and 12,000 jointly for the evaluation purposes. A precision of 99.98% and an accuracy of 96.65% is achieved using the above approaches.

Ahmed and Fatma [73] applied a triple edge strategy to develop a Hybrid Intrusion detection system in which the Naïve Bayes feature selection (NBFS) technique has been applied for dimensionality reduction, Optimized Support Vector Machines (OSVM) is applied for outlier rejection and Prioritized K-Nearest Neighbours (PKNN) classifier is employed for classifying the input attacks. Different datasets are applied such as the KDD Cup-'99 dataset, NSL-KDD dataset, and Kyoto 2006 + dataset. 18 effective features are selected from the KDD Cup-'99 dataset having a detection rate of 90.28%. 24 features are selected from Kyoto 2006 + dataset with a detection ratio of 91.60%. The author has compared with previous work and has a best overall detection ratio of 94.6%. Tirtharaj Dash et al. [74] report two new hybrid intrusion detection methods that are gravitational search (GS) and sequence of gravitational search and particle swarm optimization called (GSPSO). Gravitational search is a population-based heuristic algorithm that is based on mass interactions and the law of gravity. It comprises of search agent who interacts among each other with heavier masses by the gravitational force and its performance is evaluated by its mass. The slow movement of the search agent assures the good solution of the algorithm. The combinational approach has been carried out to train ANN with models such as GS-ANN and GSPSO-ANN. The random selection of 10% features is selected for training purposes, while 15% is used for testing purposes is applied successfully for intrusion detection purposes. The KDD Cup-'99 dataset was applied as a metric for calculation. Normalisation of dataset was done for uniform distribution by MATLAB. An average detection ratio of 95.26% was achieved.

Yao and Wang [75] introduced a hybrid framework for IDS, K-Means algorithm is employed for clustering purposes. In the classification phase, many machine learning algorithms (SVM, ANN, DT, and RF) which are all supervised learning algorithms are compared on different parameters. The supervised learning algorithm has various parameters for different kinds of attacks (DoS, U2R, Probe, and R2L). The proposed hybrid algorithm has achieved an accuracy rate reaching 96.70% on the KDD Cup-'99 dataset. Alauthaman et al. [76] proposed a technique of peer to peer, bot detection based on an adaptive multilayer feed forward neural network in cooperation with decision trees. CART is used as a feature selection technique to prefer the relevant features. With all those features, a multilayer feed-forward NN training model is built applying the resilient back-propagation learning algorithm. Network traffic reduction techniques were applied by using six rules to pick the most relevant features. Twenty-nine features are selected from six rules. The proposed approach has an accuracy rate of 99.20% and a detection rate of 99.08% respectively on ISOT and ISCX datasets.

Sidra and Faheel [77] introduce a genetic algorithm, which is based on vectors. In this technique vector chromosomes are applied. The novelty of this algorithm is to show the chromosomes as a vector and training data as metrics. It grants multiple pathways to have the fitness function. Three feature selection techniques are chosen, linear correlation-based feature (LCFS), modified mutual information-based feature selection (MMIFS), and forward feature selection algorithm (FFSA). The novel algorithm is tested in two datasets (KDD Cup-'99 and CDU-13). Performance metrics show that vector-based genetic algorithm has a high DR of 99.8% and a low FPR of 0.17%. Saud and Fadl [78] introduced an IDS model applying the machine learning algorithm to a big data environment. This paper employs a Spark-Chi-SVM model. ChisqSelector is applied as a feature selection method and constructed an IDS model by applying the SVM as a classifier. The comparison is done with Spark Chi-SVM classifier and Chi- Logistic regression classifier. KDD Cup-'99 dataset is applied for the metrics evaluation process. The result shows Spark Chi-SVM model has good performance having an AUROC of 99.55% and AUPR of 96.24%. Venkataraman and Selvaraj [79] report the hybrid feature selection structure for the efficient classification of data. The symmetrical uncertainty is employed to find the finest features for classification. The Genetic algorithm is applied to find the finest subset of the features with greater accuracy.

The author combined (SU-GA) as a hybrid feature selector. Matlab and Weka tools are applied to develop the proposed hybrid feature selector. Different classification algorithm (J48, NB, SMO, DT, JRIP, Kstar, Rand Frst, Multi Perptn,) are used to classify different attacks. The average learning accuracy with Multi Perpn and SU-GA is highest having 83.83% on the UCI dataset. Neeraj et al. [80] applied knowledge computational intelligence in network intrusion detection systems. This paper introduces an intelligent and Hybrid NIDS model. This model, then integrates the fuzzy logic controller, multilayer perception, adaptive Neuro-Fuzzy interference system, and a Neuro fuzzy genetic. The proposed system has trio modules collector, analyser, and predictor modules for gather and filter network traffic, classifying the data, and to prepare the final decision in assuming knowledge on the accurate attack. The experiment is performed on the KDD Cup-'99 dataset and achieved a false alarm detection accuracy of 99%. Akash and Khushboo [81] implemented a hybrid technique. DBSCAN is applied for choosing the enhanced features for IDS and eliminating the noise from data. Based on Euclidean distance and minimum number of points DBSCAN groups together the points that are close to cach other. For grouping data, K means clustering is proposed. SMO (sequential minimal optimization) classifier is implemented for classification purposes. The experimental setup is performed on the KDD Cup-'99 dataset with certain attributes. The approach DBKSMO achieved an accuracy of about 98.1%. MATLAB and WEKA tools are applied to execute the full process.

Rajesh and Manimegalai [82] introduced an Effective IDS applying flawless feature selection outlier detection and classification. The author employs a feature selection technique named intelligent flawless feature selection algorithm (IFLFSA) to select the finest number of features that are effective for identifying the attacks. To identify the outliers from the dataset, an entropy-based weighted outlier detection (EWOD) approach is applied. An intelligent layered technique is employed for efficient classification. The KDD Cup-'99 dataset is applied for experimental purposes. The comprehensive detection accuracy wraps the detection accuracy on four categories of attacks namely DoS, Probe, U2R, and R2L. The detection accuracy of the suggested system has achieved the rate of 99.45%. Unal et al. [83] implemented a hybrid approach for IDS using machine learning approaches. Naive bayes, K-Nearest neighbor is used for classification purposes, while random forest and J-48 algorithm are used as a decision tree to train the algorithm on different types of attack. The author proposed a CfsSubsetEval approach and WrapperSubsetEval as the two feature selection techniques. NSL-KDD dataset is used as an evaluation process. The overall accuracy of 99.86% is obtained on all types of attacks. Pragma et al. [84] introduce a hybrid IDS to the hierarchical filtration of anomalies (HFA). This work proposes an ID3, a decision tree, which is employed for classifying data into corresponding classes. K-NN is used to assign the class label to an unknown data point that is based on its class labels of the k nearest neighbors. Isolation forest is introduced to isolate an anomaly against normal instances. The proposed algorithm (HFA) is performed on the NSL-KDD dataset and KDD Cup-'99 dataset. The overall performance on the KDD Cup-'99 dataset, has an accuracy rate of 96.92%, detection ratio of 97.20% and FPR of 7.49%. The proposed algorithm performance with the NSL-KDD dataset has an accuracy rate of 93.95% has a detection ratio of 95.5%, and an FPR of 10.34%. Sushruta and Chandrakanta [85] report a new approach named BFS-NB hybrid model in IDS. This paper proposes the best first search technique for dimensionality reduction and was employed for an attribute selection technique. The Naïve Bayes classifier is implemented as a classification technique and to improve the accuracy of identifying the intrusions. The BFS-NB algorithm is performed on the KDD dataset gathered from the US air force. The classification accuracy of the BS-NFB is 92.12% while sensitivity and specificity analysis of 97 and 97.5% is achieved. Table 1 depicts the Taxonomy of the latest hybrid intrusion detection methods.

5 Evaluations of Intrusion Detection System

This section describes the features of KDD Cup-'99, NSL-KDD, and Kyoto 2006 + datasets and different metrics to classify the performance of the IDS.

Table 1 Taxonomy of latest hybrid intrusion detection methods

Year	Authors/research papers	Algorithms	Techniques	Dataset	Evaluation criteria	Feature selection	Results
2017	Neha Acharya, Shailendra Singh: An IWD-based feature selection method for intrusion detection system. Soft Computing Springer 2017. pp: 4407–4416. DOI 10.1007/s00500-017-2635-2	IWD, SVM	Feature reduction using IWD (Intelligent water drops), SVM, is used as a classifier	KDD Cup-'99 Dataset	Precision rate, Detection rate, False Alarm rate, Accuracy rate	Yes	Achieves a precision rate of 99.10%, Detection rate of 99.40%, Accuracy rate of 99.05%, False alarm of 1.40%
2017	Arif Jamal Malik, Farrukh Aslam Khan: A hybrid technique using binary particle swarm optimization and decision tree pruning for network intrusion detection. Cluster Comput Springer 2017. pp: 667–680. DOI 10.1007/s10586-017-0971-8	Particle swarm optimization (PSO)	Particle swarm Optimization (PSO) algorithm/PSO are used for node pruning of Decision Trees and the pruned Decision Trees is used for classification of the network intrusions	KDD Cup-'99 Dataset	Precision rate, Accuracy rate, IDR, FPR. Time	Yes	A precision of 99.98%, Accuracy of 96.65%, IDR of 92.71%, FPR of 0.136 and Evaluation performance of 383.58 sec. is achieved
2017	Ahmed I. Saleh, FatmaM. Talaat, LabibM. Labib: A hybrid intrusion detection system (HIDS) based on prioritized k-nearest neighbors and optimized SVM Classifiers. Artif Intell Springer 2017. pp: 403–443. DOI 10.1007/s10 462-017-9567-1	Naïve based feature selection, Optimized SVM Algorithm, Prioritized KNN Algorithm	Hybrid HIDS strategy (Based on Naïve based feature selection), Naïve based is used as feature selection, OSVM is used for outlier rejection. PKNN is used for classifying input attacks	KDD Cup-'99 NSL-KDD and Kyoto 2006 + Dataset	Detection Rate, Sensitivity, Specificity, Precision	Yes	An overall detection rate of 94.6%, Sensitivity rate of 53.24%, Specificity of 98.21%, Precision of 56.62%, is achieved on all the datasets

(continued)

Table 1 (continued)

Year	Authors/research papers	Algorithms	Techniques	Dataset	Evaluation criteria	Feature selection	Results
2017	Tirtharaj Dash: A study on intrusion detection using neural networks trained with evolutionary algorithms. Soft Comput Springer 2017. pp: 2687–2700. DOI 10.1007/s00500-015-1967-z	Artificial Neural Network	Gravitational search (GS), and combination of GS and particle Swarm optimization (GSPSO) are implemented to train ANN	NSL-KDD Dataset	Detection rate, Mean Squared Error MSE, Time	No	A detection ratio of 95.26%, MSE of 0.4527% and training time of 103.70 seconds is achieved
2017	Haipeng Yao, Qiyi Wang: An Intrusion Detection Framework Based on Hybrid Multi-Level Data Mining, Int J Parallel Prog. Springer 2017. pp: 1-19. DOI 10.1007/s10766-017-0537-7	Hybrid Multilevel data mining algorithm	MH-DE (Multilevel Hybrid Data Engineering), MH-ML (Multilevel Hybrid Machine Learning, MEM (Micro Expert Module) to train on KDD Cup-'99 dataset	KDD Cup-'99 Dataset	Accuracy rate, Detection rate, Precision rate, Recall, F-Value	Yes	Accuracy of 96.70%, Precision of 96.55%, F-Value of 96.60%, Recall of 96.70% and on all types of attack a detection rate of 66.69% is achieved
2018	Mohammad Alauthaman, Nauman Aslam, Li Zhang, Rafe Alasem, M. A. Hossain: A P2P Botnet detection scheme based on decision tree and adaptive multilayer neural networks. Neural Comput & Applic 29, Springer 2018. pp. 991–1004. DOI 0.1007/s00521-016-25645	Neural Network with resilient back propagation algorithm, CART	Feature reduction is done by CART, Neural Network with resilient back propagation algorithm for updating the weights	ISOT & ISCX dataset.	Accuracy rate, Detection rate, FPR	Yes	Detection of 99.08%, An accuracy rate of 99.20%, and FPR of 0.75% is achieved

(continued)

Table 1 (continued)

Year	Authors/research papers	Algorithms	Techniques	Dataset	Evaluation criteria	Feature selection	Results
2018	Sidra Ijaz, Faheel A. Hashmi: Vector Based Genetic Algorithm to optimize predictive analysis in network security: Applied intelligence, Volume 48, issue 5, Springer 2018. pp: 1086-1096. DOI 10.1007/s10489-017-1026-9	Vector based genetic algorithm	Chromosomes as vector and training data as metrics'	CTU-13 dataset and KDD Cup-'99 dataset	Accuracy rate, FPR	Yes	Accuracy rate for DOS is 99.8% and FPR of 0.17% is achieved
2018	Suad Mohammed. Fadl Mutahe: Intrusion detection model using machine learning algorithm on Big Data environment: In proc. journal of big data. Springer 2018. pp: 1-12. DOI 10.1186/s40537-018-0145-4	Support Vector Machine (SVM)	Chisqselector using SVM classifier for feature reduction	KDD Cup-'99 dataset	AUROC, AUPR, Time	Yes	Area under curve (AUROC) of 99.55%, Area under precision recall curve (AUPR) of 96.24% and training time of 10.79 se
2018	Sivakumar Venkataraman, Rajalakshmi Selvaraj: Optimal and Novel Hybrid Feature Selection Framework for Effective Data Classification: Advances in systems, control and application. Springer 2018. pp: 499–514. DOI 10.1007/978-981-10-4762-6_48	Symmetrical Uncertainty and genetic algorithm (SU-GA) is used as classification algorithm	Symmetric uncertainty is used to obtain best features/genetic algorithm is applied on selected features	UCI dataset.	Accuracy rate	Yes	The average learning accuracy of 83.83% is achieved, and overall evaluation time of 0.23 seconds is achieved on all methods

(continued)

Table 1 (continued)

Year	Authors/research papers	Algorithms	Techniques	Dataset	Evaluation criteria	Feature selection	Results
2018	Neeraj Kumar, Upendra Kumar: Knowledge Computational Intelligence in Network Intrusion Detection Systems. In: Knowledge Computing and Its Applications. Springer 2018. pp: 161-176. DOI 10.1007/978-981-10-6680-1_8	Genetic algorithm	Fuzzy Logic Controller, Neuro- fuzzy inference system, Multi layer perception, Neural Fuzzy genetic for classification of attacks	KDD Cup-'99 dataset	Accuracy rate	Yes	A detection accuracy of 99% is achieved
2019	Akash Saxena, Khushboo Saxena: Hybrid Technique Based on DBSCAN for Selection of Improved Features for Intrusion Detection System: Emerging Trends in Expert Applications and Security. pp: 365377. Springer 2019. DOI 10.1007/978-981-13-2285-3_43	DBSCAN, SMO, K-Means clustering	DBSCAN is used to eliminate noise in data, K-Means is used for grouping of data, and SMO is used for detection of intrusions.	KDD Cup-'99 dataset.	Accuracy rate, Detection rate	Yes	An approx accuracy of 98.1% and approx Detection rate of 70% is achieved

(continued)

Table 1 (continued)

Year	Authors/research papers	Algorithms	Techniques	Dataset	Evaluation criteria	Feature selection	Results
2019	Rajesh Kambattar, Manimegalai Rajkumar: An Effective Intrusion Detection System Using Flawless Feature Selection, Outlier Detection and Classification: Progress in Advanced Computing and Intelligent Engineering. Springer 2019. pp: 203–213. DOI 10.1007/978-981-13-1708-8_19	Intelligent flawless feature selection algorithm (IFLFSA), Entropy Based weighted outlier rejection (EWOD), Intelligent layered classification algorithm	IFLFSA is used as feature selection; EWOD is used to detect outliers in data. Intelligent layered classification algorithm is applied to classify the data	KDD Cup-'99 dataset	Accuracy rate	Yes	Overall accuracy of 99.45% is achieved
2019	Unal Cavusoglu: A new hybrid approach for intrusion detection using machine learning methods: Applied Intelligence (2019) 49. pp: 2735–2761. Springer 2019. DOI 10.1007/s10489-018-01408-x	Naïve bayes, Random forest, J-48, K-Nearest Neighbor (KNN) algorithm	CfsSubsetEval and WrapperSubsetEval as a two feature selection techniques, while Naïve bayes, Random forest, J-48, K-Nearest Neighbor (KNN) algorithm is used as a classifier	NSL-KDD Dataset	Accuracy rate, Detection Rate, FP rate, F-measure, MCC, TP rate, Time	Yes	Overall detection ratio of 0.9828%, Overall accuracy rate of 99.86%, Overall TPR of 0.929%, Overall FPR of 0.00042%, Overall F-Measure of 0.9625 %, MCC of 0.955 % and total time to evaluate is 10.625 seconds (Performed on NSL KDD, 25 attributes on all types of attacks)

(continued)

Table 1 (continued)

Year	Authors/research papers	Algorithms	Techniques	Dataset	Evaluation criteria	Feature selection	Results
2019	Pragma Kar, Soumaya Banerjee, Kartick Chandra Mondal, Gautam Mahapatra, Samiran Chattopadhay: A Hybrid Intrusion Detection System for Hierarchical Filtration of Anomalies. In: Information and Communication Technology for Intelligent Systems. Smart Innovation, Systems and Technologies, vol 106. Springer 2019, pp: 417–426. DOI 10.1007/	ID3, K-Nearest neighbor, Isolation Forest,	ID3 is applied as feature selector, K-nearest neighbor is used to apply a class to unknown data point, Isolation forest is applied to segregate normal data from anomaly	KDD Cup-'99 dataset, NSL-KDD Dataset	Accuracy rate, Detection rate, False alarm rate	Yes	Performance on NSL-KDD dataset has accuracy of 93.95% has a detection rate of 95.5% and a false alarm rate of 10.34%. The performance with KDD Cup-'99 dataset has an accuracy of 96.92%, detection rate of 97.20% and false alarm rate of 7.49%
2019	Sushruta Mishra, Chandrakanta Mahanty, Shreela Dash, Brojo Kishore Mishra.: Implementation of BFS-NB Hybrid Model in Intrusion Detection System: Recent Developments in Machine Learning and Data Analytics. Advances in Intelligent Systems and Computing, vol 740. Springer. 2019 pp: 167–175. DOI 10.1007/978-981-13-1280-9_17	Best first search and Naïve bayes, (BFS-NB) algorithm.	Naïve bayes is used as classifier and best search is used as attribute optimization method.	KDD dataset from US Air force	Accuracy, Specificity, Sensitivity	Yes	The classification accuracy of BS-NFB is 92.12%, Sensitivity analysis of 97% and specificity of 97.5% is achieved

5.1 KDD Cup-'99 Dataset

The KDD Cup-'99 is an extensively used dataset in the field IDS. KDD Cup-'99 data is captured from the DARPA 98, IDS evaluation program. The entire KDD Cup-'99 dataset comprises 4,898,431 single connection records; each contains 41 features classified as normal or attacks. KDD Cup-'99 dataset classifies the attacks into four types, namely Probe, User to Root, Denial of Service, and Remote to Local [86].

- *Probe.* The intruder gathers the knowledge from the networks or hosts scans the whole networks or hosts that are prone to attacks. An intruder then exploits the system vulnerabilities by looking at the known security breaches so that the whole system is compromised for malicious purposes.
- *Denial of Service (DOS).* This kind of attack results in the unavailability of computing resources to legitimate users. The intruder overloads the resources, by making the computing resources busy, so that genuine users are unable to utilize the full resources of the computer.
- *User to Root (U2R).* The intruder tries to acquire the root access of the system or the administrator privileges by sniffing the passwords. The attacker then looks for the vulnerabilities in the system, to acquire the gain of the administrator authorization.
- *Root to Local (R2L).* The intruder attempts to gain access to the remote machine, which does not have the necessary and legal privilege to access that machine. The attacker then exploits the susceptibility of the remote system, tries gaining access rights to the remote machine. Table 2 describes the categories of attack and attack names.

The KDD Cup-'99 intrusion datasets mainly consist of three components: The entire KDD Cup-'99 dataset consists of the examples of normal and attack connections, 10% KDD dataset consists of the data, which aims for training the classifiers, and the KDD test dataset for the testing of classifiers. Table 3 describes the dataset features.

The connection features can be classified into four categories namely:

- *Basic features.* It is accessed from the header of the packet, without inspecting the packet contents, such as (protocol type, duration, flag, service, and the number of bytes that are sent from the source to the destination and contrarily).

Table 2 Four categories of attack

Category of attack	Attack name
Probe	satan, portsweep, nmap, ipsweep
DoS	neptune, back, smurf, land, teardrop, pod
U2R (User to Root)	perl, loadmodule, buffer_overflow, rootkit
R2L (Remote to Local)	warezlient, spy, warezmaster, multihop, guesspasswd, phf, imap, ftp_write

Table 3 KDD Cup-'99 Dataset Features

Index	Feature name	Description
1	Duration	Connection length
2	protocol_type	Protocol types (TCP, ICMP, and UDP)
3	Service	Mapping of destination port to the service (ftp, telnet, http…)
4	Flag	Connection status
5	src_bytes	Source to destination. (Total number of bytes)
6	dst_bytes	Destination to source. (Total number of bytes)
7	Land	If the address of the source and destination are same from the place, then it is = 1/otherwise 0
8	wrong_fragment	Indicates the number of the wrong fragments
9	Urgent	Indicates the number of the urgent packets
10	Hot	Indicates the number of the hot indicators
11	num_failed_logins	No. of failed attempts at login
12	logged_in	If logged in = 1/if login failed = 0
13	num_compromised	It indicates the compromised states
14	root_shell	With a root account a command interpreter is running then root shell = 1/otherwise its 0
15	su_attempted	su is attempted = 1/otherwise its 0
16	num_root	Number of times the root is accessed
17	num_file_creations	To create new files, the number of operations used
18	num_shells	Number of command interpreters that are active
19	num_access_files	Number of operations to create the file
20	num_outbound_cmds	In the ftp sessions the number of outgoing commands
21	is_host_login	If the login is on the host login file list then, is host login = 1 /otherwise it is 0
22	is_guest_login	If a guest logs into system then, is guest login = 1/ otherwise it is 0
23	Count	At a given interval, the number of connections to the same host as the current connection
24	srv_count	At a given interval, the number of connections to the same service as the current connection
25	serror_rate	SYN errors with proportion of connections
26	srv_serror_rate	SYN errors with proportion of connections
27	rerror_rate	REJ errors with proportion of connections
28	srv_rerror_rate	REJ errors with proportion of connections
29	same_srv_rate	Same services with proportion of connections
30	diff_srv_rate	Different services with proportion of connections
31	srv_diff_host_rate	Different hosts with proportion of connections

(continued)

Table 3 (continued)

Index	Feature name	Description
32	dst_host_count	Same destination having the number of connections
33	dst_host_srv_count	Same destination that apply the same service having the number of connections
34	dst_host_same_srv_rate	Same destination that apply the same service having the number of connections
35	dst_host_diff_srv_rate	Different services having same host, has the number of connections
36	dst_host_same_src_port_rate	System having same source port has the number of connections
37	dst_host_srv_diff_host_rate	Same service incoming from different hosts having the number of connections
38	dst_host_serror_rate	Host having a S0 error has the number of connections
39	dst_host_srv_serror_rate	The host and the specified service having S0 error have the number of connections
40	dst_host_rerror_rate	The host having an RST error has the number of connections
41	dst_host_srv_rerror_rate	The host and the specified service having an RST error have the number of connections

- *Content features.* It is decided by investigating the details of the TCP packet (number of failed attempts to login the computer).
- *Traffic features.* The features are calculated in proportionate to window intervals. It is further divided into the same service features and same host features. The same service features it analysis the connections in last two seconds having the same service as a present connection. The same host features, it analyses the connections in the last two seconds of the host having the same destination as its present connection and then computes the stats of the protocol behavior.
- *Time features.* The same service features and same host features and are also called time-based features. The time features decide the length of time from the source IP address to the destination IP address.

5.2   NSL-KDD Dataset

KDD Cup-'99 datasets are having several redundant records and duplicate records. The redundant record is having an overall 78% and duplicate records have an overall 75%. This redundancy and duplicity prevent it from classifying the other records [87].

A contemporary NSL-KDD dataset was proposed [88] that does not include the redundant and duplicate records in training and testing data [89], which helped in overcoming the redundancy and duplicity issue. There is a total 37 attacks in

Table 4 NSL-KDD: Number of instances in training and test data

Attack Classes	Total number of instances in the training set	Total number of instances in the test set
DoS	land (18), teardrop (892), pod (201), neptune (41,214), smurf (2,646), back (956)	neptune (4,657), pod (41), smurf (665), apache2 (737), worm (2), mailbomb (293), processtable (685), udpstorm (2), teardrop (12), land (7), back (359)
Probe	portsweep (2,931), ipsweep (3,599), nmap (1,493), satan (3,633),	satan (735), ipsweep (141), nmap(73), portsweep (157), mscan (996), saint (319)
R2L	guess_passwd (53), ftp_write (8), imap (11), phf (4), multihop (7), warezmaster (20), warezclient(890), spy(2)	ftp_write (3), phf (2), warezmaster (944), snmpguess (331), httptunnel (133), named (17), guess_passwd (1,231), imap (1), multihop (18), xlock (9), snmpgetattack (178), xsnoop (4), sendmail (14)
U2R	buffer_overflow (30), rootkit (10), perl (3), loadmodule (9)	rootkit (13), loadmodule (2), perl (2), sqlattack (2), xterm (13), ps (15), buffer_overflow(20)
Total	58,630	12,833

the testing dataset out of which 21 different attacks are of training dataset and the remaining attacks are available only for testing the data.

The attack classes are categorized into DoS, Probe, R2L, and U2R categories. In the training data, the normal traffic consists of 67,343 instances having an overall of 125,973 instances. In test data the normal traffic consists of 9711 instances, having an overall of 22,544 instances. Table 4 depicts the total stats of instances in training and test data.

5.3   Kyoto 2006 + Dataset

The Kyoto 2006 + dataset consists of original data that is captured from network traffic data between the years 2006 and 2009. The new version consists of further data collected from the year 2006 to 2015. This dataset is taken by applying the email server, honey pots, dark net sensors, and WebCrawler [90, 91]. The authors have employed various types of darknet, honeypots on five of the networks in indoors and outdoors of the Kyoto University and from the honeypots, the traffic data is possessed.

There are 50,033,015 normal session, 43,043,225 attack sessions and 425,719 sessions relevant to unknown attack. The author's extracted 14 statistical features from the honeypots of 41 features from the KDD Cup-'99 dataset. Table 5 shows the statistical features of the Kyoto 2006 + dataset, which is taken from the dataset of KDD Cup-'99.

Besides, the authors also obtained ten features extra, which facilitate the authors to further examine, what kind of attacks appear on computer networks. The redundant

Table 5 Statistical features in the Kyoto 2006 + dataset derived from the KDD Cup-'99 dataset

Index	Feature name	Description
1	Duration	Connection length (in number of seconds)
2	Service	Connection service types such as (http, telnet, etc.)
3	Source bytes	The source IP address that sends the number of data bytes
4	Destination bytes	The destination IP address that sends the number of data bytes
5	Count	For past two seconds, the number of connections having same for Source and destination IP address for current connection
6	Same_srv_rate	Identical service having the count feature with proportion of connections
7	Serror_rate	'SYN' errors having the count feature with proportion of connections
8	Srv_serror_rate	'SYN' errors having the service count with proportion of connections
9	Dst_host_count	Between the past 100 connections whose destination IP address is same to the current connection, the number of connections having source IP address is same to that of the current connection
10	Dst_host_srv_count	Between the past 100 connections whose destination IP address is same to that of the current connection, the number of connections having service type is same to that of the current connection
11	Dst_host_same_src_port_rate	Source port having the same of the current connection in Dst_host_count feature with proportion of connections
12	Dst_host_serror_rate	'SYN' errors having Dst_host_count feature with proportion of connections
13	Dst_host_srv_serror_rate	'SYN' errors having Dst_host_srv_count feature with proportion of connections
14	Flag	The connections state at the time of which connection was written

and insignificant features, content features are ignored as they are unsuitable and time consuming for NIDS to obtain without domain knowledge. Table 6 depicts the additional features of the Kyoto 2006 + dataset.

5.4 Performance Metrics

This section describes various performance metrics used to validate the results in IDS. Those metrics indicate all the measurements that are used by researchers to

Table 6 Additional features in Kyoto 2006 + dataset

Index	Feature name	Description
1	IDS_detection	It indicates that if IDS is started an alarm for the connection. '0' indicates that alarms are not triggered and an Arabic numeral indicates various kinds of alarms. Parenthesis implies the same amount of alarm
2	Malware_detection	It specifies that if malware was noticed at the connection. '0' indicates that malware was not noticed, and string implies the identical malware noticed at the connection. Parenthesis implies the amount of similar malware
3	Ashula_detection	It indicates if exploit codes and shellcode were applied in connection. '0' indicates no exploit code and shellcode were observed, and an Arabic numeral implies the distinctive types of the exploit codes or shellcodes. Parentheses indicate the same figure of exploit code or shellcodes
4	Label	It means that, there was attack in the session or not '1' indicates as normal. '−1' means familiar attack was observed in the session, and '−2' indicates unidentified attack was observed in the session
5	Source_IP_Address	It indicates that source IP address is utilized in session
6	Source_Port_Number	It implies that the source port number is utilized in the session
7	Destination_IP_Address	It implies that the destination IP address is utilized in the session
8	Destination_Port_Number	It implies that the destination port number is utilized in the session
9	Start_Time	It implies that at which moment the session was started
10	Duration	It indicates the total time duration when the session is set

prove the results obtained in any of: True positive (TP), True negative (TN), False negative (FN), False positive (FP) [92].

5.4.1 Confusion Matrix

It is known as error metrics, allowing the correlation between predicted and actual classes. It is important for measuring the accuracy, recall, precision, specificity, and ROC, AUC curve. It helps in the visualisation performance of the algorithms and it is often applied to depict the performance of the classifier on the testing dataset.

Class	Predicted positive class	Predicted negative class
Actual positive class	FN	TP
Actual negative class	TN	FP

5.4.2 Accuracy

The ratio of true predictions, precisely classified instances, calculated by:

$$\text{Accuracy} = \frac{\text{TN} + \text{TP}}{(\text{TN} + \text{TP} + \text{FN} + \text{FP})}$$

5.4.3 Error Rate

The ratio of all the predictions made, which are maliciously classified: It is calculated by:

$$\text{ER} = 1 - \text{Acc.}$$

5.4.4 True Positive Rate

It is termed as the intrusions that are accurately classified as an attack by the IDS. It is also called recall, sensitivity, or detection rate. It is calculated by:

$$\text{TPR} = \frac{\text{TP}}{\text{FN} + \text{TP}}$$

5.4.5 False Positive Rate

It is the regular patterns that are wrongly identified as an attack. It is also called a false alarm rate is given by:

$$\text{FPR} = \frac{\text{FP}}{\text{TN} + \text{FP}}$$

5.4.6 True Negative Rate

It is the regular patterns that are correctly identified as normal. It is also known as specificity given by:

$$\text{TNR} = 1 - \text{FPR.}$$

5.4.7 False Negative Rate

It is termed as the intrusions that are wrongly identified as normal. It is calculated by:

$$FNR = 1 - TPR.$$

5.4.8 Precision

It is the amount of behavior precisely classified as an attack. It is calculated by:

$$Precision = \frac{TP}{(FP + TP)}$$

5.4.9 F-measure

It is the measurement of accuracy tests. It is described as the weighted harmonic mean of the precision and recall and it also called as f-value or f-score:

$$FM = 2 \times \frac{recall \times precision}{recall + precision}$$

5.4.10 Matthews's Correlation Coefficient

It is used only in the binary intrusion detection system in which it computes the observed and predicated binary classification. It is calculated by:

$$MCC = \frac{(TP \times TN) - (FP \times FN)}{\sqrt{(TP + FN)(TP + FP)(TN + FN)(TN + FP)}}$$

The detection achievement of IDS is classified using the area under curve (AUC) and receiver operating characteristic (ROC). ROC reveals the changes in detection ratio with contrast in the internal verge to produce a low or high false positive rate. If AUC values are larger, then the classifier performance will be better. Another important metric that are considered in the evaluation of IDS is CPU consumption, its throughput, and power consumption which may run on different hardware configuration on high speed networks with or without the limited number of the resources.

6   Research Opportunities

The inception of Intrusion Detection Systems has started in the year 1980, numerous literature studies have presented on this topic to date, there exists a vide varieties of research opportunities in this field. Some of the research opportunities are listed below.

- *Trusted IDS.* There is a demand to develop a dependable intrusion detection systems that accurately classifies the novel intrusions, such a degree that it can be efficiently trusted without or little human interventions as possible. The robustness of IDS still needs further investigation to new evasion techniques.
- *Nature of Attacks.* The knowledge information about the nature of new intrusion attacks needs to be continuously studied and updated frequently, as the attackers are motivated enough and try different ways to intrude the IDS systems. Updating the knowledge of new attacks helps in preventing the new attacks that penetrate the IDS systems.
- *The efficiency of IDS.* One of the major concerns about IDS that exists today is its run time efficiency. There is a concern about IDS, that it consumes more resources results in degrading the IDS performance. The need of the hour is to develop efficient IDS that is computationally inexpensive and consumes fewer resources without affecting the performance factors of IDS.
- *Datasets.* NSL-KDD, Kyoto 2006 +, and KDD Cup-'99 are considered as the benchmark datasets in the field of IDS, still one needs to evaluate the performance of IDS on other datasets such as CAIDA, ISCX, ADFA-WD, ADFA-LD, CICIDS, BoT-IoT, etc. Evaluating the performance of IDS in different datasets helps us to know the actual performance of IDS on different environments.

7   Conclusion

In this study, an analysis of various intrusion detection systems that applies the various machine learning techniques was observed. Different hybrid machine learning-based intrusion detection techniques, proposed by various researchers have been analyzed. The outcome of the results and observing various parameters such as accuracy, detection rate, and false positive alarm rates are the most important contributions of this study. The study of the KDD Cup-'99, NSL-KDD, and Kyoto 2006 + dataset, with its different parameters, are discussed in this article. Finally, observation of the various performance metrics used for evaluation of results in intrusion detection systems is analyzed. By observing various parameters in this article it is concluded that "A Network Hybrid Model for Intrusion Detection System" was an efficient one.

Acknowledgements Sincere thanks to Dr. Philippe Fournier-Viger, the founder of the SPMF data mining library for providing us his deep insights into our work.

References

1. https://cybersecurityventures.com/research/
2. Liao, H.-J., Richard, C.-H.: Intrusion detection system a comprehensive review. J. Netw. Appl. 16–24. Elsevier
3. Mohammed, M., Khan, M.B.: Machine Learning Algorithms and Applications. CRC Press Taylor and Francis Group
4. Tsai, C.-F., Hsu, Y.-F.: Intrusion detection by machine learning, a review. Expert Systems with Applications, pp. 11994–12000. Elsevier
5. Kang, M., Jameson, N.J.: Machine learning Fundamentals Prognostics and health management in electronics. Fundamentals. Machine Learning, and Internet of Things. Willey Online Library
6. Quinlan, J.R.: Machine Learning, vol. 1, no. 1
7. Quinlan, J.R.: C4.5: Programs for Machine Learning, vol. 16, pp. 235–240. Morgan Kaufmann Publishers, Inc.
8. Littlestone, N., Warmuth, M.K.: The weighted majority algorithm. Inf. Comput. **108**(2), 212–261
9. McCulloch, W., Pitts, W.: A logical calculus of ideas immanent in nervous activity. Bull. Math. Biophys. **5**(4), 115–133
10. Freund, Y., Schapire, R. E.: Large margin classification using the perceptron algorithm. Mach. Learn. **37**(3), 277–296
11. Pearl, J.: Bayesian networks. A model of self-activated memory for evidential reasoning. In: Proceedings of the 7th Conference of the Cognitive Science Society, University of California, Irvine, CA, pp. 329–334. Accessed 01 May 2009
12. Rish, I.: An empirical study of the Naive Bayes classifier. IJCAI Workshop on Empirical Methods in AI
13. Altman, N. S.: An introduction to kernel and nearest-neighbor nonparametric regression (PDF). The American Statistician, 46 (3), pp. 175–185
14. Yuan, G.-X., Ho, C.-H.: Recent advances of large-scale linear classification. Proceedings of the IEEE, pp. 2584–2603
15. Cortes, C., Vapnik, V.N.: Support-vector networks. Mach. Learn. **20**(3), 273–297
16. Platt, J.C.: Probabilistic outputs for support vector machines and comparisons to regularized likelihood methods, pp. 61–74. Advances in Large Margin Classifiers, MIT Press
17. MacQueen, J. B.: Some methods for classification and analysis of multivariate observations. In: Proceedings of 5th Berkeley Symposium on Mathematical Statistics and Probability, pp. 281–297. University of California Press
18. Kaufman, L, Rousseeuw, P.J.: Clustering by means of Medoids. In: Statistical Data Analysis Based on the Norm and Related Methods, pp. 405–416. North-Holland
19. Duda, R.O, Hart, P.E.: Pattern Classification and Scene Analysis. Wiley
20. Dempster, A.P., Laird, N.M., Rubin, D.B.: Maximum likelihood from incomplete data via the EM algorithm. J. R. Stat. Soc. 1–38
21. Baum, L.E., Petrie, T.: Statistical inference for probabilistic functions of finite State Markov Chains. Ann. Math. Stat. 1554–1563
22. Kohonen, T.: The self-organizing map. In: Proceedings of IEEE, pp. 1464–1480
23. Ng, A., Jordan, M.: On discriminative versus generative classifiers. A comparison of logistic regression and naive bayes. Adv. Neural Inf. Process Syst.
24. Blum, A., Chawla, S.: Learning from labeled and unlabeled data using graph mincuts. In: Proceedings of the 18th International Conference on Machine Learning
25. Joachims, T.: Transductive inference for text classification using support vector machines. In: Proceeding of the 16th International Conference on Machine Learning (ICML), pp. 200–209. Morgan Kaufmann, San Francisco (1999)
26. Chapelle, O., Schölkopf, B., Zien, A.: Semi-supervised Learning. MIT Press
27. Zhu, X.: Semi-supervised Learning Literature Survey. University of Wisconsin, Madison
28. Nigam, K., Mccallum, A.K.: Text classification from labeled and unlabeled documents using EM. Machine Learning, vol. 39, pp. 103–134. Springer

29. Szummer, M., Jaakkola, T.: Partially labeled classification with Markov random walks. Advances in Neural Information Processing Systems
30. Yu, N.: Domain adaptation for opinion classification, a self training approach. J. Inf. Sci. Theory Pract
31. Blum, A., Mitchell, T.: Combining labeled and unlabeled data with co-training. In: COLT: Proceedings of the Workshop on Computational Learning Theory
32. Barto, A.G, Sutton, R.S., Anderson, C.W.: Neuron like adaptive element that can solve difficult learning control problems. IEEE Trans. Syst. Man Cybern. 834–846
33. Sutton, R.S.: Learning to predict by the method of temporal differences. Mach. Learn. 9–44
34. Watkins, C.J, C.H, Dayan, P.: Q learning. Mach. Learn. 279–292
35. Kumar, P.R, Variya, P.P.: Stochastic System: Estimation, Identification and adaptive control. Prentice Hall, Englewood Cliffs, NJ
36. Sutton, R.S.: Integrated architectures for learning and planning and reacting based on the approximating dynamic programming. In: Proceedings on Seventh International Conference on Machine Learning, Austin, T.X Morgan Kaufmann
37. Moore, A.W., Atkeson, C.G.: Prioritized sweeping. Reinforcement learning with less data and less time. Mach. Learn.
38. Barto, A.G, Bradke, S.J, Singh S.P.: Learning to act using real time dynamic programming. Artif. Intell. 81–138
39. Lior, R.: Ensemble learning. Pattern classification using ensemble methods. Ser. Mach. Perception Artif. Intell. **85**
40. Schapire, R.E.: The strength of weak learnability. Mach. Learn. 197–227
41. Breiman, L.: Bagging predictors. Mach. Learn. **24**(2), 123–140 (1996d)
42. Efron, B., Tibshirani, R.: An Introduction to the Bootstrap. Chapman & Hall, New York, NY (1993)
43. Smyth, P., Wolpert, D.: Stacked density estimation. In: Jordan, M.I, Kearns, M.J., Solla, S.A. (eds.), Advances in Neural Information ProcessingSystems, vol. 10, pp. 668–674. MIT Press, Cambridge, MA (1998)
44. Wolpert, D.H.: Stacked generalization. Neural Netw. **5**(2), 241–260 (1992)
45. Kohavi, R., John, G.: Wrappers for feature subset selection. Artif. Intell. Spec. Issue Relev. 273–324
46. Langley, P.: Selection of relevant features in machine learning. In: Proceedings of the AAAI Fall Symposium on Relevance. AAAI Press
47. Miller, A. J.: Subset Selection in Regression. Chapman and Hall, New York
48. Kohavi, R.: Wrappers for Performance Enhancement and Oblivious Decision Graphs. Ph.D. thesis, Stanford University
49. Langley, P.: Selection of relevant features in machine learning. In: Proceedings of the AAAI Fall Symposium on Relevance. AAAI Press
50. Pearson, K.: On Lines and Planes of Closest Fit to Systems of Points in Space, pp. 559–572, Philosophical Magazines
51. Breiman, L., Friedman, J.H., Olshen, R.A., Stone, C.J.: Classification and Regression Trees, Monterey, CA, Wadsworth & Brooks/Cole Advanced books & Software
52. Anderson, J.P.: Computer society threat monitoring and surveillance. Fort Washington, PA Computer Security Research Centre
53. Halme, L., R.: AIN'T misbehaving-A taxonomy of anti-intrusion techniques. Comput. Secur. **14**(7), 606 (1995)
54. Nisioti, A., Mylonas, A.: From intrusion detection to attacker attribution. Comprehensive survey of unsupervised methods. IEEE Commun. Surv. Tutor. **20**, 3369–3388
55. Sebyala, AA, Olukemi T, Sacks L.: Active platform security through intrusion detection using Naive Bayesian network for anomaly detection. In: The London Communications Symposium. Citeseer, London
56. Fan, W, Miller, M, Stolfo, S, Lee, W, Chan P.: Using artificial anomalies to detect unknown and known network intrusions. Knowl. Inf. Syst. **6**(5), 507–527
57. Vapnik, V.: The Nature of Statistical Learning Theory, 2nd edn. Springer, New York

58. Williams, G., Baxter, R., He, H., Hawkins, S., Gu, L.: A comparative study of ANN for outlier detection in data mining. In: Proceedings of IEEE International Conference on Data Mining (ICDM'02), Maebashi City, Japan, pp. 709–712. IEEE

59. Liao, Y, Vemuri V,R.: Use of K-nearest neighbor classifier for intrusion detection. Comput. Secur. **21**(5), 439–448

60. Gruschke, B.: Integrated event management. Event correlation using dependency graphs. In: Proc. of the 9th IFIP/IEEE International Workshop on Distributed Systems, pp. 130–141. Operations & Management (DSOM 98), Newark, DE, USA

61. Chandola, V., Banerjee, A., Kumar, V.: Anomaly detection. A survey ACM Computure Surveys, vol. 41, no. 3, pp. 1–72 (2009)

62. Tylman, W.: Anomaly-based intrusion detection using bayesian networks. In: Third International Conference on Dependability of Computer Systems Szklarska, Poreba, Poland, pp. 211–218, 26–28 June 2008

63. Botha, M, Von, Solms, R.: Utilising fuzzy logic and trend analysis for effective intrusion detection. Comput. Secur. **22**(5), 423–434

64. Cha, B.R., Vaidya, B., Han, S.: Anomaly intrusion detection for system calls using the soundex algorithm and neural networks. In: 10th IEEE Symposium on Computers and Communications (ISCC'05), Cartagena, Spain, pp. 427–433. IEEE

65. Eskin, E., Arnold, A., Prerau, M., Portnoy, L., Stolfo, S.: A geometric framework for unsupervised anomaly detection. Detecting intrusions in unlabeled data. In: Proceedings of the Conference on Applications of Data Mining in Computer Security, pp. 78–100. Kluwer Academics

66. Fangfei, W., Qingshan, J., Lifei, C., Zhiling, H.: Clustering ensemble based on the fuzzy KNN algorithm. In: Eighth ACIS International Conference on Software Engineering, Artificial Intelligence, Networking, and Parallel/Distributed Computing (SNPD'07), Qingdao, July 30, 2007–Aug 1, 2007, vol 3, pp. 1001–1006 (2007)

67. Idris, NB., Shanmugam, B.: Artificial intelligence techniques applied to intrusion detection. In: IEEE India Conference Indicon (INDICON'05), Chennai, India, pp. 52–55, 11–13 Dec 2005

68. Mukkamala, S., Sung, AH., Abraham, A.: Intrusion detection using an ensemble of intelligent paradigms. J. Netw. Comput. Appl. **28**(2), 167–182

69. Agrawal, S., Agrawal, J.: Survey on anomaly detection using data mining techniques. In: Proceeding with 19th International Conference on Knowledge Based and Intelligent Information and Engineering Systems, vol. 60, pp. 708–713. Elsevier (2015)

70. Acharya, N., Singh, S.: An IWD-based feature selection method for intrusion detection system. Soft Computing, pp. 4407–4416, Springer (2017). https://doi.org/10.1007/s00500-017-2635-2

71. Shah-Hosseini, H.: Optimization with the nature-inspired intelligent water drops algorithm. Dos Santos, W.P. (ed.) Evolutionary computation. I-Tech, Vienna, pp. 298–320. ISBN 978–953-307-008-7

72. Malik, A.J., Khan, F.A.: A hybrid technique using binary particle swarm optimization and decision tree pruning for network intrusion detection. Cluster Computing, pp. 667–680, Springer (2017). https://doi.org/10.1007/s10586-017-0971-8

73. Saleh1, A.I., Talaat1, F.M.: A hybrid intrusion detection system (HIDS) based on prioritized k-nearest neighbors and optimized SVM classifiers. Artificial Intelligence Review, pp. 403–443. Springer (2017). https://doi.org/10.1007/s10462-017-9567-1

74. Dash, T.: A study on intrusion detection using neural networks trained with evolutionary algorithms. Soft Computing, pp. 2687–2700. Springer (2017). https://doi.org/10.1007/s00500-015-1967-z

75. Yao, H., Wang: An Intrusion Detection Framework Based on Hybrid Multi-Level Data Mining, International Journal of Parallel Programming, pp. 1–19. Springer (2017). https://doi.org/10.1007/s10766-017-0537-7

76. Alauthaman, M., Aslam, N.: A P2P Botnet detection scheme based on decision tree and adaptive multilayer neural networks. Neural Computing & Applications, pp. 991–1004. Springer (2018). https://doi.org/10.1007/s00521-016-2564-5.

77. Ijaz, S., Hashmi, F.A.: Vector based genetic algorithm to optimize predictive analysis in network security. Applied intelligence, vol. 48, issue 5, pp. 1086–1096. Springer (2018). https://doi.org/10.1007/s10489-017-1026-9

78. Mohammed, S., Mutaheb, F.: Intrusion detection model using machine learning algorithm on Big Data environment, proceedings. J. Big Data 1–12. Springer (2018). https://doi.org/10.1186/s40537-018-0145-4

79. Venkataraman, S., Selvaraj, R.: Optimal and Novel Hybrid Feature Selection Framework for Effective Data Classification, Proceedings with: Advances in Systems, Control and Application, pp. 499–514. Springer (2018). https://doi.org/10.1007/978-981-10-4762-6_48

80. Kumar, N., Kumar, U.: Knowledge Computational Intelligence in Network Intrusion Detection Systems, Knowledge Computing and Its Applications, pp.161–176. Springer (2018). https://doi.org/10.1007/978-981-10-6680-1_8

81. Saxena, A., Saxena, K.: Hybrid Technique Based on DBSCAN for Selection of Improved Features for Intrusion Detection System, Emerging Trends in Expert Applications and Security, pp. 365–377. Springer (2019). https://doi.org/10.1007/978-981-13-2285-3_43

82. Kambattan, R., Rajkumar, M.: An effective intrusion detection system using flawless feature selection, outlier detection and classification. Progress in Advanced Computing and Intelligent Engineering, pp. 203–213. Springer (2019). https://doi.org/10.1007/978-981-13-1708-8_19

83. Cavusoglu, U.: A new hybrid approach for intrusion detection using machine learning methods. Applied Intelligence, pp. 2735–2761. Springer 2019. https://doi.org/10.1007/s10489-018-01408-x

84. Kar, P., Banerjee, S., Mondal, K.C., Mahapatra G., Chattopadhyay S.: A hybrid intrusion detection system for hierarchical filtration of anomalies. Information and Communication Technology for Intelligent Systems, Smart Innovation Systems and Technologies, vol. 106, pp. 417–426. Springer (2019). https://doi.org/10.1007/978-981-13-1742-2_41

85. Mishra, S., Mahanty, C., Dash, S., Mishra, B.K.: Implementation of BFS-NB hybrid model in intrusion detection system, recent developments in machine learning and data analytics. Advances in Intelligent Systems and Computing, vol. 740, pp. 167–175. Springer (2019). https://doi.org/10.1007/978-981-13-1280-9_17

86. Al-Dhafian, B., Ahmad, I, Al-Ghamid, A.: An Overview of the Current Classification Techniques, International Conference on Security and Management, Las Vegas, USA, pp. 82–88, July 27–30

87. Revathi, S., Malathi, A.: A Detailed Analysis on NSL-KDD Dataset Using Various Machine Learning Techniques for Intrusion Detection. Int. J. Eng. Res. Technol. 2(12), 1848–1853

88. Tavallaee, M., Bagheri, E., Lu, W., Ghorbani, Ali, A.: A detailed analysis of the KDD Cup-'99 data set. In: Proceedings of the 2009 IEEE Symposium on Computational Intelligence in Security and Defense Applications, Ottwa, Canada, July 8–10

89. Kavitha, P., Usha, M.: Anomaly based intrusion detection in WLAN using discrimination algorithm combined with Naïve Bayesian classifier. J. Theor. Appl. Inf. Technol. 62(1), 77–84

90. Singh, R., Kumar, H., Singla, R.K.: An intrusion detection system using network traffic profiling and online sequential extreme learning machine. Expert Syst. Appl. 42(22), 8609–8624

91. Song, J., Takakura, H., Okabe, Y., Eto, M., Inoue, D., Nakao, K.: Statistical analysis of honeypot data and building of Kyoto 2006+ dataset for NIDS evaluation. In: Proceedings of the 1st Workshop on Building Anal. Datasets and Gathering Experience Returns for Security, Salzburg, pp. 29–36 (2006)

92. Hindy, H., Brosset, D.: A Taxonomy and Survey of Intrusion Detection System Design Techniques, Network Threats and Datasets, pp. 1–35, 9, June 2018. arXiv. 1806.03517v1 [cs.CR]

93. Fournier-Viger, P., Lin, C.W., Gomariz, A., Gueniche, T., Soltani, A., Deng, Z., Lam, H. T.: The SPMF open-source data mining library version 2. In: Proceedings of the 19th European Conference on Principles of Data Mining and Knowledge Discovery (PKDD 2016) Part III, pp. 36–40. Springer LNCS 9853 (2016). https://www.philippe-fournier-viger.com/spmf/

Indexing in Big Data Mining and Analytics

Ali Usman Abdullahi⑩, Rohiza Ahmad, and Nordin M. Zakaria

Abstract Big data analytics is one of the best ways of extracting values and benefits from the hugely accumulated data. The rate at which the global data is accumulating and the rapid and continuous interconnecting of people and devices is overwhelming. This further poses additional challenge to finding even faster techniques of analyzing and mining the big data despite the emergence of specific big data tools. Indexing and indexing data structures have played an important role in providing faster and improved ways of achieving data processing, mining and retrieval in relational database management systems. In doing so, index has aided in data mining by taking less time to process and retrieve data. The indexing techniques and data structures have the potential of bringing the same benefits to big data analytics if properly integrated into the big data analytical platforms. A lot of researches have been conducted in that direction, and this paper attempts to bring forward how the indexing techniques have been used to benefit the big data mining and analytics. Hence, this can bring the impact that indexing has on RDBMS to the folds of big data mining and analytics.

Keywords Big data · Big data analytics · Data mining · Indexing techniques · Index

A. U. Abdullahi (✉)
Computer Science Education Department, Federal College of Education (Tech), Gombe, Nigeria
e-mail: usmanali@fcetgombe.edu.ng
URL: http://www.fcetgombe.edu.ng

R. Ahmad · N. M. Zakaria
Computer and Information Sciences Department, Universiti Teknologi PETRONAS, Seri Iskandar, Perak, Malaysia
e-mail: rohizaahmad@utp.edu.my

N. M. Zakaria
e-mail: nordinzakaria@utp.edu.my

1 Introduction

The exponential growing nature of global data was collated and presented by the Statista in their report titled—Volume of data/information created worldwide from 2010 to 2025. The report indicated that the world's overall volume of both created and copied data as of that 2020 year would be 50.5ZB. Also, this volume is expected to rise three-fold within five years and is estimated to be 175 by 2025 [38]. This fact gives a sense of hugeness of data size that is termed as big data [9, 10]. This global data is accumulated from various forms, comprising structured, semi-structured and unstructured masses of data that need analysis so dearly. Since the data are collated and stored into datasets, then those enormous datasets are also referred to as big data/datasets [5, 34, 44].

According to Chen et al. [9, 10], the increase in volume indicates how big the generation, the collection and the scaling of data masses have become. While increase in velocity indicates the need for timely and rapid collection and analysis of big data in order to maximally utilize its commercial value, the increase in variety indicates that big data comprises various forms which may include unstructured and semi-structured, besides the usual structured data [4].

The IDC on its part presented a different opinion on big data. In its 2011 report, IDC viewed big data from its technological and architectural angle as the data is designed for extraction of economic value. The economic value comes from the very large volume and widely varied data, which have high velocity of discovery, capture and analysis [44].

The IDC definition added another 'V' to make it 4Vs model, with the fourth 'V' being value. Therefore, a broader definition of the big data could be derived from all earlier ones as the term is used in describing enormous datasets, whose contents are characterized by the 4Vs: (i) Volume—very large amount of data; (ii) Variety—different forms of the data, i.e., structured, semi-structured and unstructured gathered from different sources including images, documents and complex record; (iii) Velocity—the data has been constantly changing contents, which come from complementary data collection, archived and streamed data [6, 10, 45]; and (iv) Value—very huge value and very low density [10].

In addition, recent literatures include veracity as the fifth 'V', the characteristic of the big data after being convinced that none of the earlier described characteristics of big data have covered that. By veracity, it means that there are uncertainty and effect of accuracy on the quality of collected data [4, 5, 42, 44].

The value in big data is extracted by proper and efficient retrieval of the data from the big datasets. Thus, fast processing of the retrieved data is determinant to faster and timely analysis and access to the required data. Indexing is one of the most useful techniques for faster data retrieval during processing and accessing. Therefore, the industrial 4.0 data-driven processes are in need of such faster data retrieval. Once the retrieval and access processes involving big data usage are made faster, a lot of benefits are achieved. These benefits include energy/power saving and improving hardware durability and reduce heat generation.

1.1 Objective of the Chapter

The main objective of this chapter was to present indexing techniques that can be used in searching and effective retrieval of data during data mining. The strategy of all the indexing techniques is to restrict the amount of input data to be processed during the mining of any dataset. The chapter highlights those indexing techniques that have the potentials of working better with big data.

1.2 Taxonomy of the Chapter

The focus of the chapter is to include in it all relevant literatures that are searched and downloaded from the major publication databases. The databases include IEEEXplore, DBLP, Scopus, Springer and others. Then, titles, abstracts, introductions and conclusions of papers that covered application of indexing in big data mining and analytics were described. This was done using selection criteria in order to pick the papers that matched and have highlighted the structure of various indexing approaches used for big data mining and analytics. In addition, the chapter identifies the indexing approaches that have potential for big data mining and those that have less potential. Figure 1 depicts the diagrammatic sketch of the taxonomy used for the chapter.

The chapter's perspective is to center the discussion of literatures on the indexing approaches, their data structure, their mode of application, their pros and cons, and prospect for big data mining. In addition, the chapter attempts to cover indexing from its application on RDBMS up to it's use in big data mining and analytics. This may enable specialist and practitioners of big data mining and analytics to have wider view on the indexing approaches, and when, how and where to apply each of the approaches for better results.

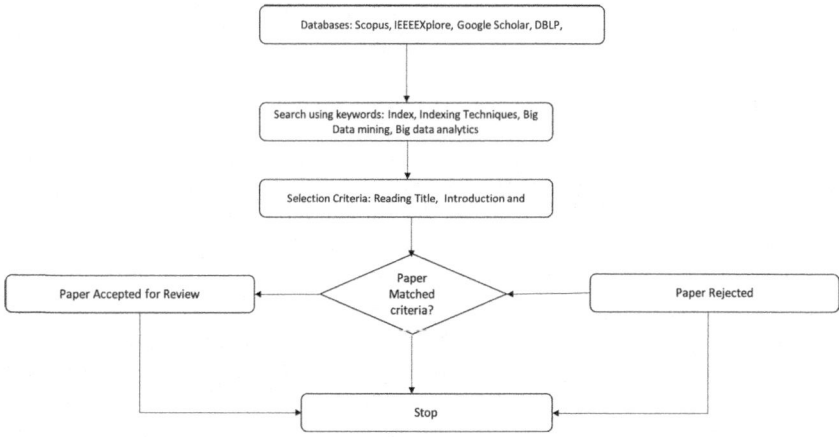

Fig. 1 A sketch of the taxonomy used for the chapter

The remaining part of the chapter is organized as follows. Sect. 2 presents the basic types of the indexing, Sect. 3 discusses the online indexing which is improvement of the basic indexing. Section 4 enumerates the indexes inbuilt to MapReduce process, which is a dedicated process of big data processing. Then, Sect. 5 presents the user-defined indexes, and finally, Sect. 6 concludes the paper.

2   Index and Indexing

Indexing, as information retrieval technique, is the process of generating all the suitable data structures that allow for efficient retrieval of stored information [35]. The term index refers to the suitable data structure needed, to allow for the efficient information retrieval [26]. Usually, the data structures used for indexing in most cases do not store the information itself. Rather, it uses other data structures and pointers to locate where the data is actually being stored. A meta index is also used to store additional information about stored data like the external name of a document, its length, which can be ranked as the output of the index.

In RDBMS, the index is a data structure that speed up the operation of data retrieval from database tables. The index comes with additional cost for writing/reading it into/from the index table. This leads to more storage usage in order for the extra copy of data created by the index to be maintained. The indexes are used in locating data quickly by going straight to the data location without having to search all the rows in the database table every time that table is accessed. Indexes are the bases for providing rapid random lookups and efficient access of ordered recorded.

While in big data, if index is to be implemented the same way as it works with the RDBMS, there is no doubt that it will be so big to the extent that the intended fast record retrieval may not be achieved. Therefore, there is a need for a closer look at the indexing technique in a way that will be reasonably small and maintain the targeted goal of faster record retrieval.

2.1   Index Architecture and Indexing Types

The index has two architectures: non-clustered and clustered index, as shown in Fig. 3. The non-clustered architecture presents data in an arbitrary order, but maintains a logical ordering in which rows may be spread out in a file/table without considering the indexed column expression. This architecture uses a tree-based indexing that has a sorted index keys and pointers to record at its leaf node level. The non-clustered index architecture is characterized by having the order of the physical rows of the indexed data differing with the order in the index [32]. The non-clustered index sketch is displayed in Fig. 3 (Fig. 2).

On the other hand, clustered index architecture changes the blocks of data into a certain distinct order to match the index. This results in the ordering of the row

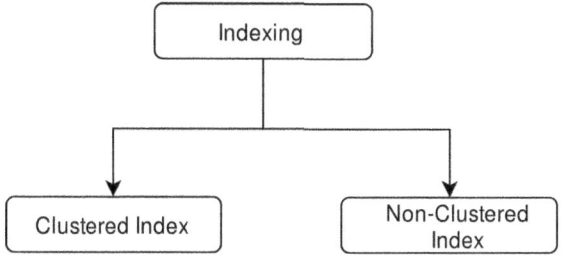

Fig. 2 A sketch of index types

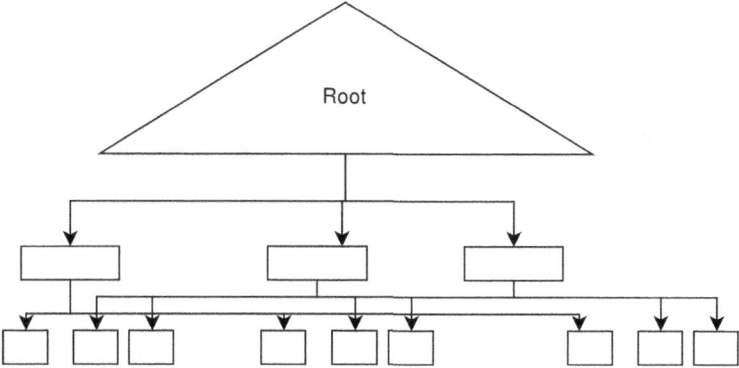

Fig. 3 A sketch of non-clustered index

data. It can greatly increase the overall speed of retrieval in a sequential accessed data or reverse order of the index or among a selected range of items. The major characteristic of the clustered index architecture is that the ordering of the physical data rows is in accordance with that of the index blocks that points to them [29]. The clustered index sketch is displayed in Fig. 4.

There are many types of indexing in existence, and below is the overview of some of them, as well as some of the studies involved each of them and their applicability in big data situation.

2.2 Bitmap Index

The bitmap index uses bit array called bitmaps to store the bulk of its data. Bitmap is a special type of index that uses bitwise logical operation on the bitmaps to answer most of the queries run against it. The bitmap index works basically in situations where index values are repeated very frequently unlike other index types commonly used. The other types are most efficient when indexed values are not repeated at all or they are repeated smaller number of times. The gender field of a database table is

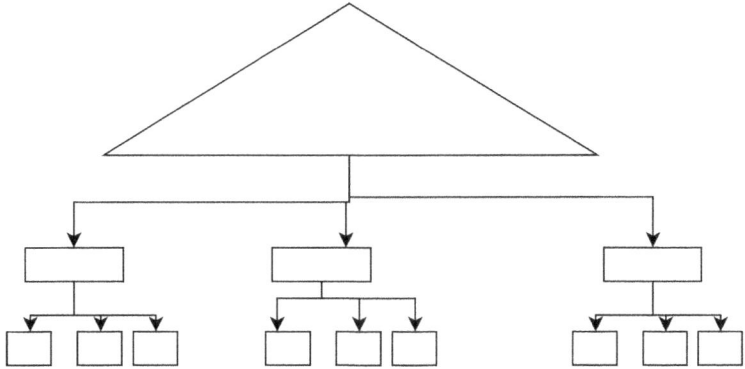

Fig. 4 A sketch of clustered index

a good example for a bitmap index. This is so because no matter how many tuples there are in a database table, the field will only have two possible values: male or female. A typical bitmap index data structure is shown in Fig. 5.

The bitmap index has been used in wide variety of areas and in big data analytics. For instance, bitmap index application was used in the aspect of index compression. The approach helps the compression to work better with high cardinality attribute data. Wu et al. [39] present a study and analysis of some of the compression techniques that use bitmap indexing, namely, byte-aligned bitmap compression (BBC) and word-aligned hybrid (WAH). These techniques were able to reduce compressed data sizes and improve their performance. The authors' motivation was the fact that most of the empirical researches do not include comparative analysis among these different techniques, and the result of their own work showed that compressed bitmap indexes appeared to be smaller in size compared to that of B+-Tree, with WAH occupying half of the space of B+-Tree while the BBC occupies half the space of WAH.

Table

Record	Gender
Row 1	M
Row 2	F
Row 3	M
Row 4	F
Row 5	M
Row 6	F
Row 7	M

Bitmap Index

Gender	Row 1	Row 2	Row 3	Row 4	Row 5	Row 6	Row 7
Male	1	0	1	0	1	0	1
Female	0	1	0	1	0	1	0

Fig. 5 A bitmap index form from gender table

Another work was done by Fusco et al. [13] using bitmap index as a compression approach to minimize CPU workload and consumption rate of disk. The platform used for the work was a streaming network data, which requires a real-time indexing. The authors introduced what they called COMPAX, a variant of compressed bitmap index that supersedes the word-aligned hybrid (WAH) in terms of throughput of indexing, shorter retrieval time and higher compression rate. The NETwork Flow Index (NET-FLI), which highly optimizes real-time indexing as well as data retrieval from larger-scale repositories of network, was used. The NET-FLI synergies COMPAX and locality-sensitive hashing (LSH) are used for streaming reordering in an online setup to achieve the target of the research. This combination results in higher insertion rates of up to 1 million flows per second many folds over what is obtainable in typical commercial network flow. The ISPB also allows the performance of complex analysis jobs by administrators.

In addition, an effort was made to automate indexing process as well as resolving the index selection problem (ISP) using bitmap by [27]. They came up with a technique that merges the features of index selection techniques (IST) and those of linear programming for optimization to minimize cost. The result was a new method that solves ISP externally and uses optimizer for choosing the set of indexers to be used. The clustering data mining technique was used and when benchmarked it outperformed Microsoft SQL Server Index Selection Tool (IST) in terms of speed of selection and suggestion of indexes. Even though the bitmap index was designed for RDBMS, it has also been used on some HLQLs on big data to improve information retrieval. For example, a group of students incorporated bitmap index into Hive to improve the retrieval of Facebook users information using gender field.

The bitmap technique as a candidate of indexing in big data has been tried, and in some specific situation it gives some improved results. However, the bitmap index does generate large volume of data as its data structure alongside the volume of the big data itself. Hence, it cannot be used for general data retrieval in big data because big data contains variety of data values.

2.3 Dense Index

The dense indexing approach uses a file/table with pair of keys and pointers to each record in the data file/table, which is sorted. Every key in the dense index is associated to a specific pointer to one of such records. If the underlining architecture of the indexes is a clustered one with duplicate keys, dense index just points to the first record with the said key [15]. Figure 6 displayed a sample of the dense index. In the dense index, each index entry consists of a search key and a pointer. The search key holds the value to be searched while the pointer stored the identifier to the disk location containing the corresponding record and the offset that identifies the point where the record starts within the block.

The dense index as an ordered index either stores index entry for all records when the approach it is using is non-clustered, or it just stores the index entry to the first

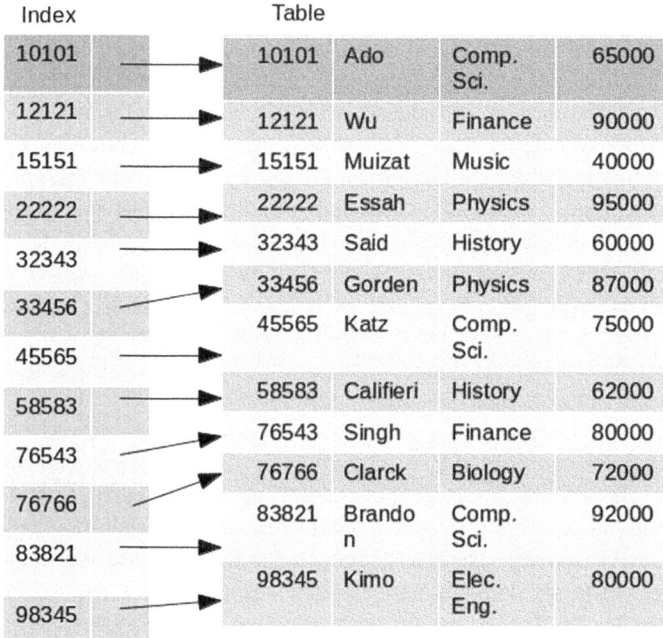

Fig. 6 A dense index on employee table

search key when using the clustered index approach [35]. The dense index uses file in storing its index data structure and there is a dedicated pointer to each record and the data has to be ordered. These two facts suggested that the index is going to grow so big that it will be close to the size to that of the stored data. Hence, there are no studies using dense index in big data retrieval.

2.4 Sparse Index

The sparse indexing method also uses a file/table that contains pair of search keys and pointers. The pointers are pointing to blocks instead of individual records in the data file/table, which is sorted in the order of the search keys. In sparse index, index entries appeared for some of search key values. An index entry is associated to a specific pointer to one of such blocks. If the underlining architecture of the indexes is a clustered one with duplicate key, then the index just points to the lowest search key in each block [15]. To locate any search key's record, an index entry is searched with the largest value that is less than or equal to the given search key value. Then, the searching starts at the record pointed to by the index entry and moves down until the target is found [32, 35]. Figure 7 depicts the explanation given above for sparse index.

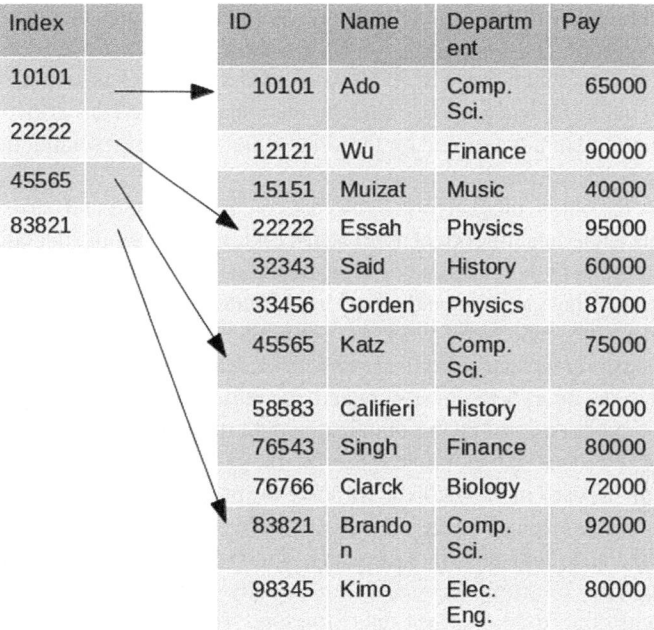

Fig. 7 A sparse index on employee table

The dense and sparse indexes are the most common types of ordered index that most relational databases use for generating query execution plans. However, for both the dense and sparse index types, the use of files/tables to keep the pairs of search keys and pointers may make them very unsuitable for big data indexing due to two reasons: (1) The records and block of data file for big data will be distributed over different clusters, which will make it very difficult to maintain such files. (2) The volume of the big data will make the size of such index files to be unnecessarily very large, which may lead to unreasonable costs of space and maintenance time.

3 Online Indexes

3.1 Online Indexing

Most techniques of indexing are having one drawback or the other. Thus, the big data indexing requires the study of other modified indexing techniques. Some of these modified indexes include the work of Chaudhuri et al. [8], which introduced the online indexing. The online index is an extension to the use of external tuning tools to optimize the physical design of a database through the analysis of representative workload set.

This technique works by monitoring the steps of an actual workload without knowing it upfront, and to create an index automatically when one of the query's execution plans is generated. The index is created at the background of a running workload in one complete go. The online indexing has lessened the work on the side of the database administrator and on the side of the system by avoiding the need of creating the index externally.

However, the online indexing usage has not covered the multi and partial indexes, which are necessary and important approaches for query processing in both OTP and OLAP. The results of comparison between the online indexing and the conventional indexing showed that the online index performed better than the conventional one due to the following: knowledge of workload is not required before creating the index. Since the index is created as side effect of query execution, its entries cover only what is specified in the query's predicates; hence, the reason for better performance. The online index can use any of the basic indexing data structure for its implementation, be it B-Tree or ordered tables-based index [8, 16, 21].

Amir et al. [2] used online indexing to solve the problem of managing unbounded length keys that are found in XLM paths, IP addresses, multi-dimensional points, multi-key data and multi-precision numbers. This type of data is categorized as big data. These types of string-based keys are usually atomic and indivisible; hence requires a customized comparison data structure. Their proposed online indexing happens to work with any data structure with a worst-case complexity of $O(\log n)$, which they reported to be the best based on their experiment. A suffix tree was cited as one of the application area of the proposed online indexing [2].

3.2 Database Cracking

Another effort in the direction of 'on-the-fly' indexing was made by Idreos et al. [21]. The authors introduced a combination of automatic index selection and partial indexes called database cracking. This approach uses the side effect of running current query and future ones to refine the structure of its index. The database cracking substituted the normal scanning of stored data for index creation and query processing, with pivoted partitioning of the data using the predicates from the query.

By doing that, one of the partitions will be containing only the tuples that answers the query. The underlining data structure used by the cracker index could be quick sort generated B-Tree or a hash, whose partitions expected to be beneficial to arriving queries.

In database cracking, no external index or external tuning tools are required, but just the monitoring of queries. The present and arriving queries are used to build and optimize the index. Database cracking proves to be advantageous compared to online indexing, and by extension the ordinary query processing. For the fact that the partitioning has to be carried out in small number for every predicate (fan-out of 2–3), it takes a lot of queries to get the index fully optimized [17, 18, 41].

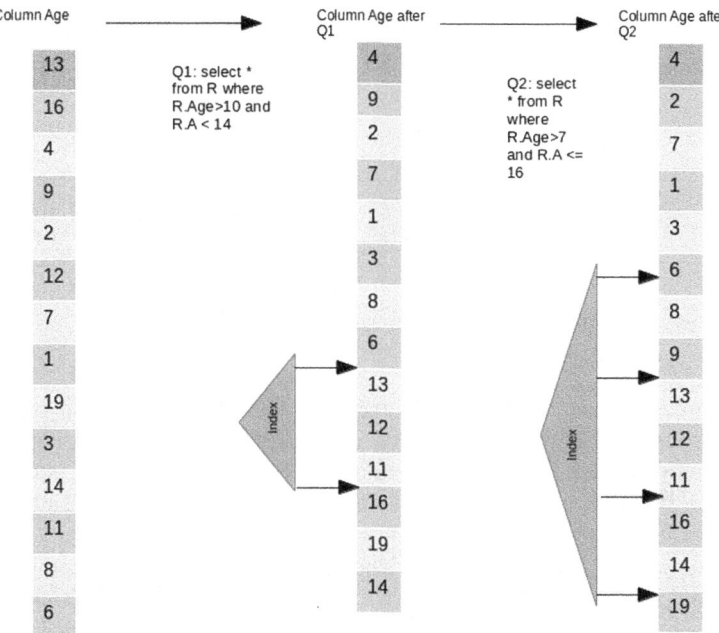

Fig. 8 An example of database cracking

Figure 8 displays an example of database cracking. The database cracking was implemented in one 'MonetDB' as a successful alternative to index scanning [7, 28].

However, the database cracking was observed to be CPU-intensive rather than I/O bound. Pirk et al. [28] proposed an enhancement that has taken the database cracking from being CPU bound to I/O bound. Input/output bound operation is best suited for big data processing. The authors used approaches such as predication, vectorization, data parallelism and CPU multi-thread on SIMD instructions to achieve the method. The results of their work showed that it is 25 times faster than the first database cracking.

3.3 Adaptive Merge

Graefe, Goetz, Kuno and Harumi [18] have introduced a modification of database cracking called adaptive merge. The adaptive merge also creates index by using the predicates specified in a query and used the partial index to answer that given query. The process also increments and optimizes the index with the help of the underlining data structure, the partitioned B-Tree. Adaptive merge uses merge sort for the index optimization and works based on 'monitor queries then build index' approach.

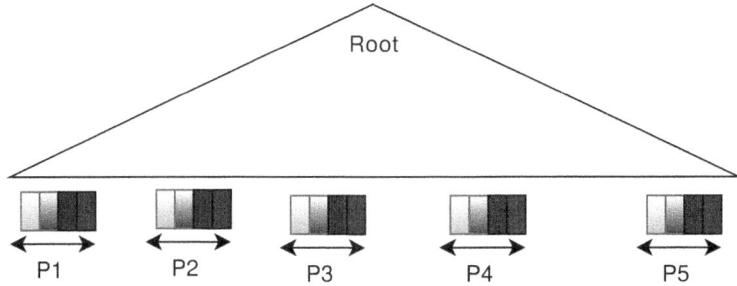

Fig. 9 An adaptive merge index is based of B-Tree

Unlike database cracking that uses partitioning of its data structure and works only on in-memory database, the adaptive merge works on external block access like flash memory and disc as well as in-memory. In terms of performance, the adaptive merge indexing optimizes with far less number of queries compared to database cracking. This is due to the fact that merge process's fan-in is unlimited as against limited fan-out of partitioning [18, 22]. Then Idreos et al. [22] made a further attempt to bring the two approaches into one, by replacing the index initial creation stage of the adaptive merge. The authors suggested a data structure that does not require much resource to build such as array, but optimizes by merging. This attempt was believed to work well with bug datasets. Figure 9 presents the partitioned B-Tree that was generated by adaptive merge index.

The last three indexing methods also referred to as oblivious indexes work by creating the index on-the-fly and automatically. The indexes use the concept of monitoring any issued query and then build the index as the side effect of the query's execution. However, all the above-discussed indexing techniques work with RDBMS and in an OLTP approach. The OLTP usually has short queries that are frequently posed to the database as the main operational system. So, there are number of arriving queries to increment and/or optimize an index if the need arises. On the contrary, the situation is different in the case of OLAP, especially in batch-oriented situations, which are the most common analysis when it comes to big data. Also, the mentioned cases are mostly different when MapReduce is to be used for such analysis.

The drawbacks of the RDBMS in processing big data imply that their corresponding index types have the same drawbacks. This prompted researchers of big data to customize and, in some cases, develop different indexing strategies for the big data analysis as mentioned earlier. These developed indexing strategies for big data information retrieval systems are being used by big data analytics.

3.4   Big Data Analytics Platforms

Upon the realization of the drawbacks of the traditional database systems, research efforts continue with more focus in the direction of fast processing of big data, especially with OLAP queries (big data analytics). The analytical challenge of big data indicates that it has outgrown that capacity of traditional DBMS resources. This further prompts for the development of new technologies to solve the challenge. As a result of the above, Dean and Ghemawat [11] had presented the notable new technologies of Google File System (GFS), which is a distributed file system used for data storage across clusters. The authors also introduced the Google MapReduce programming model, which is used for data analysis and indexing [9, 10, 14, 23].

The MapReduce is a programming paradigm that uses divide and conquer approach to problem-solving. MapReduce uses functional programming approach with only two functions: Map and Reduce. The functions are deployed to the chunks of data that are stored on the distributed file system. The tasks to be carried out by the functions are to be defined by the user and they are executed in parallel on the various blocks of data. The MapReduce works in a key/values filtering and aggregation manner to solve the problem in question [19, 23, 24, 36, 43].

The MapReduce remained the most powerful and the most accepted approach for the big data mining and analytics [24, 37]. One of the major reasons for its popularity among researchers and industry experts is its flexibility. Users can intuitively write code to solve virtually any problem. Besides that, MapReduce also supports very high parallel programming, even though at low level. MapReduce design nature was to eradicate input/output (I/O) problem associated to extract, load and transfer (ELT) [23].

The design migrates the computation to various computing units instead of moving data into memory for computation. Furthermore, the Hadoop MapReduce implementation remains the fastest, one reported, in handling very large data analysis. In addition, larger percentage of research works' experiment involving big data analytics that are found in the literatures are done using the Hadoop MapReduce or they are related to it or its associated tools. Lastly, reports and releases coming from the big IT companies indicate that most of them use tools that are supported by the Hadoop/MapReduce in performing their big data analysis [24, 31, 33, 37]. Table 1 highlights some of the features of three attempted solutions for big data analytics.

4   Inherent Indexes in MapReduce

A simple inverted index is said to be inherently and trivially implemented in MapReduce as one of the effective tasks for textual data retrieval. This is highlighted by Dean and Ghemawat [11] in their original paper on MapReduce and cited by Graef and kuno [18]. When performing inverted index with MapReduce, the map function parses the split covering certain input, and emits sequence of <key, value> pairs.

Table 1 Comparison between big data analytics approaches

S. no.	Big data analytic	Parallel programming approach	Dynamic flexibility	Schema less	SQL support	Extract, load and transfer
1	MapReduce	Yes	Yes	Yes	No	Not directly
2	Algebraic workflow [12]	Yes	No	Yes	Yes	No
3	AterixDB [1]	Yes	No	No SQL Like	Semi-structured	Yes

The reduce function accepts all pairs of the same key, sort the corresponding values and emits <key, list(value)> pair. The set of all output pairs form a simple inverted index. McCreadie [25] deducted that two interpretations of the above scenario can be a per-token indexing or a per-term indexing in relation to the indexing of corpus datasets.

The per-token indexing strategy involves emitting of <term, doc-ID> pairs for each token in a document by the map function. However, the reduce function does the aggregation of each unique term with its corresponding doc-ID to obtain the term frequencies (tf), after which the completed posting list for that term is written to disk. So, if a term appears tf times it will be indicated as such. The advantage of this strategy is that it makes the map phase very simple. However, it has the potential of costing more memory and time due to storage of large intermediate results, network traffic and prolonging sort phase (Fig. 10).

On the other hand, the per-term indexing uses the map function to emit tuple in the form <term, (doc-ID, tf)>, and this reduces the number of emit operation as only unique term per document is emitted. The reduce function in this inter-operation only sorts the instances by document to obtain the final posting list sorted by ascending doc-ID. Ivory information retrieval system uses this approach. Also, a combiner function can be used to generate tfs by performing a localized merge on each map task's output (Fig. 11).

4.1   Per Document Indexing

This is the inverted indexing technique used by Nutch platform on the Hadoop to index document for faster search. Nutch tokenized the document during map phase and the map function emits tuples in the form of <document, doc-ID >, while the reduce phase writes all index structures. Though this strategy emits less, the value of each emit used to have more data and have reduced intermediate results, thus achieving higher levels of compression than single terms. Documents are indexed on same reduce task easily due to the sorting of document names [25].

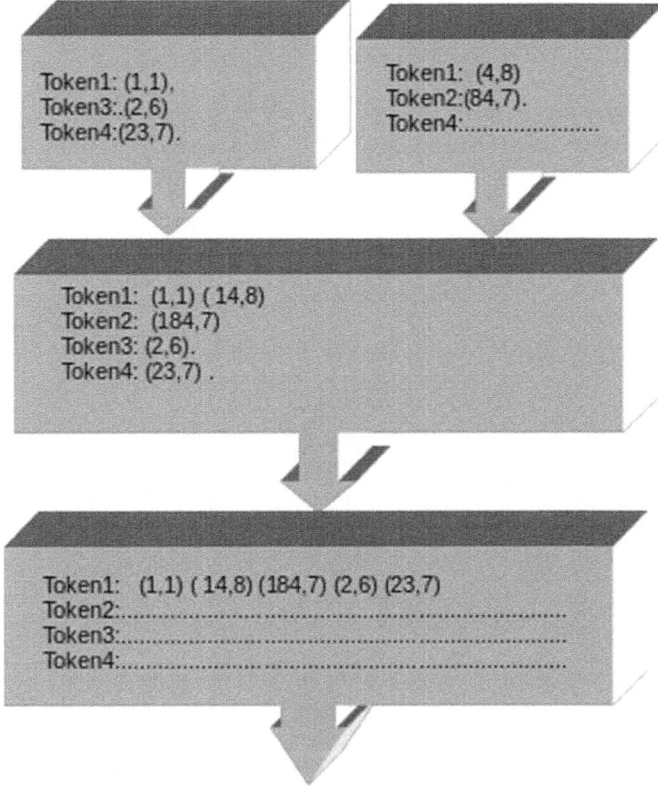

Fig. 10 A sample of per token index using MapReduce

4.2 Per-Posting List Indexing

This indexing technique is based on a single-pass indexing, which splits data onto multiple map tasks, with each operating on its own data sub-set. The map task serves as the scanning phase of the single-pass indexing. As this process run on the document, compressed posting lists are built in memory for each term. This partial index is flushed from the map task when the memory run low or when all the documents are processed. The flushing is done by emitting a set in the form of <term, posting list> pairs for all terms present in the memory. Before taking up of intermediate results by reduce task, the flushed partial indexes are sorted and stored on disk first by map number and then by flushed numbers. In order to achieve globally correct ordering of posting list for each term, the posting lists are merged by map number and flushed number. The term, posting lists, is merged together by the reduce

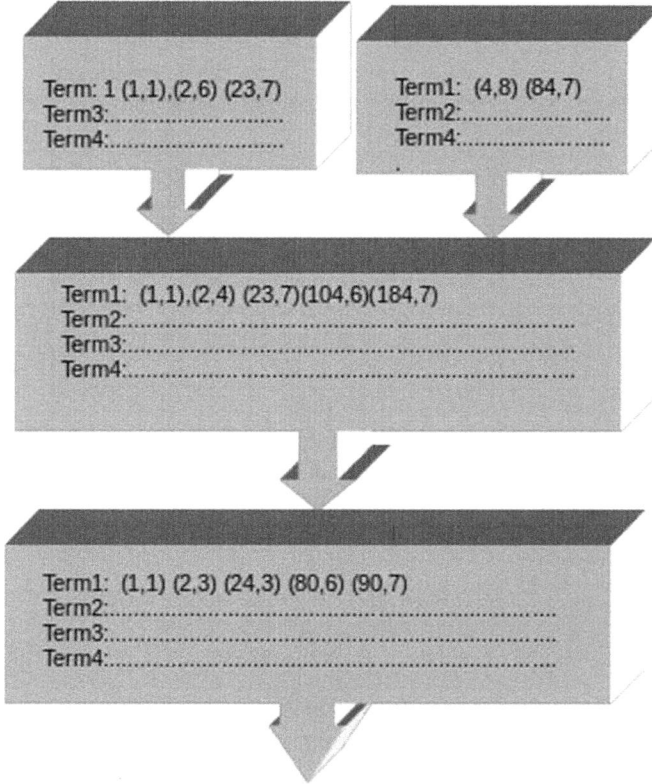

Fig. 11 A sample of per term index using MapReduce

function to form the standard index comprising full posting lists. The standard index is compressed using EliasGamma technique by storing only the distance doc-IDs [25] (Fig. 12).

All of the four strategies of inverted indexing in MapReduce discussed above are the main task focused on and carried out by the MapReduce job. This is against the primary function of indexing in RDBMS, which is to speed up access of stored data in order to improve the performance of other processes. Thus, the aim of indexing is not only to use but also to improve parallel processing of the document contents. Rather, the primary aim of indexing is to improve the performance of parallel processor itself. This improvement is to be achieved in addition to the underlining parallel processor that MapReduce is programmed to accomplish. Thus, there is a need for additional indexing scheme that works with the parallel processor and serves the same purpose with what index does in RDBMS.

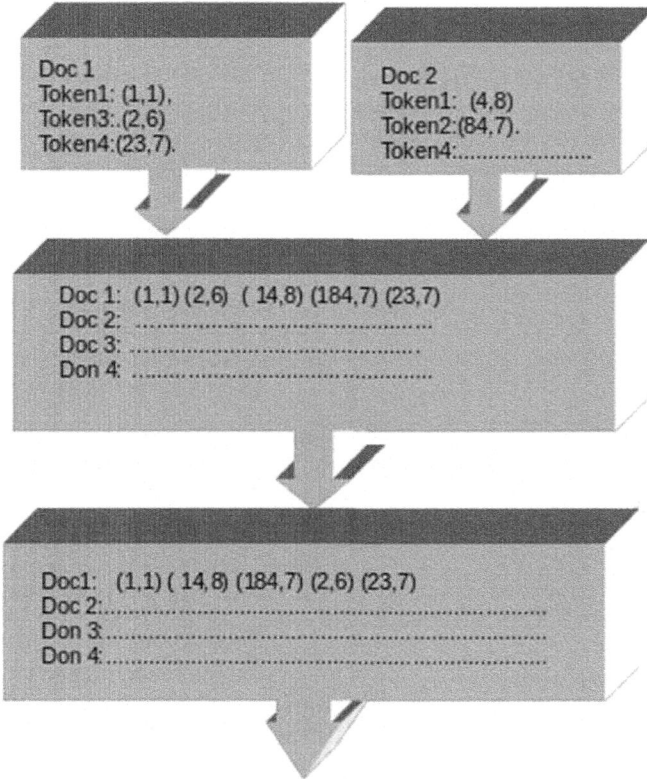

Fig. 12 A sample of per document index using MapReduce

5 User-Defined Indexing in MapReduce

For the user-defined indexing used in MapReduce and big data analytics, Yang and Parker [40] have employed HDFS's file component as B-Tree nodes to achieve indexing. In their approach each file contains data and pointer to lower files in the tree hierarchy, which is considered its children. During query processing the tree is traversed to locate the required segment of data to be processed using an improved Map-Reduce-Merge-Traverse version of MapReduce. After locating the data, then the map, the reduce and the merge tasks are performed on it to return the record set that answers the given query.

Also, An et al. [3] have used blockIds from the HDFS as the search keys of their B+-Tree-based index. When a given query is to be processed, the B+-Tree-based index is first searched to determine the start and the end of contiguous blocks that formed the index, and the result of the search formed the input data to be scanned. Then, only the blockIds that are returned from such search are used by MapReduce for main query processing. Hence, through this process preventing the full scan of the input data is done. In addition, Richter et al. [30] used the copies of replica stored by HDFS, to index different data attributes, which may likely be used as incoming query's predicates. When a MapReduce query arrives, their library checks the fields contained in the query's predicates and used the clustered index built on that field to return the blockIds of the data required to answer the given query.

Furthermore, in all the mentioned studies, the authors used indexing data structure that scale logarithmically, thereby improving data processing and retrieval. This is done by preventing the MapReduce from full scan of input data by guiding the process to just scanning and processing the data that corresponds to the output of the indexes. Moreover, many other researches were conducted on indexing using different types of big data; however, those researches are closely tied to that type of big data as the indexing data structure and the index implementations are determined by the nature of the data itself [20]. Table 2 displays the summary of the index approaches, their memory requirement and big data potentials.

Table 2 Comparison between big data analytics approaches

Reference	Indexing approach	Data structure	Memory/storage	Potential for big data mining
Wu et al. [39], Fusco et al. [13]	Bitmap	Tabular	Requires large space of memory and storage	Has good potential for big data mining
Gracia et al. [15], and Silberschez et al. [35]	Dense	Tabular	Requires large space of memory and storage	Not a good approach for big data mining
Gracia et al. [15], Rys [32] and Silberschez et al. [35]	Sparse	Tabular	Requires large space of memory and storage	Not a good approach for big data mining
Chaudhuri et al. [8]	Online indexing	Vectors	Requires small space of memory and storage	A good approach for big data mining
Idreos et al. [21]	Database cracking	Arrays	Requires small space of memory and storage	A good approach for big data mining
Graefe et al. [18]	Adaptive merge	Tree-based	Requires average space of memory and storage	A very good approach for big data mining
Yang and Parker [40]	B-Tree	Tree-based	Requires average space of memory and storage	A good approach for big data mining
An et al. [3]	B+-Tree	Tree-based	Requires small space for of memory and storage	A good approach for big data mining
Richter et al. [30]	HAIL	File-based	Requires large space of memory and storage	A good approach for big data mining

6 Conclusion

Moreover, it can be simply deducted from the above reviews that user-defined index and content indexing as tool for optimizing the performance of information retrieval has been successful. It can also be added that there is high potential for improving big data analytics through the use of advanced indexing techniques and data structures. Particularly, if these techniques and data structures are hybridized, improved and customized to work with MapReduce, they will surely improve the performance of big data analytics.

It has been highlighted that MapReduce is one of the most popular tools for big data analytics. However, the low-level nature of its implementations has given rise to the development of HLQLs. The HLQLs ease the programmer's task of handling the analysis. The review also highlighted the different types of index in both RDBMS and those used in big data analytics using different big data analytics platforms, including MapReduce and its index approaches. The improvement can be achieved by changing the indexing approach or using more efficient data structure.

References

1. Alsubaiee, S., Altowim, Y., Altwaijry, H., Behm, A., Borkar, V., Bu, Y., Carey, M., Cetindil, I., Cheelangi, M., Faraaz, K., et al.: Asterixdb: a scalable, open source bdms. Proc. VLDB Endow. **7**(14), 1905–1916 (2014)
2. Amir, A., Franceschini, G., Grossi, R., Kopelowitz, T., Lewenstein, M., Lewenstein, N.: Managing unbounded-length keys in comparison-driven data structures with applications to online indexing. SIAM J. Comput. **43**(4), 1396–1416 (2014)
3. An, M., Wang, Y., Wang, W.: Using index in the mapreduce framework. In: Web Conference (APWEB), 2010 12th International Asia-Pacific, pp. 52–58. IEEE (2010)
4. Bachlechner, D., Leimbach, T.: Big data challenges: Impact, potential responsesand research needs. In: IEEE International Conference on Emerging Technologies and Innovative Business Practices for the Transformation of Societies (EmergiTech), pp. 257–264. IEEE (2016)
5. Bajaber, F., Elshawi, R., Batarfi, O., Altalhi, A., Barnawi, A., Sakr, S.: Big data 2.0 processing systems: Taxonomy and open challenges. J. Grid Comput. **14**(3), 379–405 (2016)
6. Berman, J.J.: Principles of big data: preparing, sharing, and analyzing complexinformation. Newnes (2013)
7. Boncz, P.A., Zukowski, M., Nes, N.: Monetdb/x100: hyper-pipelining query execution. Cidr. **5**, 225–237 (2005)
8. Chaudhuri, S., Narasayya, V.: Self-tuning database systems: a decade of progress. In: Proceedings of the 33rd International Conference on Very Large Data Bases, pp. 3–14. VLDB Endowment (2007)
9. Chen, C.P., Zhang, C.Y.: Data-intensive applications, challenges, techniques andtechnologies: a survey on big data. Inf. Sci. **275**, 314–347 (2014)
10. Chen, M., Mao, S., Liu, Y.: Big data: a survey. Mob. Netw. Appl. **19**(2), 171–209 (2014)
11. Dean, J., Ghemawat, S.: Mapreduce: a flexible data processing tool. Commun. ACM **53**(1), 72–77 (2010)
12. Dias, J., Ogasawara, E., de Oliveira, D., Porto, F., Valduriez, P., Mattoso, M.: Algebraic data flows for big data analysis. In: 2013 IEEE International Conference on Big Data, pp. 150–155. IEEE (2013)

13. Fusco, F., Vlachos, M., Stoecklin, M.P.: Real-time creation of bitmap indexes onstreaming network data. VLDB J. Int. J. Very Large Data Bases **21**(3), 287–307 (2012)
14. Gani, A., Siddiqa, A., Shamshirband, S., Hanum, F.: A survey on indexing techniques for big data: taxonomy and performance evaluation. Knowl. Inf. Syst. **46**(2), 241–284 (2016)
15. Garcia-Molina, H., Ullman, J.D., Widom, J.: Database System Implementation, vol. 654, 2nd ed edn. Prentice Hall Upper Saddle River, NJ (2014)
16. Glombiewski, N., Seeger, B., Graefe, G.: Waves of misery after index creation.BTW 2019 (2019)
17. Graefe, G., Idreos, S., Kuno, H., Manegold, S.: Benchmarking Adaptive Indexing, pp. 169–184. Springer (2011)
18. Graefe, G., Kuno, H.: Self-selecting, self-tuning, incrementally optimized indexes. In: Proceedings of the 13th International Conference on Extending Database Technology, pp. 371–381. ACM (2010)
19. Hong, Z., Xiao-Ming, W., Jie, C., Yan-Hong, M., Yi-Rong, G., Min, W.: A optimized model for mapreduce based on hadoop. TELKOMNIKA (Telecommunication Computing Electronics and Control) **14**(4) (2016)
20. Ibrahim, H., Sani, N.F.M., Yaakob, R., et al.: Analyses of indexing techniques onuncertain data with high dimensionality. IEEE Access **8**, 74101–74117 (2020)
21. Idreos, S., Kersten, M.L., Manegold, S.: Database cracking. In: CIDR, vol. 7, pp. 7–10 (2017)
22. Idreos, S., Manegold, S., Kuno, H., Graefe, G.: Merging what's cracked, crackingwhat's merged: adaptive indexing in main-memory column-stores. Proc. VLDB Endow. **4**(9), 586–597 (2011)
23. Khasawneh, T.N., AL-Sahlee, M.H., Safia, A.A.: Sql, newsql, and nosql databases: A comparative survey. In: 2020 11th International Conference on Information and Communication Systems (ICICS), pp. 013–021 (2020)
24. Lee, S., Jo, J.Y., Kim, Y.: Performance improvement of mapreduce process bypromoting deep data locality. In: 2016 IEEE International Conference on Data Science and Advanced Analytics (DSAA), pp. 292–301. IEEE (2016)
25. McCreadie, R., Macdonald, C., Ounis, I.: On single-pass indexing with mapreduce.In: Proceedings of the 32nd international ACM SIGIR conference on Research and development in information retrieval. pp. 742–743. ACM (2009)
26. McCreadie, R., Macdonald, C., Ounis, I.: Mapreduce indexing strategies: Studyingscalability and efficiency. Inf. Process. Manage. **48**(5), 873–888 (2012)
27. Nang, J., Park, J.: An efficient indexing structure for content based multimediaretrieval with relevance feedback. In: Proceedings of the 2007 ACM symposium on Applied computing, pp. 517–524. ACM (2007)
28. Pirk, H., Petraki, E., Idreos, S., Manegold, S., Kersten, M.: Database cracking: fancy scan, not poor man's sort! In: Proceedings of the Tenth International Workshop on Data Management on New Hardware, p. 4. ACM (2014)
29. Ramakrishnan, R., Gehrke, J., Gehrke, J.: Database Management Systems, vol. 3. McGraw-Hill New York (2010)
30. Richter, S., Quian´e-Ruiz, J.A., Schuh, S., Dittrich, J.: Towards zero-overhead staticand adaptive indexing in hadoop. VLDB J. **23**(3), 469–494 (2014)
31. Roy, S., Mitra, R.: A survey of data structures and algorithms used in the contextof compression upon biological sequence. Sustain. Humanosphere **16**(1), 1951–1963 (2020)
32. Rys, M.: Xml and relational database management systems: inside microsoft sqlserver 2005. In: Proceedings of the 2005 ACM SIGMOD International Conference on Management of Data, pp. 958–962. ACM (2005)
33. Sevugan, P., Shankar, K.: Spatial data indexing and query processing in geocloud. J. Test. Eval. **47**(6) (2019)
34. Shireesha, R., Bhutada, S.: A study of tools, techniques, and trends for big dataanalytics. IJACTA **4**(1), 152–158 (2016)
35. Silberschatz, A., Korth, H.F., Sudarshan, S., et al.: Database System Concepts, vol. 4. McGraw-Hill New York (1997)

36. Silva, Y.N., Almeida, I., Queiroz, M.: Sql: From traditional databases to big data. In: Proceedings of the 47th ACM Technical Symposium on Computing Science Education, pp. 413–418. ACM (2016)
37. Sozykin, A., Epanchintsev, T.: Mipr-a framework for distributed image processing using hadoop. In: 2015 9th International Conference on Application of Information and Communication Technologies (AICT), pp. 35–39. IEEE (2015)
38. Statista: Volume of data worldwide from 2010–2025. https://www.statista.com/statistics/871 513/worldwide-data-created/ (2020)
39. Wu, K., Otoo, E., Shoshani, A.: On the performance of bitmap indices for highcardinality attributes. In: Proceedings of the Thirtieth international conference on Very large data bases-Volume 30. pp. 24–35. VLDB Endowment (2004)
40. Yang, H.C., Parker, D.S.: Traverse: simplified indexing on large map-reduce-mergeclusters. In: International Conference on Database Systems for Advanced Applications. pp. 308–322. Springer (2009)
41. Ydraios, E., et al.: Database cracking: towards auto-tunning database kernels. SIKS (2010))
42. Zakir, J., Seymour, T., Berg, K.: Big data analytics. Issues Inf. Syst. **16**(2), 81–90 (2015)
43. Zhang, Q., He, A., Liu, C., Lo, E.: Closest interval join using mapreduce. In: 2016 IEEE International Conference on Data Science and Advanced Analytics (DSAA), pp. 302–311. IEEE (2016)
44. Zhang, Y., Ren, J., Liu, J., Xu, C., Guo, H., Liu, Y.: A survey on emerging computing paradigms for big data. Chinese J. Electron. **26**(1) (2017)
45. Zikopoulos, P., Eaton, C.: Understanding big data: Analytics for enterprise classhadoop and streaming data. McGraw-Hill Osborne Media (2011)

Two-Steps Wrapper-Based Feature Selection in Classification: A Comparison Between Continuous and Binary Variants of Cuckoo Optimisation Algorithm

Ali Muhammad Usman, Umi Kalsom Yusof, and Syibrah Naim

Abstract Feature selection (FS) is the process of eliminating irrelevant features and improving classification performance while maintaining the standard of the data. Based on evaluation criteria, FS can be either filter or wrapper. Wrappers are computationally expensive due to many feature interactions in the search space. In this study, cuckoo optimisation algorithm (COA) along with its binary (BCOA) are used as wrapper-based FS for the first time to explore the promising regions in the search space, which obtained improved classification accuracy and selected better quality subsets of features within a shorter time. Based on that we developed two different fitness functions. The first one (BCOA-FS) and (BCOA-FS) adopt the standard wrapper-based evaluation with emphasis mainly on the classification performance. Whereas in the second one (BCOA-2S) and (COA2S) combine the first one in another evaluation process with a focus on both the number of features and classification accuracy. The results obtained indicate that COA-FS and BCOA-FS can select fewer features with better accuracy on both categorical and continuous label data, with BCOA-FS better than COA-FS. Similarly, COA-2S performed better than BCOA-FS and COA-2S and is comparable to the existing works. BCOA-2S outperformed the three of the existing studies on the majority of the datasets with almost 10 and 5% on both classification accuracy and number of selected features, respectively.

Keywords Feature selection · Wrapper-based · Cuckoo optimisation algorithm · Binary Cuckoo optimisation algorithm and Classification

A. M. Usman · U. K. Yusof (✉)
School of Computer Sciences, Universiti Sains Malaysia, 11800 Pulau Penang, Malaysia
e-mail: umiyusof@usm.my

A. M. Usman
e-mail: alimuhammad@fcetgombe.edu.ngs

A. M. Usman
Department of Computer Sciences, Federal College of Education (Technical), P.M.B 60, Gombe, Nigeria

S. Naim
Technology Department, Endicott College of International Studies (ECIS), Woosong University, Daejeon, Korea
e-mail: syibrah@wsu.ac.kr

© The Author(s), under exclusive license to Springer Nature Switzerland AG 2021
H. Chiroma et al. (eds.), *Machine Learning and Data Mining for Emerging Trend in Cyber Dynamics*, https://doi.org/10.1007/978-3-030-66288-2_6

1 Introduction

In classification, feature selection (FS) is mainly used to minimise the features in a data set while maintaining its standard [23, 68]. The aim is to select the most relevant subsets that are sufficient enough to describe the target class [62]. FS can be supervised [51], non-supervised [12] and semi-supervised [61]. In dealing with the supervised type, the class label of the data set is defined already in contrast with the non-supervised that the class label is unaware. Whereas, semi-supervised combine both supervised and non-supervised (i.e. with both label and unlabelled data) [32].

The supervised FS is classified further as a filter, wrapper and hybrid approach depending on their evaluation criteria [64]. In the filter-based approach, the features are evaluated without considering any classification algorithm, which makes them computationally fast. However, the filter ignores feature dependence or relationship among selected or ranked features, which subsequently affects the classification performance (i.e. either error rate or accuracy) [15].

In the wrapper-based method, a classification algorithm is used to evaluate each subset of features selected, and hence, it achieves better classification accuracy or error rate [26, 32, 64]. The major shortcomings of the wrapper-based approach are computationally expensive and not favourable on high-dimensional data sets. The most common examples of the wrapper-based approach are sequential forward selection (SFS) [60], sequential backward selection (SBS) [36] plus q take away r [16] and genetic algorithm (GA) [37] among others.

In FS, the classification performance of any of the approach (filter or wrapper) is measured in according to feature size (number of selected features), error rate (or accuracy) and computational time. Machine learning algorithms are commonly used to measure or evaluate the goodness of the selected subset of features in terms of accuracy or error rates [64]. Examples of the most widely used ones include support vector machine (SVM) [57], K-nearest neighbour (KNN) [24] and Gaussian Naïve Bayes (GNB) [34] among others.

The wrapper-based approach requires one to determine a classification algorithm and uses its performance as the evaluation standard. It searches for features that are suitable to the machine learning algorithm that increases the accuracy [39, 40]. Classification accuracy, as well as the selected subsets of features, are used to determine the prediction performance in wrapper model [54, 66]. As such, prediction accuracy and less prone to local optima are the critical advantage of wrapper over the filter. Hence, the outcomes are mostly more encouraging than the findings of the filter models.

However, the shortcomings of the wrapper-based approach include a high risk of overfitting data, classifier dependency, highly computationally intensive and not favourable for astronomical dimensional data [41]. Examples of wrapper techniques are a sequential forward selection (SFS), sequential backward elimination (SBE), plus q take away r and beam search. Others include simulated annealing, randomised hill-climbing, genetic algorithm and estimation of distribution algorithm among others [40, 46]—the steps on how the wrapper-based approach is illustrated in Fig. 1.

Fig. 1 Wrapper-Based
feature selection procedure

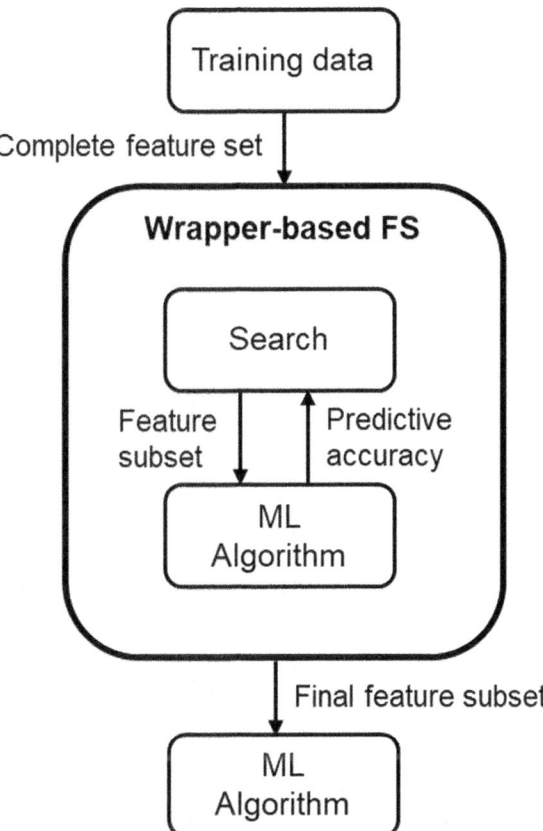

Finding the optimal subsets of features with the less computational cost is quite a demanding task because of the search space, and the number of features needed to search in the solutions are too many. Hence, FS is considered as an NP-hard problem [33, 39, 52, 64]. In searching for the best subsets of features, [26, 39] identified three search strategies: complete search, sequential or heuristic search and the random search.

The complete search works by finding all the possible feature subsets while evaluating them one after the other to select the best subset of features with the highest classification performance. There is a guarantee of getting the optimal results based on the laid down criteria in it. However, for a data set with N number of features, there will be 2^N subsets to be generated and evaluated, which is almost impractical for a considerable value of N. More so, [41] argued that 'search is complete does not mean that it must be exhaustive'.

The heuristic or sequential search add or remove features in sequential order; the remaining features that are not selected are considered later for selection in that manner. By doing so, the choice of the features may likely end up with the same pattern as complete search. However, there is no guarantee of finding the target solution [52].

Random search is the most popularly used among all the search strategies [42]. It aims at creating stability between the heuristic and complete search by combining the advantages of both. It begins with a randomly selected subset of features and progress in two ways. Either to follow the heuristic search and insert some randomness or to generate the next subset in a completely random manner [39, 42].

Out of all these search strategies, the random search is the only one that can escape from local optimum in the vast search space due to the involvement of randomness and mostly finish within the shortest time [26, 40, 64].

There are some works on the wrapper-based FS that applies different search strategies on different meta-heuristic algorithms. For example, [9, 10] used an artificial immune system for the FS. Also a particle swarm optimisation (PSO) is reported in [31, 44, 52–54, 63, 65].

Recently, differential evolution (DE) in [27], cuckoo search in [13, 18], grasshopper optimisation algorithm in [43], genetic algorithm (GA) in [3, 19, 30, 59, 66] are reported. In addition to that, genetic programming is used for the wrapper-based FS in [48], and recently a flower pollination algorithm for FS as also in [56]. It clearly shows that the meta-heuristic algorithms are suitable for addressing these problems. Despite the attempt to solve the lingering issues of the wrapper-based FS still, the existing works cannot successfully evolve the best subset of features with improving accuracy on some of the data sets [23].

The cuckoo optimisation algorithm (COA) presented in [49] is among the evolutionary algorithms that show promising results in handling different combinatorial optimisation problem including NP-hard, despite its proven records, especially in dealing with filter-based FS in [55]. Its application, specifically for the wrapper-based FS, is not fully investigated.

This study aimed to find the best subsets of features with lesser feature size and yet maintain the same or even better classification accuracy compared to using full-length features within a short period. Also, investigate the difference between COA and BCOA in wrapper-based FS.

To accomplish this goal, a pair of two FS frameworks are developed based on BCOA and COA. These proposed algorithms were studied and compared with other FS algorithms presented in other works on benchmark problems of varying difficulties.

Precisely, this study will examine

1. Whether adopted COA wrapper-based FS algorithms would choose the best subsets of feature, that has least feature size, less computational and accomplish the best error rate compared to full-length features, and would outpace the adopted BCOA wrapper-based single objective algorithms;
2. Whether adopted BCOA wrapper-based FS algorithms would choose the best subsets of feature and can attain the best performance than the adopted COA wrapper-based algorithms above;
3. Whether COA wrapper-based algorithm with two steps evaluation would choose sets of best features subsets and would outpace the two steps BCOA wrapper-based algorithm, and other existing works; and

4. Whether BCAO wrapper-based algorithm with two steps evaluation would choose sets of best features subsets and would outpace all other approaches stated directly above.

The rest of the paper is prearranged as follows: Part 2 is the background containing the details about the adopted COA and BCOA along with related works. The proposed wrapper-based feature selection approaches are presented in Part 3, while Part 4 is the experimental design, data sets used along with benchmark approaches. Then Part 5 is the presentation of the results while Part 6 concludes the entire work and suggests future work areas.

2 Background

2.1 Cuckoo Optimisation Algorithm

The original Cuckoo Optimisation Algorithm (COA) is strictly made for a continuous optimisation problem. At the same time, the binary version (BCOA) can be applied to solve problems that are in binary or discrete form. COA used for FS is very scarce in the literature. The size or dimension of the search space (i.e. the full-length features in every data set) is n. Every habitat in the COA is assigned by using a vector of n decimal numbers. The location of habitat i in dth length is xid normally in the range [0, 1]. To know in case if a feature is selected or otherwise, a verge $0 < \theta < 1$ is mandatory to equate it with the decimal numbers in the habitat position. If eventually, $xid > \theta$, then feature d is chosen else d is not be chosen.

COA developed by [49] is adopted, and the detail of how it works is

1. An array called "habitat" is used for the optimisation problem as show in Eq. 1.

$$habitat = [x_1, x_2, \ldots, x_{Nvar}] \tag{1}$$

2. Five and twenty eggs are used as the lower and upper limits, respectively, for every iteration.
3. They lay their eggs within a maximum range distance from their habitat in Equation.

$$ELR = \alpha \times \frac{number\ of\ current\ cuckoos}{total\ number\ of\ eggs} \times e_{new} \tag{2}$$

An α represent an integer number.
4. P% (those without any profit value) of the laid eggs are killed.
5. A k-means (K = 3 or 5) clustering is used for the grouping.
6. All cuckoos deviate φ radians while flying λ% to the goal, as shown in Eq. 3.

$$\lambda \sim U(0, 1) \qquad \varphi \sim (-\omega, \omega) \tag{3}$$

where $\lambda\ U(0, 1)$ means that λ is a uniformly distributed random within range of 0 and 1. ω is limits an aberration from goal habitat.

2.2   *Binary Cuckoo Optimisation Algorithm*

Binary Cuckoo Optimisation Algorithm (BCOA) is mostly used to solve FS problem; meanwhile, the representation of the habitat is in the form of a binary string, where the position of every habitat is a boolean 1 which signifies that a feature is chosen and 0 otherwise. Assuming X_G and X_C represent the respective goal and current habitat. Then, Eq. 4 computes the X_{NH} next habitat as follows:

$$X_{NH} = X_C + rand(X_G - X_C) \tag{4}$$

A sigmoid function is applied in Eq. 5 to use X_{NH} as binary to record it within [0, 1]. Then Eq. 6 alters the values to either 0 or 1.

$$S = \frac{1}{(1 + e^{-X_{NH}})} \tag{5}$$

$$IF\ (S > rand)\ THEN\ X_{NH} = 1\ AND\ IF\ (S < rand)\ THEN\ X_{NH} = 0 \tag{6}$$

2.3   *Related Works*

This part reviews some related works on wrapper-based FS. Both the traditional and meta-heuristic ones, as shown in the subsequent parts. However, this study focuses mostly on the evolutionary algorithms; for more details on the swarm intelligence based approaches refer to [7].

2.3.1   Classical Wrapper-Based Feature Selection

As mentioned earlier, wrapper-based FS algorithms are highly computationally cost compared to the filter-based FS algorithms [15, 48]. Perhaps, this is due to the longer evaluation processes involved in the training and testing of the classifier. Furthermore, since the search space of the FS problem is exponential to the number of features. Therefore, searching for the entire search space is impractical. Based on that the existing wrapper-based techniques used stochastic or greedy search [21, 42].

The most common FS techniques that practice the greedy hill-climbing are sequential feature selection (SFS) [1] and sequential backward selection (SBS) [41].

In SFS, it begins with an empty set of features and keeps on adding one feature at a time in an iterative manner until adding another feature will not enhance the existing classification performance then it stops. Unlike, in SBS where it starts with a full set of features and keeps on looping to remove one feature at a time until removal of a feature cannot improve the existing classification accuracy (error rate). Apart from the computational cost incurred on a large number of data sets, another major drawback of both SFS and SBS is the nesting effect, since any feature that is added or removed cannot be undone. Thus, they both are trapped into the local optima easily [40, 41].

Although, [38] developed a "plus q take away r" technique that will escape the nesting effect, SBS was applied r times in a back-tracking order while SFS is applied q times in forwarding step order. Determining better numbers for q and r is required, to solve this problem of having fixed values for both q and r. Then, [47] enhanced it by introducing a floating-point in both SFS (sequential forward floating selection (SFFS)) and SBS (sequential backwards floating selection (SBFS)) that automatically determine the value of q and r. Although, both SFFS and SBFS proved to be useful in some cases, [67] argued that they could likely trap into local optimal even if the benchmark function is monotonic (neither decrease nor increase) and yet is a small-scale problem.

Inline spectral frequencies (LFS), the number of features to be used for evaluation in every step are limited. As such, the computational efficiency of the sequential forward's methods was enhanced by the LFS and sustained an analogous accuracy of the selected subset of features. But, LSF ranks all features without taking into consideration whether some features are present or not, and this restricts the performance of the LSF algorithm particularly the interaction between features.

2.3.2 Wrapper-Based Feature Selection with Meta-Heuristic Algorithms

As mentioned earlier, meta-heuristic algorithms have become more robust in handling NP-hard problems, including FS. Huang and Wang [30] employed GA for both FS and SVM parameter optimisation on a real-world data set. The results obtained are in favour of the GA in terms of classification accuracy and fewer number of features compared with the grid algorithm reported in work. Also, [59] proposed another GA for FS and SVM for parameter selection in the detection of diabetic retinopathy. A promising result was obtained on 60 images of data sets. An enhanced GA (EGA) was proposed in [19] to reduce text dimensionality. It is incorporated with six filter FS methods to create a hybrid one. Finally, experimental results showed that the hybrid outperformed the single approach as well as the traditional GA. Recently in [3], the highest accuracy of 99.48% was attained on two different Wisconsin breast cancer data sets. GA was used for FS before applying the five different classifiers. The results obtained are better than the others.

Unler and Murat [53] present a discrete PSO for FS in binary classification problems. The proposed approach incorporates an adaptive FS technique which dynamically takes into consideration the relevance and dependence of the features included in the feature subset. The experimental results indicated that the proposed discrete PSO algorithm is competitive in terms of both classification accuracy and computational performance compared with the scatter search and tabu search algorithms on openly available data sets.

Vieira et al. [58] proposed a modified binary PSO (MBPSO) for FS with simultaneous optimisation of SVM parameters to predict the outcome of patients with septic shock. The results indicated that MBPSO performed very well compared with the standard PSO both in terms of accuracy and features selected. However, when compared to GA, the same accuracy was recorded, but the MBPSO select fewer features.

Similarly, [31] developed a supervised PSO-based rough set FS for medical data diagnosis. Two different algorithms PSO-based relative reduct (PSO-RR) and PSO-based quick reduct (PSO-QR) are presented. The results obtained showed that the proposed algorithms performed better in terms of the fewer number of features, classification accuracy and computational time compared to the standard PSO and other methods reported.

Recently, PSO initialisation and updating mechanism are changed to suit better FS problems in [52]. The discretisation is applied before the FS since discretisation is considered an essential task of FS. A potential particle swarm optimisation (PPSO) is proposed which employs a modern illustration that can minimise the search space of the problem and an advanced fitness function to assess candidate solutions better and direct the search process. The results of the experiments on the ten high-dimensional data sets disclosed that PPSO chooses fewer than 5% of the number of features for all data sets. Compared with the two-stage method which uses bare-bone PSO (BBPSO) for FS on the discretised data, PPSO attains a better accuracy on seven data sets. Furthermore, PPSO gains improved classification accuracy than evolve PSO (EPSO) on eight data sets with a reduced feature size on six data sets. Moreover, PPSO also performs better than the three compared approaches and achieves similar to one approach on majority data sets in terms of both learning capacity as well as generalisation ability.

To predict heart disease among patients, [18] used cuckoo search and rough set for FS, and the disease prediction is made using fuzzy. A better result was achieved in four different benchmark data sets.

Recently, [13] used a modified cuckoo search along with rough set to build the fitness function that takes several features into the reduce set and classification into consideration. SVM and KNN are used to evaluate the performance of the proposed approach. The results obtained indicate the superiority of the method used and can significantly improve performance.

Despite the attempt to solve the lingering issues of the wrapper-based FS, still the existing works cannot successfully evolve the best subset of features with improving accuracy on some of the data sets [23]. COA presented in [49] is among the evolutionary algorithms that show promising results in handling different combinatorial optimisation problem, including NP-hard; However, despite its proven records, especially in dealing with filter-based FS in [55].

COA has been applied to solve different kinds of problems. Recently, it is used with harmony search for optimum tuning of fuzzy PID controller for LFC of interconnected power systems in [20]. Energy-aware clustering in wireless sensor networks in [35], accelerated COA was proposed in [22] where simulated annealing algorithm was used in place of the k-means clustering of the standard COA in vehicle routing problems.

Compared to GA, the imperialist competitive algorithm (ICA), CSA and PSO. COA is simpler to implement and can converge rapidly [6, 29]. Its application, specifically for the wrapper-based FS, is not fully investigated.

3 Proposed Wrapper-Based Feature Selection Approaches

Thus, in this part first, both BCOA and COA are adopted and used for wrapper-based FS. The detail of how each of the experiments was carried out can be seen in the subsequent parts.

3.1 BCOA and COA for Feature Selection

Two wrapper-based FS are proposed, namely, BCOA-FS and COA-FS. Throughout the evolutionary training process, Eq. 7 is applied as the fitness evaluation function to estimate and evaluate the best cuckoo habitat i, where the position x_i signifies the subsets of features.

$$Fitness(x_{(i)}) = Error\,Rate \tag{7}$$

where *ErrorRate* is calculated based on Eq. 8:

$$Error\,Rate = \frac{(FP + FN)}{(TP + TN + FP + FN)} \tag{8}$$

where FP, FN, TP, and TN, are the respective false positives, false negatives, true positives, and true negatives.

3.2 A Combined Fitness Function for BCOA and COA Feature Selection

The subset of feature selected by both BCOA-FS and COA-FS may probably comprise some redundancy since the fitness function in Eq. 7 does not reduce the features. However, it hypothesises that the same or less accuracy might be realised using a smaller subset of features. To additionally minimise the feature size deprived of affecting the classification error rate, a two-step FS method (BCOA-2S and COA-2S) is introduced, where the entire evolutionary procedure is separated into two steps.

In step 1, both BCOA-FS and COA-FS emphasises on improving accuracy. Whereas, in step 2, the features are involved in the fitness function. Furthermore, step 2 begins with the solutions realised in step 1, which certifies the reductions in the features according to the subsets of the features with the best accuracy.

The proposed two-step fitness function employed in both BCOA-2S and COA-2S is shown in Eq. 9:

$$
Fitness_2(x_i) = \begin{cases} Step\ 1, & Error\,Rate \\ Step\ 2, & Error\,Rate\ \beta * \dfrac{M}{n} + (1 - \beta) * \frac{M\,Eror\,Rate}{n\,Eror\,Rate} \end{cases} \quad (9)
$$

where Error Rate is the classification error rate attained by the selected subset of features. $\beta \in [0, 1]$ is a constant number within the range [63]. M denotes the size of selected features and n is the total feature size. $n\,Error\,Rate$ is the error rate obtained by using the total feature size for classification on the training set. In step 2, the fitness function considers both the feature size as well as the error rate. It guarantees that these two components are in a similar array, i.e. [0, 1], and the of feature size is normalised and represented by M/n.

The classification performance is represented by $(M\,Error\,Rate)/(n\,Error\,Rate)$ rather than Error Rate alone to circumvent the circumstances, whereby Error Rate is too insignificant (for instance, < 0.005), and M/n plays a significant role inside the fitness function. In a situation like this, the feature size considers most compared to the error rate, which might have a subset of features with high error rate compared to using the full-length feature size. Meanwhile Error Rate would be lesser than $n\,Error\,Rate$ at the end of the step 1, $(M\,Error\,Rate)/(n\,Error\,Rate)$ is in the similar array as M/n, i.e. [0, 1].

As soon as they are joined into a single fitness function, β is employed to display the comparative significance of the chosen features and $(1 - \beta)$ displays the outstanding significance of the error rate. The $Error\,rate$ is expected to be more significance compared with feature size, thus β is assign to be lesser than $(1 - \beta)$ (i.e. $\beta < 0.5$). The pseudocode of (BCOA-FS and BCOA-2S) along with (COA-FS and COA-2S) can be seen in Algorithm 1 and Algorithm 2, respectively. The main difference between BCOA-FS, COA-FS and BCOA-2S depend on the fitness evaluation function, that is illustrated mostly in the grey lines of algorithms.

The detailed of the proposed wrapper-based BCOA is depicted in Algorithm 1, whereby Eqs. 7 and 8 have been used as the respective fitness functions. The grey colour signifies the areas where the equations and initialisation as per feature selection problems are used in the proposed algorithms.

Algorithm 1 Proposed BCOA-FS and BCOA-2S
1: **Start**
2:    **Initialise each habitat with some features from a dataset**
3:    **Collect the features in their respective habitats**
4:    **Explain ELR for every single cuckoo using Equation 5 and Equation 6**
5:    **Allow the cuckoos to lay their eggs in their matching ELR**
6:    **Destroy those cuckoos familiar by the multitude birds**
7:    **Allow egg to incubate and baby chicken raise**
8:    **Estimate the environment of every recently grownup cuckoo**
9:    **Limits cuckoos' highest number in location and abolish those that exist in poorer environments**
10:    **Group cuckoos and discover finest cluster and choose goal line environment**
11:    **Allow the new cuckoo populace to settle at the goal line environment**
12:    **Return the optimum solution (selected features)**
13:    **Evaluate the fitness function according to Equation 8 in BCOA-FS and Equation 9 in BCOA-2S**
14:    **If the stop condition is satisfied to stop, else go to 3**
15: **Stop**

Algorithm 2 Proposed COA-FS and COA-2S
1: **Start**
2:    **Initialise each habitat with some features from a dataset**
3:    **Collect the features in their respective habitats**
4:    **Explain ELR for every single cuckoo using Equation 2 and Equation 3**
5:    **Allow the cuckoos to lay their eggs in their matching ELR**
6:    **Destroy those cuckoos familiar by the multitude birds**
7:    **Allow egg to incubate and baby chicken raise**
8:    **Estimate the environment of every recently grownup cuckoo**
9:    **Limits cuckoos' highest number in location and abolish those that exist in poorer environments**
10:    **Group cuckoos and discover finest cluster and choose goal line environment**
11:    **Allow the new cuckoo populace to settle at the goal line environment**
12:    **Return the optimum solution (selected features)**
13:    **(8) in COA-FS and (9) in COA-2S**
14:    **If the stop condition is satisfied to stop, else go to 3**
15: **Stop**

Figure 2 shows that each cuckoo is initialised according to Eq. 1 for each of the data sets. Then, the number of features in each habitat are collected while those eggs detected in the habitat are killed. At this juncture, the fitness function for both BCOA-FS and COA-FS are evaluated using Eqs. 7 and 8, respectively. Whereas, the fitness function for the combined FS (BCOA-2S and COA-2S) are evaluated using

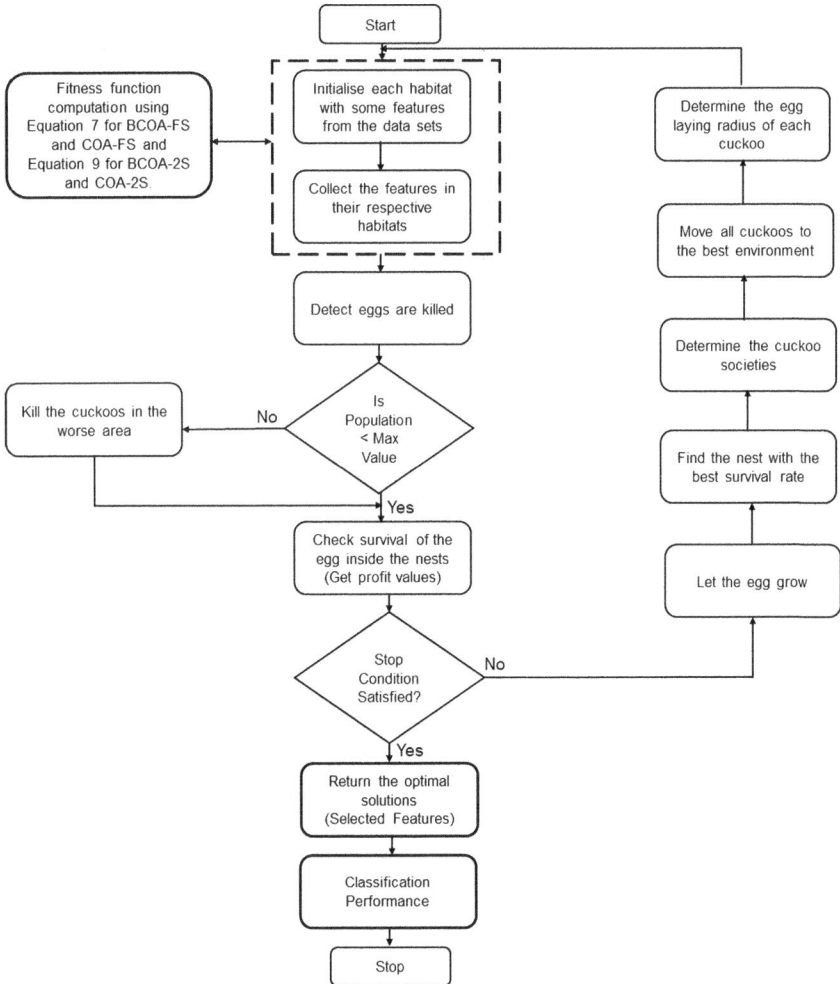

Fig. 2 Flowchart of the (BCOA-FS and BCOA-2S) and (COA-FS and COA-2S)

the fitness function in Eq. 9. The population is compared with maximum value, and if the population is less than the maximum value, then the cuckoo in the worst area would be killed, otherwise it gets profit values (check the survival of the egg inside the nest). Then stop condition evaluated; if yes, it leads the eggs to grow. However, the nest found with the best survival rate among the cuckoo societies is transferred to the best society according to Eqs. 5 and 6 for the BCOA. Whereas, Eqs. 2 and 3 are for the standard COA. Based on Eq. 2, one can find the best ELR and repeat all the steps. Otherwise, return the optimum solution of the highest ranked features. Then, finally, the best, along with the average of them, are selected using a classifier.

The time complexity of both BCOA-FS and COA-FS based on the fitness function in Eq. 7 is $O(\frac{1}{m}) + O(\frac{1}{n})$. The term m and n represent the number of selected features and the population size, respectively. The binary search for using BCOA runs in $O(n)$ time, while the COA search in $O(log_2 n)$ time. Thus, the computational complexity of BCOA-FS is $O(\frac{1}{m}) + O(\frac{1}{n}) + O(m)$ and COA-FS is $O(\frac{1}{m}) + O(\frac{1}{n}) + O(log_2 n)$.

Based on the fitness function in Eq. 8, the complexity is $O(\frac{1}{m^2}) + O(\frac{1}{n^2})$. Therefore, the total complexity of the BCOA-2S is $O(\frac{1}{m^2}) + O(\frac{1}{n^2}) + O(n)$ and that of COA-2S is $O(\frac{1}{m^2}) + O(\frac{1}{n^2}) + O(log_2 n)$. Therefore, BCOA-FS and COA-FS can complete its process within a shorter time in most cases compared to its BCOA-2S and COA-2S counterpart.

4 Experimental Design

This part describes the data sets used in conducting the experiments. Parameters settings, as well as benchmark, approaches are used to test the performance of the proposed methods.

4.1 Experimental Datasets

The data sets used in this study are the 26 well-known University of California Irvine (UCI) Machine learning data sets with distinct features. It contains a different number of features ranging from 9 to 500, 14 categorical and 12 continuous data type, 72–5000 instances, 13 binary classes and 13 multi-classes. These different appearances of the data set, especially on the number of features that contain smaller, medium and large features are the motives behind the selection of the data sets as shown in Table 1. The data sets can be found in [17] or can be downloaded freely at https://www.ics.uci.edu.ml/~earn.

Furthermore, most of the data sets have been used recently in the works of [2, 15, 43, 44], which clearly show that the data sets are goods for benchmarking FS problems. The data sets contain both categorical and continuous data that can be useful in demonstrating the comparison between the categorical discrete and the continuous data. Continuous data have infinite values in the form of decimal numbers, while the categorical discrete values are mostly finite values in groups.

4.2 Experimental Parameter Settings

The parameters employed for the experiments were set as follows: initial and maximum population are set to 5 and 20, respectively. Moreover, the proposed algorithms were run 40 independent times on each data set. The parameter settings used for the

Table 1 List of data sets

S/N	Data set	Features	Classes	Instances	Data type
1	Wine	13	3	178	Continuous
2	Australian	14	2	690	Continuous
3	Zoo	17	7	101	Continuous
4	Vehicle	18	4	846	Continuous
5	Lymphography (Lymph)	18	4	148	Categorical
6	Mushroom	24	2	5644	Categorical
7	Spect	22	2	267	Categorical
8	German	24	2	1000	Continuous
9	Leddisplay	24	10	1000	Categorical
10	WBCD/BreastEW	30	2	569	Continuous
11	Ionosphere (Ionosp)	34	2	351	Continuous
12	Dermatology	34	6	366	Categorical
13	Soybean Large	35	19	307	Categorical
14	Chess (KrvskpEW)	36	2	3196	Categorical
15	Connect4	42	3	44473	Categorical
16	LungCancer (Lung)	56	3	32	Continuous
17	Promoter	57	2	106	Categorical
18	Sonar	60	2	208	Continuous
19	Splice	60	3	3190	Categorical
20	Optic	64	10	5620	Categorical
21	Audiology	68	24	226	Categorical
22	Coil2000	85	2	9000	Categorical
23	Hillvalley	100	2	606	Continuous
24	Musk1 (Clean1)	166	2	476	Continuous
25	DNA	180	3	3186	Categorical
26	Madelon	500	2	4400	Continuous

proposed $COA - FS, COA - 2S, BCOA - FS$ and $BCOA - 2S$ algorithms are chosen based on the work of [45, 49]. The maximum number of iterations was set to 100.

Also, similar to the work of [63] and [65]. In the experiments, all the rows in each of the data sets were partition into two groups: a training group and a test group. The most partitioning approach is that 2/3 (about 66%) of the rows in the data sets are in the training group and 1/3 (almost 33%) of the rows are in the test group [11]. To simplify the process, we divide 70% of the rows into each data set as the training group and the remaining 30% as the test group. The rows are chosen so that the percentage of rows from various classes are equal in both the training group. The proposed wrapper-based methods need a classifier to estimate the suitability of the

Table 2 Existing wrapper-based approaches

References	Type	Acronym	Year
[8]	Single-objective	GSBS	1994
[9]	Single-objective	BAIS	2009
[63]	Single-objective	ErFS and 2SFS	2012
[27]	Single-objective	ABC-ER, ABC-Fit2C	2018

selected subsets of features. A KNN (with K = 5) was used in the experiments, to reduce the wrapper-based computational time [4].

The experiments of GSBS and LFS are carried out using the popularly known Waikato Environment for Knowledge Analysis (WEKA) [28]. The entire settings in LFS along with GSBS are saved to the defaults since they can obtain better results. Also, a 5NN was used in both LFS and GSBS, which generate a unique solution (feature subset) for each data set.

4.3 Benchmark Approaches

Scrutinise the concert of the proposed wrapper-based approaches in this chapter. The results found are related to the previous works, as shown in Table 2. From (Table 2), two traditionally known wrapper-based FS methods, namely linear forward selection (LFS) [25] and greedy stepwise backward selection (GSBS) [8] are used as benchmark methods. Both LFS, together with GSBS, were consequential of SFS and SBS, respectively. LFS [25] limits the number of features that are selected in each step of the forward selection, which can reduce the number of evaluations. As such, the LFS is computationally less expensive compared to the SFS and will get better results. More details about the LFS is in [25].

On the other hand, the greedy stepwise based FS algorithm mostly shifts either forward or backwards in the search space [8]. Provided that the LFS makes a forward selection, a backward search is selected in the greedy stepwise search to create a greedy stepwise backward selection (GSBS). GSBS begins with all the feature size and halts if the removal of any outstanding feature results in a reduction in evaluation measure, i.e. the error rate of the classifier. Also, the work in [63] was used as a benchmark method for both single and multi-objective wrapper-based approach, due to the similarities in the data sets. The detail explanation of the results obtained and the analysis is presented in the subsequent sections.

The details of the results obtained are offered in the subsequent section.

5 Results and Discussions

This part deliberates on the results of the proposed methods, comparison between them and other existing works that their work coincide with the data sets apply in this study.

5.1 Results of the Proposed BCOA-FS and COA-FS

The results of both the categorical and continuous data sets for BCOA-FS and COA-FS are displayed in Tables 4 and 3, respectively. The results showed a comparison between all the proposed wrapper-based methods. From the tables, "BCOA-FS" and "COA-FS" represent the proposed wrapper-based methods that adopt both BCOA and COA, respectively. "All" stands for all features used for each of the data sets. Besides, "Ave Size", "Ave Acc" and "Best Acc" represents average feature size, average accuracy and best accuracy attained by each of the data sets for the 40 independent runs, respectively.

The results proposed BCOA-FS outperformed its COA-FS counterpart on the continuous data sets. Out of the 12 data sets in the table (Table 3), they recorded similar feature size, best accuracy and average accuracy on WineEW, Australian, Zoo and to some extents on Vehicle data sets. However, as the number of features increases, BCOA-FS outperformed COA-FS on the remaining eight data sets. In addition to that a similar performance was slightly noticed between BCOOA-FS and COA-FS on HillValley datasets. On the average, it is clear that BCOA-FS outperformed its COA-FS counterpart on the majority of the data sets, and hence considered the best wrapper-based feature selection.

Alternatively, a comparison between BCOA-FS and COA-FS was made on categorical data sets, as shown in the results Table 4. Similar to the continuous data sets, the categorical data sets also recorded similarities in terms of the mean of selected features, best accuracy and average accuracy on data sets with fewer feature size as such as Lymph, Mushroom, Spect and Leddisplay. However, from Dermatology that has 34 total number of available features, there is a change in performance between BCOA-FS and COA-FS. The results also imply that as the feature size increase the BCOA-FS perform better than the COA-FS in all the data sets except in Coil2000. Perhaps due to a large number of instances in the Coil2000 data set.

5.2 Results of the Proposed BCOA-2S and COA-2S

The results of both the categorical and continuous data sets for BCOA-2S and COA-2S are also displayed in Tables 4 and 3, respectively. The terms "BCOA-2S" and "COA-2S" represents the proposed combined accuracy and selected features into a

Table 3 Results of the BCOA-FS, COA-FS, BCOA-2S and COA-2S for continuos data sets

Datasets	Approach	Ave-Size	Best-Acc	Ave-Acc	Time(s)	Datasets	Approach	Ave-Size	Best-Acc	Ave-Acc	Time(s)
WineEW	All	13	77.25			Australian	All	14	71.25		
	LFS	7	74.57				LFS	4	71.25		
	GSBS	8	86.21				GSBS	12	70.45		
	BCOA-FS	8	86.21	96.75	6120		BCOA-FS	3.42	89.25	86.25	10767.6
	COA-FS	8	86.21	96.75	6217		COA-FS	3.42	89.25	86.25	10871.8
	COA-2S	7	86.21	97.25	6516		COA-2S	3.32	90.56	87.58	11170.8
	BCOA-2S	7	100	86.21	6417		BCOA-2S	3.32	90.58	87.58	11071.8
Zoo	All	16	86.21			Vehicle	All	18	84.25		
	LFS	8	86.21				LFS	9	84.05		
	GSBS	7	86.21				GSBS	15	78.51		
	BCOA-FS	9	86.21	96.25	4559.4		BCOA-FS	9.1	89.56	85.22	10143.4
	COA-FS	9	86.21	96.25	4665.4		COA-FS	9.1	89.56	85.22	10261.6
	COA-2S	8	86.21	98.56	4964.4		COA-2S	9	91.25	91.25	10560.6
	BCOA-2S	8	86.21	98.56	4865.4		BCOA-2S	9	91.56	91.56	10461.6
Germany	All	24	86.21			WBCD	All	30	95.20		
	LFS	3	86.21				LFS	10	84.11		
	GSBS	17	86.21				GSBS	24	85.62		
	BCOA-FS	12.5	86.21	72.48	17044.2		BCOA-FS	13.41	95.21	91.28	8393.4
	COA-FS	12.1	86.21	74.25	17180.8		COA-FS	13.52	94.75	92.45	8813.07
	COA-2S	11.5	86.21	75.45	17479.8		COA-2S	13.21	98.22	98.22	9804.96
	BCOA-2S	11.2	86.21	75.70	17380.8		BCOA-2S	12.45	100.00	100.00	9448.416
IonosphereEW	All	34	86.21			LungCancer	All	56	70.00		
	LFS	5	86.21				LFS	6	90.00		
	GSBS	30	86.21				GSBS	33	90.00		
	BCOA-FS	12.56	88.82	86.21	7835.4		BCOA-FS	27	90.65	88.65	76995
	COA-FS	12.12	92.46	86.21	8227.17		COA-FS	27.6	90.31	89.56	80844.75
	COA-2S	11.85	94.26	86.21	9038.7		COA-2S	25.2	92.25	92.25	116816
	BCOA-2S	10.65	98.25	86.21	8710.02		BCOA-2S	25	95.50	95.50	112568.2

(continued)

Table 3 (continued)

Datasets	Approach	Ave-Size	Best-Acc	Ave- Acc	Time(s)
SonarEW	All	60	86.21		
	LFS	5	86.21		
	GSBS	45	86.21		
	BCOA-FS	23.42	90.65	86.21	67260.6
	COA-FS	24.5	86.45	86.21	70623.6
	COA-2S	21.9	96.58	86.21	107956
	BCOA-2S	19.8	98.20	86.21	104030
Musk1(Clean1)	All	166	86.21		
	LFS	10	86.21		
	GSBS	122	86.21		
	BCOA-FS	84.25	88.88	86.21	70106.4
	COA-FS	85.6	88.81	86.21	73611.7
	COA-2S	41.25	88.97	86.21	120893
	BCOA-2S	50.1	90.50	86.21	116497

Datasets	Approach	Ave-Size	Best-Acc	Ave- Acc	Time(s)
Hillvalley	All	100	56.75		
	LFS	8	57.69		
	GSBS	90	49.45		
	BCOA-FS	45.25	68.52	65.22	72903.6
	COA-FS	46.12	64.26	59.62	76548.78
	COA-2S	44.6	70.65	68.56	118503
	BCOA-2S	42.4	74.25	70.35	114193.8
Madelon	All	500	70.90		
	LFS	7	64.62		
	GSBS	489	51.28		
	BCOA-FS	255.5	80.00	80.00	143303.4
	COA-FS	259.2	79.89	77.47	150468.6
	COA-2S	250.6	81.45	79.45	218918.7
	BCOA-2S	248.52	83.45	81.65	210958

single fitness function for both BCOA and COA, respectively. All other headings in the table are the same as explained in the previous subsection.

There are 14 categorical data sets and 12 continuous data sets that make a total of the 26 data sets used in this research. Out of all the 14 categorical data sets, BCOA-2S accomplished better results than COA-2S in terms of the average number of selected features, best accuracy and average accuracy on almost all the data sets. Although in Leddisplay data set, it has similar performance and same best accuracy on Mushroom data set.

On the other hand, the results of the continuous data sets also are in favour of BCOA-2S compared to the COA-2S in the majority of the data sets. A similar performance was obtained on some few data sets such as WineEW, Australian, Zoo and Vehicle. However, as the feature size increases, the BCO-2S also performed better than COA-2S. It is in contrast with categorical data sets no matter the feature size, BCOA-2S performed better than its COA-2S counterpart in almost all the data sets regardless of the number of features in the data sets.

5.3 Comparison Between Proposed Methods and Classical Methods

A result of LSB and GSBS was reported to further compare with the proposed methods. The results clearly show that LFS could select fewer number of features than GSBS in the majority of the data sets. However, GSBS achieve the best classification results in most of the data sets. Although on some data sets with a fewer number of features, they recorded similar performance. But as the number of features increases, LFS select the smallest feature and GSBS obtained the best accuracy.

Comparing LSF and GSBS with the proposed wrapper-based FS, one can observe that our proposed approaches outperformed both LSF and GSBS in terms of the number of selected features, best accuracy, and average accuracy in all most all the data sets, both continuous (Table 3) and categorical (Table 4).

5.4 Comparison Between Proposed Methods and Other Existing Methods

To further evaluate the performance of the proposed methods and consequently be fair in assessing the proposed wrapper-based multi-objective. Some related works with similar datasets were used for comparison, as shown in Table 2. The details of the comparison are enumerated below:

1. **Comparison with ErFS and 2SFS**

   The results of the proposed wrapper-based feature selection are compared with the one in work [63], where ErFS and 2SFS represent the BCOA-FS and BCOA-

Table 4 Results of the BCOA-FS, COA-FS, BCOA-2S and COA-2S for categorical data sets

Datasets	Approach	Ave-Size	Best-Acc	Ave-Acc	Time(s)	Datasets	Approach	Ave-Size	Best-Acc	Ave-Acc	Time(s)
Lymph	All	18	87.25			Mushroom	All	24	99.20		
	LFS	9	93.48				LFS	8	89.51		
	GSBS	14	84.21				GSBS	18	91.25		
	BCOA-FS	9.2	93.56	92.48	9072		BCOA-FS	12.5	98.98	95.65	15597
	COA-FS	10.5	92.65	90.45	9525.6		COA-FS	13.5	98.98	94.75	16376.85
	COA-2S	8	95.00	95.00	10311.84		COA-2S	8	100.00	98.60	24017.4
	BCOA-2S	7	96.55	94.65	9936.864		BCOA-2S	7	100.00	99.80	23144.04
Spect	All	22	85.10			Leddisplay	All	24	100.00		
	LFS	10	80.00				LFS	7	100.00		
	GSBS	18	82.50				GSBS	5	100.00		
	BCOA-FS	10.8	88.98	86.65	9424.8		BCOA-FS	12	100.00	100.00	14328
	COA-FS	11	87.75	84.68	9896.04		COA-FS	12	100.00	100.00	15044.4
	COA-2S	9	91.75	88.95	10733.58		COA-2S	9	100.00	100.00	19118.88
	BCOA-2S	8.2	94.75	91.25	10343.27		BCOA-2S	9	100.00	100.00	18423.65
Dermatology	All	34	94.85			Soybean Large	All	35	87.25		
	LFS	12	83.25				LFS	11	85.56		
	GSBS	26	89.65				GSBS	24	84.21		
	BCOA-FS	12.5	90.25	89.65	15102		BCOA-FS	12.3	88.65	84.25	17188.2
	COA-FS	13.6	90.12	89.62	17557		COA-FS	12.9	86.75	85.25	18129.3
	COA-2S	10.2	93.12	92.52	17856		COA-2S	7.1	92.75	90.35	20340
	BCOA-2S	9.5	97.56	95.63	17757		BCOA-2S	13.2	95.65	91.58	20241
Chess (KrvskpEW)	All	36	92.00			Connect4	All	36	79.78		
	LFS	11	86.21				LFS	6	70.73		
	GSBS	24.2	85.65				GSBS	41	71.68		
	BCOA-FS	12.5	87.75	86.25	43833.6		BCOA-FS	15	94.25	93.15	43833.6
	COA-FS	13	86.44	85.98	48733.4		COA-FS	18	92.25	89.65	46733.4
	COA-2S	8	95.50	93.25	54032.4		COA-2S	12	96.50	93.50	54032.4
	BCOA-2S	9	100.00	98.56	53933.4		BCOA-2S	10	100.00	100.00	53933.4

(continued)

Table 4 (continued)

Datasets	Approach	Ave-Size	Best-Acc	Ave- Acc	Time(s)
Promoter	All	57	90.00		
	LFS	6	90.00		
	GSBS	50	90.00		
	BCOA-FS	20.5	92.11	92.11	77416.2
	COA-FS	25	91.89	89.68	77796.4
	COA-2S	13.6	91.25	90.65	108095.4
	BCOA-2S	12.5	94.50	91.25	107996.4
Optic	All	64	98.88		
	LFS	36	98.69		
	GSBS	38	98.75		
	BCOA-FS	30	100.00	98.98	74176.2
	COA-FS	32	100.00	98.98	75937
	COA-2S	26.5	100	100.00	106236
	BCOA-2S	20.45	100	100.00	106137
Coil2000	All	85	94.55		
	LFS	10	93.58		
	GSBS	31	93.98		
	BCOA-FS	36	92.65		74725.2
	COA-FS	37.5	93.21	92.01	78721.6
	COA-2S	20.6	93.65	92.52	109020.6
	BCOA-2S	19.6	97.75	97.75	108921.6

Datasets	Approach	Ave-Size	Best-Acc	Ave- Acc	Time(s)
Splice	All	60	79.65		
	LFS	28	78.83		
	GSBS	47	68.65		
	BCOA-FS	28	80.65	79.95	92320.2
	COA-FS	30.5	80.05	78.45	93060.4
	COA-2S	15.6	84.65	83.20	123359.4
	BCOA-2S	12.4	89.88	84.56	123260.4
Audiology	All	68	65.56		
	LFS	9	64.63		
	GSBS	24	64.63		
	BCOA-FS	23	76.62	70.46	73080
	COA-FS	22.4	75.48	69.98	75020.8
	COA-2S	18.54	79.68	73.71	105319.8
	BCOA-2S	20.9	82.56	80.54	105220.8
DNA	All	180	83.01		
	LFS	34	85.08		
	GSBS	173	82.63		
	BCOA-FS	88.5	90.55	89.25	99032.4
	COA-FS	90.5	89.25	78.65	94025
	COA-2S	48.56	91.25	90.65	144324
	BCOA-2S	52.55	92.01	90.93	144225

2S used in this research. The significant difference between the two is the use of the EC algorithm. An outstanding EA, COA, in particular, was used in this research. Whereas, the exiting works used the most common SI based algorithm (PSO). The results indicated our proposed COA and BCOA which outperformed the existing practices of PSO used in [63]. The result is not surprising because COA is reported to be more robust and can attain better results as claimed in the work of [5, 49].

From Table 5, it is clear that all the comparisons were made on the continuous data sets. Out of the 10 data sets used, it shows that in almost all cases, our proposed methods performed better than the existing one. However, in Zoo and Ionosphere data sets, for example, the existing methods performed better in terms of average accuracy. Nevertheless, the best accuracy and the number of selected features clearly show that our proposed methods performed well.

2. **Comparison with ABC-ER and ABC-Fit2C**

   The comparison between the proposed methods with ABC-ER and ABC-Fit2C in [27] is shown in Table 6. The comparison shows that our proposed methods performed better than all the seven data sets on both accuracy and number of selected features. However, even though ABC-Fit2C chooses slightly fewer features on German and Vehicle data sets than the proposed methods, but still, the proposed methods attained an improved classification accuracy compared to the ABC-Fit2C and ABC-ER. Therefore, the results displayed in Table 6 indicated that the proposed methods can effectively evolve a fewer number of features and yet achieve a better classification performance.

3. **Comparison with BAIS**

   The results obtained by the proposed methods with Bayesian and artificial immune system (BAIS) in work [10] is displayed in Table 7. The results show the superiority of the proposed methods on all the five data sets. The proposed methods outperformed the BAIS with nearly 10% of the classification accuracy on Ionosphere and Sonar data sets. Whereas, around 2–3% of improvement was realised on the proposed methods compared to the BAIS on the Mushroom, WineEW and WBCD data sets. Moreover, fewer subsets of features were selected in the proposed method than the BAIS. Therefore, both in terms of selected features and the classification accuracy, the proposed methods outperformed the BAIS in all aspects.

5.5 Comparisons Between BCOA and COA

Comparing the performance of COA and BCOA for adopted or combined objectives as shown in tables (Tables 4 and 3) for both categorical and continuous data sets, one can observe that BCOA outperformed COA in terms of number of selected features, accuracy and best accuracy for all the proposed methods.

Table 5 Comparison of (BCOA-FS, COA-FS, BCOA-2S and COA-2S) with ErFS and 2SFS

Datasets	Approach	Ave-Size	Best-Acc	Ave-Acc	Datasets	Approach	Ave-Size	Best-Acc	Ave-Acc
WineEW	BCOA-FS	8	86.21	96.75	Australian	BCOA-FS	3.42	89.25	86.25
	ErFS	8	95.96			ErFS	3.88	87.44	85.44
	COA-FS	8	86.21	96.75		COA-FS	3.42	89.25	86.25
	COA-2S	7	86.21	97.25		COA-2S	3.32	90.56	87.58
	2SFS	8	95.96			2SFS	3.42	87.44	84.24
	BCOA-2S	7	100.00	86.21		BCOA-2S	3.32	90.58	87.58
Zoo	BCOA-FS	9	86.21	96.25	Vehicle	BCOA-FS	9.1	89.56	85.22
	ErFS	9.18	97.14	95.50		ErFS	9.52	87.01	87.01
	COA-FS	9	86.21	96.25		COA-FS	9.1	89.56	85.22
	COA-2S	8	98.56	86.21		COA-2S	9	91.25	91.25
	2SFS	9.18	97.14	95.50		2SFS	8.65	87.01	84.95
	BCOA-2S	8	98.56	86.21		BCOA-2S	9	91.56	91.56
German	BCOA-FS	12.5	86.21	72.48	WBCD	BCOA-FS	13.41	95.21	91.28
	ErFS	12.58	72.00	69.41		ErFS	13.42	94.74	93.39
	COA-FS	12.1	86.21	74.25		COA-FS	13.52	94.75	92.45
	COA-2S	11.5	86.21	75.45		COA-2S	13.21	98.22	98.22
	2SFS	11.92	72.00	72.00		2SFS	5	94.74	94.74
	BCOA-2S	11.2	86.21	75.70		BCOA-2S	12.45	100.00	100.00
Ionosphere	BCOA-FS	12.56	88.82	86.21	LungCancer	BCOA-FS	27	90.65	88.65
	ErFS	12.58	93.33	88.40		ErFS	27.35	80.00	72.00
	COA-FS	12.12	92.46	86.21		COA-FS	27.6	90.31	89.56
	COA-2S	11.85	94.26	86.21		COA-2S	25.2	92.25	92.25
	2SFS	12.05	93.33	91.43		2SFS	27.38	90.00	80.00
	BCOA-2S	10.65	98.25	86.21		BCOA-2S	25	95.50	95.50

(continued)

Table 5 (continued)

Datasets	Approach	Ave-Size	Best-Acc	Ave-Acc
Hillvalyey	BCOA-FS	45.25	68.52	65.22
	ErFS	47.32	61.81	57.54
	COA-FS	46.12	64.26	59.62
	COA-2S	44.6	70.65	68.56
	2SFS	47.04	61.81	57.57
	BCOA-2S	42.4	74.25	70.35
Madelon	BCOA-FS	255.5	80.00	80.00
	ErFS	258.1	79.49	76.55
	COA-FS	259.2	79.89	77.47
	COA-2S	250.6	81.45	79.45
	2SFS	256.48	79.36	76.52
	BCOA-2S	248.52	83.45	81.65

Datasets	Approach	Ave-Size	Best-Acc	Ave-Acc
Musk1(Clean1)	ErFS	86.48	88.81	84.58
	BCOA-FS	84.25	88.88	86.21
	COA-FS	85.6	88.81	86.21
	COA-2S	41.25	88.97	86.21
	2SFS	85.58	88.81	88.88
	BCOA-2S	50.1	90.50	86.21

Table 6 Comparison of (BCOA-FS, COA-FS, BCOA-2S and COA-2S) with ABC-ER and ABC-Fit2C

Datasets	Approach	Ave-Size	Best-Acc	Ave- Acc	Datasets	Approach	Ave-Size	Best-Acc	Ave- Acc
Optical	BCOA-FS	30	100.00	98.98	Vehicle	BCOA-FS	9.1	89.56	85.22
	ABC-ER	41.13	98.10	–		ABC-ER	9.86	79.53	–
	COA-FS	32	100.00	98.98		COA-FS	9.1	89.56	85.22
	COA-2S	26.5	100	100.00		COA-2S	9	91.25	91.25
	ABC-Fit2C	37.43	98.22	–		ABC-Fit2C	7.73	77.88	–
	BCOA-2S	20.45	100	100.00		BCOA-2S	9	91.56	91.56
German	BCOA-FS	12.5	86.21	72.48	Musk1(Clean1)	BCOA-FS	86.48	88.81	84.58
	ABC-ER	10.76	70.17	–		ABC-ER	83.03	83.11	–
	COA-FS	12.1	86.21	74.25		COA-FS	85.6	88.81	86.21
	COA-2S	11.5	86.21	75.45		COA-2S	41.25	88.97	86.21
	ABC-Fit2C	9.13	70.01	–		ABC-Fit2C	80.56	82.23	–
	BCOA-2S	11.2	86.21	75.70		BCOA-2S	50.1	90.50	86.21
Ionosphere	BCOA-FS	12.56	88.82	86.21	Hillvalyey	BCOA-FS	45.25	68.52	65.22
	ABC-ER	12	92.12	–		ABC-ER	47.63	54.13	–
	COA-FS	12.12	92.46	86.21		COA-FS	46.12	64.26	59.62
	COA-2S	11.85	94.26	86.21		COA-2S	44.6	70.65	68.56
	ABC-Fit2C	11.53	91.74	–		ABC-Fit2C	44.96	54.92	–
	BCOA-2S	10.65	98.25	86.21		BCOA-2S	42.4	74.25	70.35
Madelon	BCOA-FS	255.5	80.00	80.00					
	ABC-ER	252.46	72.91						
	COA-FS	259.2	79.89	77.47					
	COA-2S	250.6	81.45	79.45					
	ABC-Fit2C	248.03	72.20	–					
	BCOA-2S	248.52	83.45	81.65					

Table 7 Comparison of (BCOA-FS, COA-FS, BCOA-2S and COA-2S) with BAIS

Datasets	Approach	Ave-Size	Best-Acc	Ave-Acc	Datasets	Approach	Ave-Size	Best-Acc	Ave-Acc
WineEW	BCOA-FS	8	86.21	96.75	Mushroom	BCOA-FS	12.5	98.98	95.65
	COA-FS	8	86.21	96.75		COA-FS	13.5	98.98	94.75
	COA-2S	7	86.21	97.25		COA-2S	8	100.00	98.60
	BAIS	7.8	98.40	–		BCOA-2S	7	100.00	99.80
	BCOA-2S	7	100.00	86.21		BAIS	11.5	98.10	–
Ionosphere	BCOA-FS	12.56	88.82	86.21	WBCD	BCOA-FS	13.41	95.21	91.28
	COA-FS	12.12	92.46	86.21		COA-FS	13.52	94.75	92.45
	COA-2S	11.85	94.26	86.21		COA-2S	13.21	98.22	98.22
	BAIS	13.4	91.20	–		BAIS	14.3	97.20	–
	BCOA-2S	10.65	98.25	86.21		BCOA-2S	12.45	100.00	100.00
Sonar	BCOA-FS	23.42	86.21	90.65					
	COA-FS	24.5	86.21	86.45					
	COA-2S	21.9	86.21	96.58					
	BAIS	23.6	77.30	–					
	BCOA-2S	19.8	86.21	98.20					

Even though BCOA is a discrete binary version of COA, however, it can be seen that it outperformed COA not only on the categorical or discrete data sets but also on the continuous data sets. Continuous or discrete data sets refer to the data sets that have their class label either as categorical or continuous.

Analysis of the computational time also shows that BCOA can complete its evolutionary process within the shortest time than the COA on the majority of the data sets. BCOA is faster than COA in around 10–5% majority of some of the data sets regardless of the continuous or categorical data sets. Meanwhile, this motivates the use of BCOA alone in the multi-objective wrapper-based feature selection. Moreover, this will avoid repetition of similar explanation of BCOA of being the best compared to its COA counterpart.

5.6 Further Discussions

The results show that both BCOA-FS and COA-FS can successfully evolve a set of features with better classification performance within a short period. However, as the number of features increase, BCOA-FS perform better than COA-FS, especially on the categorical datasets. Whereas, the COA-FS performed better mostly on the continuous class label dataset. It demonstrates that the continuous version works well on the continuous label datasets. In contrast, the binary version works well on the majority of the datasets and mostly performed better on the categorical datasets. Correspondingly, both BCOA-2S and COA-2S can successfully select the best features with better classification performance than the COA-FS and BCOA-2S on the majority of the datasets. Also, BCOA-2S outperformed COS-2S in most cases due to the use of the two-step evaluation process.

The proposed approaches used a β value of [0,1] in the evolution process. However, choosing the most appropriate value is quite a challenge. Because most of the selected features, along with their classification performance, are combined into a single fitness function. Nowadays, FS is considered as a multi-objective optimisation problem and treating the FS in that regards will solve the task much better and obtain the set of nondominated solutions.

6 Conclusions and Future Work

This paper disclosed the first study on wrapper-based feature selection using COA and BCOA. Four wrapper-based feature selections are presented. Both BCOA and COA were adopted and used as a wrapper based in the evolutionary process. Then a two-step fitness function was proposed, whereby the new classification performance obtained in the first step is combined with the number of selected features in the second step. The results obtained showed that the proposed methods performed well compared to the previous work. However, combining the two aims of the feature selection into a single fitness function cannot solve the problem better, and there will be some redundancy still among the number of selected features. Hence there is need for multi-objective feature selection that treats both numbers of selected features and classification performance simultaneously.

On the other hand, COA, especially its binary version, has performed well for FS because (1. COA representation is suitable for FS problems. The habitats in COA is N_{var}-dimensional array representing the current living position of cuckoos, which looks like the way candidate solutions are represented in the FS problem. In this case, the size of the dimensionality is the number of features. The values in any dimension/habitat display whether a feature is chosen or otherwise. (2. The search space in FS problems is too large and mostly get stuck in local optima in most of the existing methods. As such, there is a need for a global search technique. These ECs are well-known for solving problems that do not have a solution; they are robust to dynamic changes and have broad applicability [14, 49, 50]. COA is an EC; precisely an evolutionary algorithm based that has effective and efficient search operators that can search for large space to discover the optimum otherwise nearby optimum solution [14].

Acknowledgements The authors want to thank Universiti Sains Malaysia (USM) for supporting and backing the research via its Research University Grant (RUI) (1001/PKOMP/8014084) along with Woosong University, Korea.

References

1. Aha, D.W., Bankert, R.L.: A comparative evaluation of sequential feature selection algorithms. In: Learning from Data, pp. 199–206. Springer (1996)
2. Ahmad, S. et al.: Feature selection using salp swarm algorithm with chaos. In: ICFNDS '18 Proceedings of the 2nd International Conference on Intelligent Systems, Metaheuristics & Swarm Intelligence, pp. 65–69. ACM (2018)
3. Alickovic, E., Subasi, A.: Breast cancer diagnosis using GA feature selection and Rotation Forest. Neural Comput. Appl. **28.4**, 753–763 (2017). ISSN: 0941-0643
4. Alpaydin, E.: Introduction to Machine Learning. MIT Press (2014). ISBN 0262325756
5. Amiri, E., Mahmoudi, S.: Efficient protocol for data clustering by fuzzy cuckoo optimization algorithm. Appl. Soft Comput. **41**, 15–21 (2016). ISSN: 1568-4946
6. Anemangely, M. et al.: Machine learning technique for the prediction of shear wave velocity using petrophysical logs'. J. Petrol. Sci. Eng. **174**, 306–327 (2019). ISSN: 0920-4105
7. Brezocnik, L., Fister, I., Podgorelec, V.: Swarm intelligence algorithms for feature selection: a review. Appl. Sci. **8**(9), 1521 (2018)
8. Caruana, R., Freitag, D.: Greedy attribute selection. In: Machine Learning Proceedings 1994, pp. 28–36. Elsevier (1994)
9. Castro, P.A.D., Von Zuben, F.J.: Feature subset selection by means of a Bayesian artificial immune system. In: 2008 Eighth International Conference on Hybrid Intelligent Systems, pp. 561–566. IEEE (2008). ISBN: 0769533264
10. Castro, P.A.D., Von Zuben, F.J.: Multi-objective feature selection using a Bayesian artificial immune system. Int. J. Intell. Comput. Cybern. **3.2**, 235–256 (2010). ISBN: 1756-378X
11. Dobbin, K.K., Simon, R.M.: Optimally splitting cases for training and testing high dimensional classifiers. In: BMC Medical Genomics, vol. 4.1, pp. 1–8 (2011). ISSN: 1755-8794

12. Dy, J.G., Brodley, C.E.: Feature selection for unsupervised learning. J. Mach. Learn. Res. **5**(8), 845–889 (2004)
13. El Aziz, M.A., Hassanien, A.E.: Modified cuckoo search algorithm with rough sets for feature selection. Neural Comput. Appl. **29.4**, 925–934 (2018). ISSN: 0941-0643
14. Elyasigomari, V. et al.: Cancer classification using a novel gene selection approach by means of shuffling based on data clustering with optimization. Appl. Soft Comput. **35**, 43– 51 (2015). ISSN: 1568-4946
15. Faris, H. et al.: An efficient binary Salp Swarm Algorithm with crossover scheme for feature selection problems. Knowl.-Based Syst. **154**, 43–67 (2018). ISSN: 0950-7051
16. Ferri, F.J. et al.: Comparative study of techniques for large-scale feature selection. In: Machine Intelligence and Pattern Recognition, vol. 16, pp. 403–413. Elsevier (1994). ISBN: 0923-0459
17. Frank, A., Asuncion, A.: UCI Machine Learning Repository [http://archive.ics.uci.edu/ml]. In: School of Information and Computer Science, vol. 213, pp. 21–22. University of California, Irvine, CA (2010)
18. Gadekallu, T.R., Khare, N.: Cuckoo search optimized reduction and fuzzy logic classifier for heart disease and diabetes prediction. Int. J. Fuzzy Syst. Appl. (IJFSA) **6**(2), 25–42 (2017)
19. Ghareb, A.S., Bakar, A.A., Hamdan, A.R.: Hybrid feature selection based on enhanced genetic algorithm for text categorization. In: Expert Systems with Applications vol. 49, pp. 31–47 (2016). ISSN: 0957-4174
20. Gheisarnejad, M.: An effective hybrid harmony search and cuckoo optimization algorithm based fuzzy PID controller for load frequency control. Appl. Soft Comput. **65**, 121– 138 (2018). ISSN: 1568-4946
21. Gheyas, I.A., Smith, L.S. (2010). Feature subset selection in large dimensionality domains. Pattern Recogn. **43.1**, 5–13 (2010). ISSN: 0031-3203
22. Goli, A., Aazami, A., Jabbarzadeh, A.: Accelerated cuckoo optimization algorithm for capacitated vehicle routing problem in competitive conditions. Int. J. Artif. Intell. **16**(1), 88–112 (2018)
23. Gonzalez, J. et al.: A new multi-objective wrapper method for feature selection–Accuracy and stability analysis for BCI. Neurocomputing **333**, 407–418 (2019). ISSN: 0925-2312
24. Guo, G. et al.: An kNN model-based approach and its application in text categorization. In: International Conference on Intelligent Text Processing and Computational Linguistics, pp. 559–570. Springer (2004)
25. Gutlein, M. et al.: Large-scale attribute selection using wrappers. In: 2009 IEEE Symposium on Computational Intelligence and Data Mining, pp. 332–339. IEEE (2009). ISBN: 1424427657
26. Hancer, E., Xue, B., Zhang, M.: Differential evolution for filter feature selection based on information theory and feature ranking. Knowl.-Based Syst. **140**, 103– 119 (2018). ISSN: 0950-7051
27. Hancer, E. et al.: Pareto front feature selection based on artificial bee colony optimization. Inf. Sci. **422** , 462–479 (2018). ISSN: 0020-0255
28. Hastie, T., et al.: The elements of statistical learning: data mining, inference and prediction. Math. Intell. **27**(2), 83–85 (2005)
29. Hosseini-Moghari, S.-M. et al.: Optimum operation of reservoir using two evolutionary algorithms: imperialist competitive algorithm (ICA) and cuckoo optimization algorithm (COA). Water Res. Manag. **29.10**, 3749–3769 (2015). ISSN: 0920-4741
30. Huang, C.-L., Wang, C.-J.: A GA-based feature selection and parameters optimizationfor support vector machines. Expert Syst. Appl. **31.2**, 231–240 (2006). ISSN: 0957-4174
31. Hannah Inbarani, H., Azar, A.T., Jothi, G.: Supervised hybrid feature selection based on PSO and rough sets for medical diagnosis. Comput. Methods Prog. Biomed. **113.1**, 175–185 (2014). ISSN: 0169-2607
32. Hannah Inbarani, H., Bagyamathi, M., Azar, A.T.: A novel hybrid feature selection method based on rough set and improved harmony search". In: Neural Comput. Appl. **26.8**, 1859–1880 (2015). ISSN: 0941-0643
33. Jiménez, F. et al.: Multi-objective evolutionary feature selection for online sales forecasting. Neurocomputing **234**, 75–92 (2017). ISSN: 0925-2312

34. Kelemen, A. et al.: Naive Bayesian classifier for microarray data. In: Proceedings of the International Joint Conference on NeuralNetworks, vol. 2003, pp. 1769–1773. IEEE (2003). ISBN: 0780378989

35. Khabiri, M., Ghaffari, A.: Energy-Aware clustering-based routing inwireless sensor networks using cuckoo optimization algorithm. Wirel. Pers. Commun. **98.3**, 2473–2495 (2018). ISSN: 0929-6212

36. Kittler, J.: Feature selection and extraction. Handbook of Pattern recognition and image processing **1**(1), 1–37 (1986)

37. Koza, J.R. et al.: Genetic programming 1998: Proceedings of the Third Annual Conference. In: IEEE Transactions on Evolutionary Computation, vol. 3.2, pp. 159–161 (1999). ISSN: 1089-778X

38. Kuncheva, L.I., Jain, L.C.: Nearest neighbor classifier: simultaneous editing and feature selection. Pattern Recogn. Lett. **20.11**, 1149–1156 (1999). ISSN: 0167-8655

39. Liu, H., Motoda, H.: Feature extraction, construction and selection: a data mining perspective. Springer Science and Business Media (1998). ISBN: 0792381963

40. Liu, H., Motoda, H.: Feature selection for knowledge discovery and data mining. Springer Science and Business Media (2012). ISBN: 1461556899

41. Liu, H., Yu, L.: Toward integrating feature selection algorithms for classification and clustering. IEEE Trans. Knowl. Data Eng. 17.4 (2005), pp. 491–502. issn: 1041-4347

42. Liu, X.-Y. et al.: A hybrid genetic algorithm with wrapper-embedded approaches for feature selection. In: IEEE Access **6**, 22863–22874 (2018). ISSN: 2169-3536

43. Mafarja, M. et al.: Evolutionary population dynamics and grasshopper optimization approaches for feature selection problems. Knowl.-Based Syst. **145**, 25–45 (2018). ISSN: 0950-7051

44. Mafarja, M. et al.: Feature selection using binary particle swarm optimization with time varying inertiaweight strategies. In: ICFNDS'18 Proceedings of the 2nd International Conference on Future Networks and Distributed Systems, pp. 1–9. ACM (2018)

45. Mahmoudi, S., Rajabioun, R., Lotfi, S.: Binary cuckoo optimization algorithm. In: 1st National Conference on New Approaches in Computer Engineering and Information Retrieval Young Researchers And Elite Club of the Islamic Azad University, Roudsar-Amlash Branch, pp. 1–7 (2013)

46. Peng, H., Long, F., Ding, C.: Feature selection based on mutual information criteria of max-dependency, maxrelevance, and min-redundancy. IEEE Trans. Pattern Anal. Machine Intell. **27.8**, 1226–1238 (2005). ISSN: 0162-8828

47. Pudil, P., Novovicov, J., Kittler, J.: (1994). Floating search methods in feature selection. Pattern Recogn. Lett. **15.11**, 1119–1125 (1994). ISSN: 0167-8655

48. Purohit, A., Chaudhari, N.S., Tiwari, A.: Construction of classifier with feature selection based on genetic programming. In: 2010 IEEE Congress on Evolutionary Computation (CEC), pp. 1–5. IEEE (2010). ISBN: 1424469112

49. Rajabioun, R.: Cuckoo optimization algorithm. Appl. Soft Comput. **11.8**, 5508–5518 (2011). ISSN: 1568-4946

50. Sivanandam, S.N., Deepa, S.N.: Genetic algorithm optimization problems. Introduction to Genetic Algorithms, pp. 165–209. Springer (2008)

51. Song, L. et al.: Supervised feature selection via dependence estimation. In: Proceedings of the 24th International Conference on Machine Learning, pp. 823–830. ACM (2007). isbn: 1595937935

52. Tran, B., Xue, B., Zhang, M.: A newrepresentation in PSO for discretization-based feature selection. In: IEEE Transactions on Cybernetics **48.6**, 1733–1746 (2018). ISSN: 2168-2267

53. Unler, A., Murat, A.: A discrete particle swarm optimization method for feature selection in binary classification problems. Eur. J. Oper. Res. **206.3**, 528–539 (2010). ISSN: 0377-2217

54. Unler, A., Murat, A., Chinnam, R.B.: mr 2 PSO: a maximum relevance minimum redundancy feature selection method based on swarm intelligence for support vector machine classification. Inf. Sci. **181.20**, 4625–4641 (2011). ISSN: 0020-0255

55. Usman, A.M., Yusof, U.K., Naim, S.: Cuckoo inspired algorithms for feature selection in heart disease prediction. Int. J. Adv. Intell. Inf. **4.2**, 95–106 (2018). ISSN: 2548-3161

56. Usman, A.M. et al.: Comparative evaluation of nature-based optimization algorithms for feature selection on some medical datasets. I-manag. J. Image Process. **5.4**, 9 (2018). ISSN: 2349-4530
57. Vapnik, V.N.: An overview of statistical learning theory. IEEE Trans. Neural Netw. **10.5**, 988–999 (1999). ISSN: 1045-9227
58. Vieira, S.M. et al.: Modified binary PSO for feature selection using SVM applied to mortality prediction of septic patients. Appl. Soft Comput. **13.8**, 3494–3504 (2013). ISSN: 1568- 4946
59. Welikala, R.A. et al.: Genetic algorithm based feature selection combined with dual classification for the automated detection of proliferative diabetic retinopathy. Comput. Med. Imaging Graph. **43**, 64–77 (2015). ISSN: 0895–6111
60. Whitney, A.W.: A direct method of nonparametric measurement selection. IEEE Trans. Comput. **100.9**, 1100–1103 (1971). ISSN: 0018-9340
61. Xu, Z. et al.: Discriminative semi-supervised feature selection via manifold regularization. IEEE Trans. Neural Netw. **21.7**, 1033–1047 (2010). ISSN: 1045-9227
62. Xue, B., Zhang, M., Browne, W.N.: Particle swarm optimisation for feature selection in classification: novel initialisation and updating mechanisms. Appl. Soft Comput. **18**, 261–276 (2014). ISSN: 1568-4946
63. Xue, B., Zhang, M., Browne, W.N.: Particle swarmoptimization for feature selection in classification: a multi-objective approach. IEEE Trans. Cybern. **43.6**, 1656–1671 (2012). ISSN: 2168-2267
64. Xue, B. et al.: A survey on evolutionary computation approaches to feature selection. IEEE Trans. Evol. Comput. **20.4**, 606–626 (2016). ISSN: 1089-778X
65. Xue, B. et al.: Multi-objective evolutionary algorithms for filter based feature selection in classification. Int. J. Artif. Intell. Tools **22.4**, 1–31 (2013). ISSN: 0218-2130
66. Xue, X., Yao, M., Wu, Z.: A novel ensemblebased wrapper method for feature selection using extreme learning machine and genetic algorithm. Knowl. Inf. Syst. **57.389**, 389–412 (2017). ISSN: 0219-1377
67. Yusta, S.C.: Different metaheuristic strategies to solve the feature selection problem. Pattern Recogn. Lett. **30.5**, 525–534 (2009). ISSN: 0167-8655
68. Zhao, H., Sinha, A.P., Ge, W.: Effects of feature construction on classification performance: an empirical study in bank failure prediction. Expert Syst. Appl. **36.2**, 2633–2644. (2009) ISSN: 0957-4174

Malicious Uniform Resource Locator Detection Using Wolf Optimization Algorithm and Random Forest Classifier

Kayode S. Adewole⬡, Muiz O. Raheem, Oluwakemi C. Abikoye, Adeleke R. Ajiboye, Tinuke O. Oladele, Muhammed K. Jimoh, and Dayo R. Aremu

Abstract Within the multitude of security challenges facing the online community, malicious websites play a critical role in today's cybersecurity threats. Malicious URLs can be delivered to users via emails, text messages, pop-ups or advertisements. To recognize these malicious websites, blacklisting services have been created by the web security community. This method has been proven to be inefficient. This chapter proposed meta-heuristic optimization method for malicious URLs detection based on genetic algorithm (GA) and wolf optimization algorithm (WOA). Support vector machine (SVM) as well as random forest (RF) were used for classification of phishing web pages. Experimental results show that WOA reduced model complexity with comparable classification results without feature subset selection. RF classifier outperforms SVM based on the evaluation conducted. RF model without feature selection produced accuracy and ROC of 0.972 and 0.993, respectively, while RF model that is based on WOA optimization algorithm produced accuracy of 0.944 and ROC of 0.987. Hence, in view of the experiments conducted using two well-known phishing datasets, this research shows that WOA can produce promising results for phishing URLs detection task.

K. S. Adewole (✉) · M. O. Raheem · O. C. Abikoye · A. R. Ajiboye · T. O. Oladele · D. R. Aremu
Department of Computer Science, University of Ilorin, Ilorin, Nigeria
e-mail: adewole.ks@unilorin.edu.ng

M. O. Raheem
e-mail: raheem069.mo@gmail.com

O. C. Abikoye
e-mail: abikoye.o@unilorin.edu.ng

A. R. Ajiboye
e-mail: ajibabdulraheem@gmail.com

T. O. Oladele
e-mail: tinuoladele@gmail.com

D. R. Aremu
e-mail: draremu2006@gmail.com

M. K. Jimoh
Department of Educational Technology, University of Ilorin, Ilorin, Nigeria
e-mail: jimoh.km@unilorin.edu.ng

© The Author(s), under exclusive license to Springer Nature Switzerland AG 2021
H. Chiroma et al. (eds.), *Machine Learning and Data Mining for Emerging Trend in Cyber Dynamics*, https://doi.org/10.1007/978-3-030-66288-2_7

Keywords Phishing detection · Meta-heuristic · Genetic algorithm · Wolf optimization · Machine learning

1 Introduction

The significance of the world wide web (WWW) has constantly been expanding. These days the internet is an integral part of everybody's daily activities and has contributed tremendously in information sharing across the globe. Technology grows with huge speed, which makes the user to utilize it in a smarter way. As technology advances, motives for usage also grow vastly. Numerous attacks are exhibited over the network; one of them is phishing in which the attacker impersonates himself as genuine to hijack the user's credentials. Unfortunately, technological advancement combined with new sophisticated attack has increased tremendously and the number of victims is growing [1]. Such attack incorporates rogue websites that sell fake products, monetary extortion by deceiving users into uncovering delicate information, which inevitably leads to stealing of money or personality, or notwithstanding introducing malicious software on the network. There exist a broad range of methods for carrying out such assaults; for instance, social engineering, drive-by exploits, watering hole, phishing, SQL injection, man-in-the-middle, loss/theft of gadgets, denial of service and many others [2]. To engage in phishing attack, attacker can disguise uniform resource locator (URL) to appear as legitimate to the visitors of the website.

Resources and other documents on the web are accessed through URLs. URL is the worldwide location with two major elements: the protocol identifier and the resource name. The protocol identifier represents the protocol used to access the web resource, while resource name depicts the domain name or IP address where the web resource is located. These two components are separated with a colon and double forward slashes, as shown in Fig. 1.

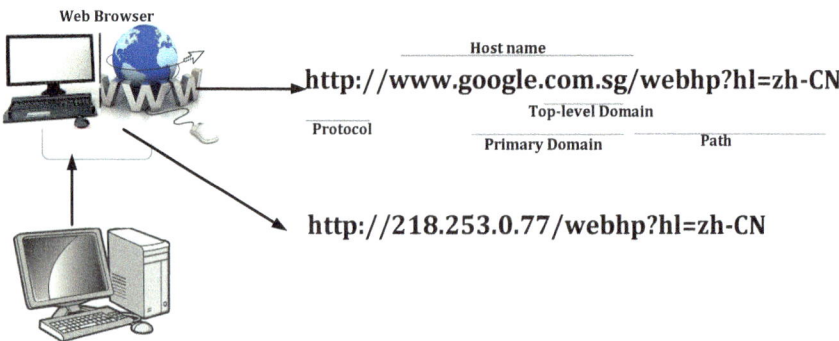

Fig. 1 Sample URL [2]

Malicious URLs have turned into a typical means to encourage cybercrimes, which include drive-by-download and spamming. Numerous attackers endeavour to utilize malicious sites to distribute malicious projects or compromise users' data. According to Kaspersky Lab, there has been increase in browser-based attacks in 2012 ranging between 946,393,693 and 1,595,587,670. However, 87.36% of these attacks were utilized maliciously. The Anti-Phishing Working Group additionally states that phishing attacks that make use of malicious links expanded from 93,462 to 123,486 during the second half of the year 2012. It has been observed by Lin et al. [3] that out of the millions of URLs utilized every day, malicious URLs are short-lived so as to avoid blacklist-based detection. Early detection method for malicious URLs utilized blacklist-based approach. This approach depends heavily on repositories of already categorized web pages. Blacklist-based method has the problem of generality owing to the fact that any URL that is not listed in the repositories might not be detected [4]. Conversely, machine learning methods have played significant roles and have also been used to build intelligent frameworks that distinguish malicious web pages from genuine ones. For example, Gupta [5] used pattern matching algorithm with word segmentation to pinpoint malicious web pages. Naïve Bayes and sequential minimal optimization (SMO) approach have been studied in the work of Aydin and Baykal [6]. Li et al. [7] proposed supervised boosting decision tree approach to detect phishing URL. Wang et al. [8] proposed a hybrid classification approach based on static and dynamic analyses for malicious website detection and Adewole et al. [9] discovered the possibility of phishing detection using hybrid rule-based method through a combination of two commonly used rule induction algorithms: JRip and Projective Adaptive Resonance Theory (PART). Notwithstanding, numerous researches have employed machine learning to deploy intelligent frameworks for malicious URLs detection, however, employing meta-heuristics approach with classification techniques to build a more accurate system to efficiently detect malicious web pages still remains an open research issue. Therefore, this chapter investigates the performance of two meta-heuristic optimization algorithms based on genetic algorithm (GA) and wolf optimization algorithm (WOA). Support vector machine (SVM) and random forest (RF) algorithms have been used as classifiers to evaluate the performance of the features selected from the two meta-heuristic algorithms. These algorithms were selected in this study based on their outstanding performances as reported in the literatures [6, 15, 16, 20].

The subsequent sections are arranged as follows: Sect. 2 focuses on related studies on malicious URLs detection, Sect. 3 discusses the methodology of the proposed framework and Sect. 4 presents the results of the various experiments conducted in this study. The last section, Sect. 5 concludes the chapter and presents the future direction of the study.

2   Related Works

Several studies have been conducted to detect malicious websites. This section discusses related studies on malicious URLs detection. For instance, Xuan and Yongzhen [10] developed malicious URLs detection system using two-dimensional barcodes and hash function. In their approach, the researchers extracted eigenvalues of malicious and benign links. The system generates black and white list library for the URLs extracted. Based on the match rules produced, the authors presented safety tips for users in accordance with these rules.

Lexicon-based approach has been studied in the literature to detect malicious websites. For instance, Darling et al. [11] employed lexical analysis of URLs to categorize websites according to the level of maliciousness. The main idea is to identify the classification model based on lexical analysis that could be employed in real time to detect malicious URLs. In addition, Lee and Kim [12] focused on detection of malicious URLs in a twitter data stream. Authors concentrated on discovering frequently distributed URLs to uncover the deviousness of associated link redirect chains. Tweets from Twitter timeline were experimented upon to develop a classification model for phishing URLs detection. Experimental result reveals that the classification approach was capable of accurately identifying suspicious links in a tweet. Gupta [5] employed algorithm for pattern matching on word segmentation to pinpoint phishing URL. Naïve Bayes and SMO approach have been studied in the work of Aydin and Baykal [6]. Li et al. [7] developed supervised boosting decision tree method to detect phishing URL. Wang et al. [8] proposed a hybrid classification approach based on static and dynamic analyses for malicious website detection.

Also, models for spam message and spam account detection on Twitter have been studied extensively in the literature [13]. In addition, Bhardwaj, Sharma and Pandit [14] proposed artificial bee colony algorithm for identifying malicious links on the web. Furthermore, a study carried out by Adewole et al. [9] applied a hybrid rule-based technique to identify malicious URLs. The research showed that PART algorithm is superior to JRip when deployed for phishing URLs detection task. Thus, their study concluded that the hybrid rule induction method that combined the rules generated from the two induction algorithms performed better than both PART and JRip in terms of accuracy.

Babagoli et al. [18] proposed a method for phishing website detection that utilizes meta-heuristic-based nonlinear regression algorithm coupled with feature selection technique. The authors evaluated the proposed approach using dataset that comprises 11055 phishing and legitimate web pages. Twenty (20) features were focused on to build the phishing detection system. Sohrabi and Karimi [19] proposed spam comments detection model from Facebook social network. Sahingoz et al. [20] proposed natural language processing (NLP) features to train classification algorithm for real-time anti-phishing detection system. Seven classification algorithms were used to evaluate the performance of the NLP-based features. The study observed that RF algorithm produced the best result with an accuracy of 97.98% for phishing URLs detection.

Although a number of machine learning models have been investigated to detect phishing URLs, however, investigation based on WOA and GA for feature optimization with SVM and RF classifiers is the main focus of this research.

3 Methodology

This section discusses the methodology employed to build the proposed model for phishing URL detection. It describes the approach for data collection as well as the optimization algorithms that were used for features optimization. Summarily, Fig. 2

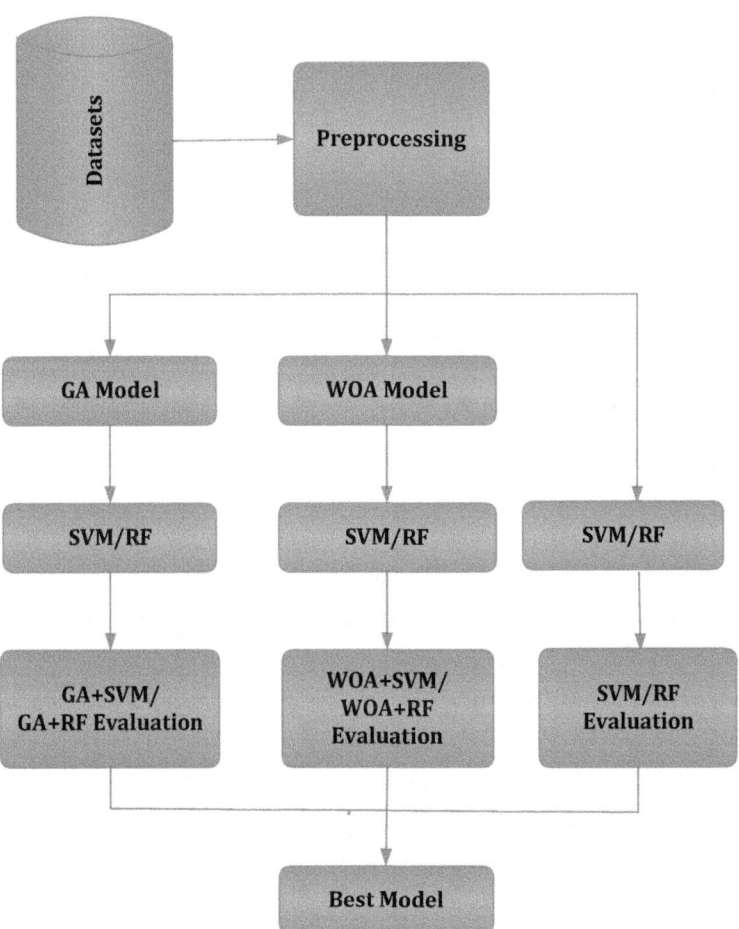

Fig. 2 Proposed framework for phishing URLs detection

Table 1 Description of the two datasets utilized in the study

Dataset name	No. of features	Attributes characteristics	No. of instances	Class distribution
Dataset1	10	Integer	1,353	Malicious (702), genuine (548), suspicious(103)
Dataset2	30	Integer	11,055	Malicious (4898), genuine (6157)

shows the proposed framework in this study to detect malicious web pages. The subsequent sections highlight the components of the proposed framework and the techniques deployed to achieve them.

3.1 Method of Data Collection and Preparation

This study examined two publicly available datasets for malicious URLs detection. The datasets were obtained from UCI machine learning repository. First dataset, henceforth referred to as Dataset1, is accessible at 'https://archive.ics.uci.edu/ml/machine-learning-databases/00379/', which has a total of 1,353 URLs instances. This dataset has ten (10) attributes for analysis. The second dataset, henceforth referred to as Dataset2, is accessible at 'https://archive.ics.uci.edu/ml/machine-learning-databases/00327/', which has a total of 11,055 URLs instances. This dataset has 30 attributes for analysis. Table 1 depicts the composition of these datasets. Detailed reports of the various features in the datasets can be found in [9].

3.2 Feature Subset Selection

Feature selection, as a data preprocessing approach, has been proven to be powerful and efficient in the preparation of data with high dimensionality for various data mining and machine learning issues. The goals of feature selection include constructing less complicated and more comprehensible models, enhancing machine mining model performance, and preparing clean and understandable data of high quality. In order to select a subset of features of high quality from the two datasets considered in this study, two meta-heuristic optimization algorithms were studied, which are GA and WOA algorithms. The subsequent sections discussed the two algorithms in detail.

3.3   Meta-Heuristics Algorithms

As discussed in the previous sections, this study investigated two meta-heuristic algorithms: GA and WOA based on their optimal performance as testified in the literature.

Genetic Algorithm. GA is a meta-heuristic search algorithm inspired by Charles Darwin's theory of natural evolution. Its operation mimics the natural selection process and the algorithm has gained wider usage for solving optimization problems over the years. GA selects fittest individuals for reproduction so as to yield offspring of the next generation. The algorithm terminates if it does not produce offspring that are significantly different from the previous generation. This means that the population has converged on this iteration and a set of solutions to the problem has been identified. In relation to feature subset selection, GA searches the best combination of features that will provide improved results in the solution space. GA has several stages which are demonstrated in Algorithm 1.

Algorithm1: Genetic Algorithm

[**Start**] Generate random population of n chromosomes (appropriate solutions for the problem)

[**Fitness**] Calculate the fitness $f(x)$ of all chromosome x in the population

[**New population**] Generate a new population by reiterating the subsequent phases till the new population is complete

    [**Selection**] Select two parent chromosomes from a population according to their fitness (the better fitness, the more chance to be selected)

    [**Crossover**] With a crossover, probability crosses over the parents to form a new offspring (children). If no crossover was achieved, offspring is an exact copy of parents.

    [**Mutation**] With a mutation, probability mutates a new offspring at each locus (position in chromosome).

    [**Accepting**] Place new offspring in a new population

[**Replace**] Use new generated population for a further run of algorithm

[**Test**] If the end condition is satisfied, **stop**, and return the best solution in current population

[**Loop**] Go to step [**Fitness**]

Wolf Optimization Algorithm. The wolf optimization algorithm (WOA) is one of the bio-inspired meta-heuristics algorithms based on wolf preying behaviour. One of its distinguishing characteristics is the concurrent possession of individual local search capability as well as flocking movement [15]. Therefore, the individual wolf in WOA searches autonomously by memorizing its particular attribute and unites with its peer when the peer is in a better location. Using this approach, long-range inter-communication amongst the wolves that characterize the searching points for candidate solutions is removed since wolves stalk their target in stillness. Also, the swarming behaviour of WOA, unlike most bio-inspired algorithms, is delegated to each individual wolf rather than to a single leader, as opposed particle swarm optimization and Firefly. WOA operates as if there are multiple leaders swarming from different directions to the best solution, instead of one direction for optimum solution by a single flock. WOA is summarized in Algorithm 2.

Algorithm 2: Wolf Optimization Algorithm

Objective function $f(x)$, $x=(x_1,x_2,...,x_d)^T$
Initialize the population of wolves, $x_i(i=1,2,...,W)$
Define and initialize parameters:
r = radius of the visual range
s = step size by which a wolf moves at a time
α = velocity factor of wolf
p_a = a user-defined threshold [0..1], it defines how commonly an enemy appears
WHILE (*t<generations && stopping criteria not met*)
   FOR i = 1 : W
    Prey_new_food_initiatively();
    Generate_new_location();
  //Check whether the next location suggested by the random generator is new.
    Otherwise, repeat generating random location.
    IF($dist(x_i,x_j) < r$ && x_j is better as $f(x_i) < f(x_j)$)
    ELSE IF
     x_i = prey_new_food_passively();
    END IF
    Generate_new_location();
    IF($rand() > p_a$)
     $x_i = x_i + rand() + v$; // escape to a new position.
    END IF
   END FOR
  END WHILE

3.4   Classification

In order to separate phishing website from legitimates ones, two classification algorithms were employed based on SVM and random forest (RF). These algorithms have been widely used in security domain and have shown significant classification performance [4, 13]. The classifiers take as input a dataset containing the class of each piece of data instance to enable the computer to learn the pattern that exists among the instances in the dataset. This pattern is then used to predict the class of new data samples. In this study, the target attribute of the Dataset1 is divided into three categories: malicious, suspicious and genuine. In the case of Dataset2, the target attribute is binary in nature, which involves malicious and genuine cases. Therefore, the goal of each classifier is to extract pattern that reveals the specific group or class each data instance is related within a given dataset.

Support Vector Machine. Support vector machine (SVM) is one of the most popular classifiers that has been employed in several domains. SVM utilizes 'kernel trick' approach to solve a nonlinearly separable problem through mapping of points into a higher-dimensional space. SVM solves the issues of overfitting that are inherent in some learning algorithms. The main idea of SVM is to compute the optimal separating hyperplane between the classes in the dataset by maximizing the margin between the classes' closest points. The maximum margin hyperplane is the one that provides the highest separation between the classes considered. In two-dimensional space, this hyperplane is a line that divides a plane into two distinctive parts corresponding to the classes in a binary classification task. SVM can solve both classification and regression tasks and can handle several types of variables. The separating hyperplanes can be defined as:

$$w_i x_i + b \geq +1, \text{ when } y_i = +1 \tag{1}$$

$$w_i x_i + b \leq -1, \text{ when } y_i = -1 \tag{2}$$

where w is the weight, x is the input, y is the output and b is bias.

Random Forest. Random forest (RF) is a type of decision tree algorithms that are based on ensemble technique. RF produces an ensemble of classifiers based on different decision trees using random feature selection and bagging technique during the training phase. The decision tree generates two categories of nodes, namely, the leaf node labelled as a class and the interior node represented with a feature. During training phase, a diverse subset of training samples is chosen with a replacement to train each decision tree. Entropy is applied to compute the information gain contributed by each feature. Let D represents the dataset with the labelled instances and C as the class such that $C = \{C_1, C_2, C_3, \ldots, C_j\}$, where j is the number of classes considered. In this study, the value of j is set to 2 or 3 depending on the specific dataset used as earlier discussed. Thus, the information needed to identify the class of an instance in the dataset D is denoted as $Info(D) = Entropy(P)$, where P is the class probability distribution such that:

$$P = \left\{ \frac{|C1|}{|D|}, \frac{|C2|}{|D|}, \frac{|C3|}{|D|}, \ldots, \frac{|Cj|}{|D|} \right\} \tag{3}$$

By partitioning D based on the value of a feature F according to subsets $\{D_1, D_2, D_3, \ldots, D_n\}$, $Info(F,D)$ according to F is computed as:

$$Info(F, D) = \sum_{i=1}^{n} \frac{|D_i|}{|D|} Info(D_i) \tag{4}$$

The corresponding information gain after obtaining the value of F is computed as:

$$\text{Gain}(F, D) = \text{Info}(D) - \text{Info}(F, D) \tag{5}$$

Then the *GainRatio* is defined as:

$$\text{GainRatio}(F, D) = \frac{\text{Gain}(F, D)}{\text{SplitInfo}(F, D)} \tag{6}$$

where *SplitInfo(F, D)* denotes the information due to the splitting of D according to the feature F. Random forest uses the majority voting of all the individual decisions to obtain the final decision of the classifier [1].

3.5 Cross-Validation

Cross-validation is a statistical technique that divides data into two parts: one used to train a model and the other used to validate the model. This method is to evaluate and compare learning algorithms [17]. Cross-validation process uses a single parameter called k. The parameter represents the number of partitions in which a given dataset can be divided. Based on this, the process is frequently called k-fold cross-validation. For instance, k = 10 becomes 10-fold cross-validation. In this study, 10-fold cross-validation method is used to train the proposed models because this approach has been widely accepted in the literature to avoid model overfitting and to produce better transparent model for phishing web page detection.

3.6 Evaluation Metric

Evaluation metrics are the methods used in determining the performance of machine learning models. For this research, the models were evaluated using parameters such as accuracy, sensitivity (recall), specificity, precision, F-measure and receiver operating characteristics (ROC). These metrics are briefly discussed as follows:

i.   Accuracy: This is the fraction of predictions that the model gets right. Mathematically, accuracy can be calculated by:

$$\text{Accuracy} = \frac{t_p + t_n}{t_p + t_n + f_p + f_n} \tag{7}$$

ii.   Sensitivity: Sensitivity is a measure of the proportion of actual positive cases that got predicted as positive or (true positive). Sensitivity is also termed as Recall. Therefore, in this case, it refers to the model ability to detect phishing URL correctly. Mathematically, sensitivity is calculated by:

$$\text{Sensitivity} = \frac{t_p}{t_{p+f_n}} \qquad (8)$$

iii. Specificity: This is defined as the proportion of actual negatives, which got predicted as the negative (or true negative). In essence, specificity measures the model ability to correctly detect web pages that are actually legitimate (i.e. not phishing web pages). It can be calculated mathematically by:

$$\text{Specificity} = \frac{t_n}{t_n + f_p} \qquad (9)$$

iv. Precision: This metric represents the proportion of the data instances that the model predicts to be relevant which is truly relevant. It is the number of true positives divided by the number of true positives plus the number of false positives. It is calculated mathematically by:

$$\text{Precision} = \frac{t_p}{t_p + f_p} \qquad (10)$$

v. F-measure: This is the weighted harmonic mean of precision and recall. It is calculated as follows:

$$F - \text{measure} = \frac{2PR}{(P + R)} \qquad (11)$$

Considering the equations above, true positive (t_p) is the number of phishing URLs that were correctly identified as phishing, false positive (f_p) is the number of legitimate URLs that were incorrectly detected as phishing. True negative (t_n) is the number of legitimate URLs that were correctly identified as legitimate, while false negative (f_n) is the number of phishing URLs that were incorrectly detected as legitimate.

4 Results and Discussion

This section explains the process employed in detecting malicious URLs using selected meta-heuristic algorithms for feature selection and SVM and RF for classification. It discusses the experiments performed, simulation tool as well as the result analysis. To evaluate the performance of each of the selected meta-heuristics algorithms, various experiments were carried out using Dataset1 and Dataset2. The experiment was performed on Windows 10 operating system, having a random-access memory (RAM) of 4 GB and 2.50 GHz Intel Core i5 CPU with 500 GB hard disk. Finally, cross-validation using 10-fold was employed to assess the performance of the selected meta-heuristic algorithms on the two phishing datasets.

4.1   Modelling and Interpretation

The experiments were conducted using WEKA 3.8.2 version, which is a simulation tool with different machine learning algorithms for predictive tasks. It has the capability for data preparation, classification, regression, clustering, associate rule mining and visualization. WEKA is a popular tool for data mining. It is an open-source and freely available platform-independent software. It has flexible facilities for scripting experiments.

Classification performance based on SVM. Table 2 presents the classification results of SVM based on all the features in the two datasets that were analysed in this study. From this result, SVM is able to achieve performance accuracy of 86.60% and 93.8% for Dataset1 and Dataset2, respectively. The false positive was 0.109 for Dataset1 and 0.066 for Dataset2. The results of other metrics considered were also promising, which shows the significance of the model for distinguishing phishing web pages from legitimate ones.

Classification performance based on RF. The results obtained based on RF algorithm without feature selection is summarized in Table 3. RF produced better results across the evaluation metrics used in this study. For instance, the accuracy of RF algorithm is 89.4% on Dataset1 and 97.2% on Dataset2. As shown in the table, ROC of the proposed RF model for two datasets was estimated at 96.3% and 99.3%, respectively. This shows the capability of the proposed framework to effectively detect phishing

Table 2 Classification based on SVM only

	Dataset1	Dataset2
Accuracy	0.860	0.938
TP rate	0.860	0.938
FP rate	0.109	0.066
Precision	0.843	0.938
Recall	0.860	0.938
F-measure	0.846	0.938
ROC area	0.900	0.936

Table 3 Classification based on RF only

	Dataset1	Dataset2
Accuracy	0.894	0.972
TP rate	0.894	0.972
FP rate	0.081	0.031
Precision	0.894	0.972
Recall	0.894	0.972
F-measure	0.894	0.971
ROC area	0.963	0.993

web pages. In addition, false alarm was reduced across the two phishing datasets. The importance of the RF model to produce relevant results is very significant as demonstrated by the results obtained with the precision metric.

Classification performance based on GA and SVM. GA was used for feature selection on Dataset1 and Dataset2, each having ten (10) and thirty (30) features, respectively, as earlier highlighted. GA selected five (5) optimal features for Dataset1 and nine (9) optimal features for Dataset2. With the reduced features for modelling, the proposed framework based on GA was able to achieve very promising results without compromising the model performance as shown in Table 4. Using the minimal number of features as selected from the GA algorithm produced accuracy of 83.7% and 93.3% for Dataset1 and Dataset2, respectively. This result reveals that the proposed approach is able to reduce model complexity while still retain better performance.

Classification performance based on GA and RF. Based on the selected features from GA, RF model produced improved results on Dataset2. The model accuracy dropped slightly on Dataset1; however, the ROC result is better than the SVM model with GA selected features. Table 5 summarized the results based on GA and RF algorithms. From this result, an accuracy of 82.3% and 94.4% was obtained on Dataset1 and Dataset2, respectively. This result implied that GA is a good feature subset selection algorithm for developing a model to detect malicious web pages.

Table 4 Classification based on GA and SVM

	Dataset1	Dataset2
Accuracy	0.837	0.933
TP rate	0.837	0.933
FP rate	0.129	0.072
Precision	0.792	0.933
Recall	0.837	0.933
F-measure	0.811	0.933
ROC area	0.866	0.931

Table 5 Classification based on GA and RF

	Dataset1	Dataset2
Accuracy	0.823	0.944
TP rate	0.823	0.944
FP rate	0.135	0.059
Precision	0.804	0.944
Recall	0.823	0.944
F-measure	0.811	0.944
ROC area	0.935	0.986

Table 6 Classification based on WOA and SVM

	Dataset1	Dataset2
Accuracy	0.837	0.933
TP rate	0.837	0.933
FP rate	0.129	0.072
Precision	0.792	0.933
Recall	0.837	0.933
F-measure	0.811	0.933
ROC area	0.866	0.931

Table 7 Classification based on WSA + RF

	Dataset1	Dataset2
Accuracy	0.823	0.944
TP rate	0.823	0.944
FP rate	0.135	0.059
Precision	0.804	0.944
Recall	0.823	0.944
F-measure	0.811	0.944
ROC area	0.935	0.987

Classification performance based on WOA and SVM. Wolf optimization algorithm (WOA) was used for feature subset selection on both datasets. The results were similar to those obtained in GA. WOA also selected five (5) optimal features for Dataset1 and nine (9) attributes for Dataset2. The results are summarized in Table 6.

Classification performance based on WOA and RF. Since the number of features selected by the two meta-heuristic algorithms considered in this study is the same, WOA with RF algorithm produced similar results when compared with GA with RF algorithm. These results are also summarized in Table 7 with an accuracy of 82.3% and 94.4% for Dataset1 and Dataset2, respectively. These results show that the two meta-heuristic algorithms considered in this study have produced promising results comparable to the results obtained without feature subset selection. This approach reduced the model complexity and guaranteed an improved prediction time when deployed in a real-life environment.

4.2 Models Comparison Based on SVM as a Classifier

Figures 3 and 4 show the comparison of the results of the models based on SVM algorithm for phishing URLs detection on Dataset1 and Dataset2, respectively. The meta-heuristic algorithms have really demonstrated comparable results considering the results obtained with the evaluation metrics.

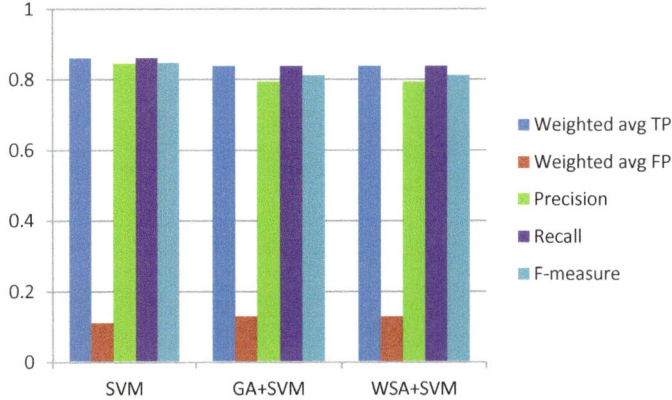

Fig. 3 Comparison of the SVM models based on Dataset1

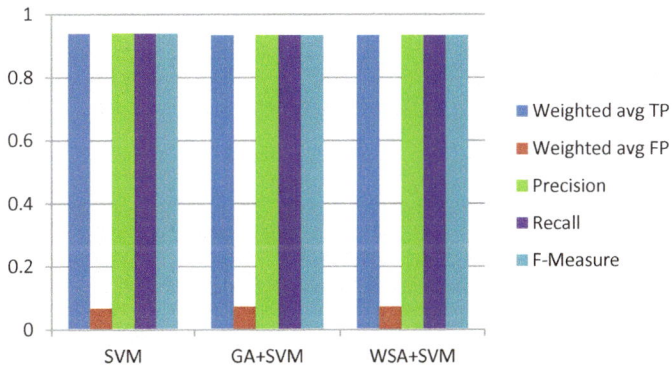

Fig. 4 Comparison of the SVM models based on Dataset2

Furthermore, the time taken by each model based on SVM classifier with and without feature subset selection revealed that the models based on meta-heuristic algorithms produced in complexity as demonstrated by their training time in seconds. However, the WOA and SVM model took the least time, having 0.16 and 8.72 s for Dataset1 and Dataset2, as shown in Figs. 5 and 6, respectively.

4.3 Models Comparison Based on RF as a Classifier

Similarly, Figs. 7 and 8 show the performance of the different RF models on Dataset1 and Dataset2, respectively. As shown in these figures, RF models based on the two meta-heuristic algorithms produced comparable results when compared with the RF model without feature subset selection.

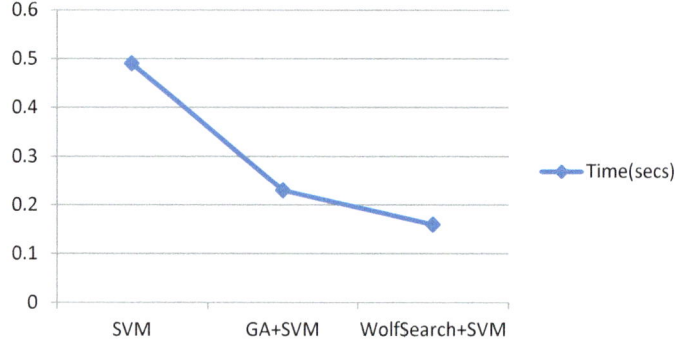

Fig. 5 Comparison of time taken for Dataset1 with SVM models

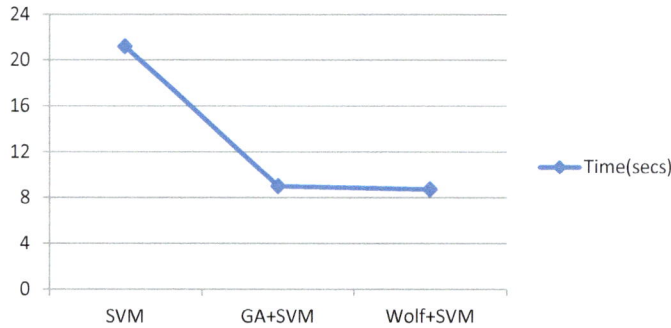

Fig. 6 Comparison of time taken for Dataset2 with SVM models

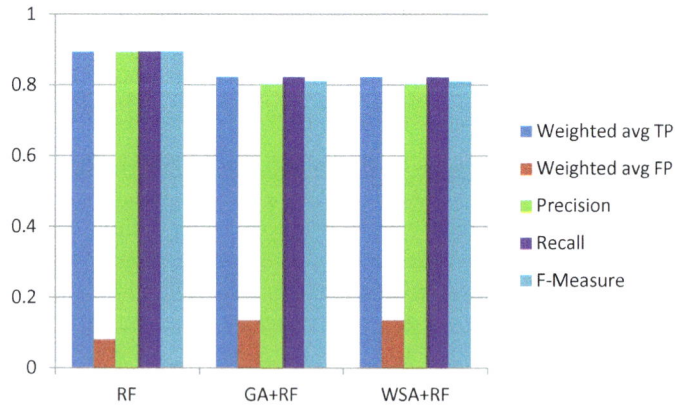

Fig. 7 RF models comparison based on Dataset1

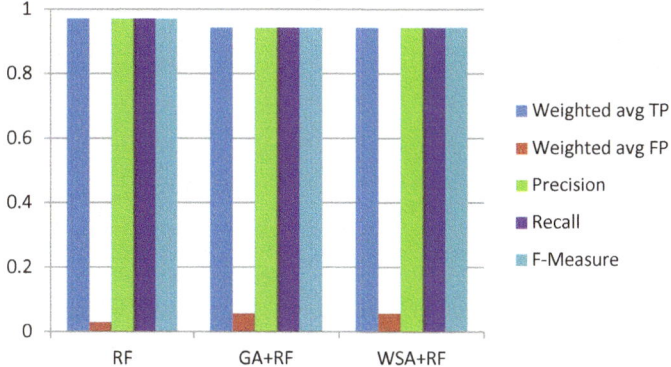

Fig. 8 RF models comparison based on Dataset2

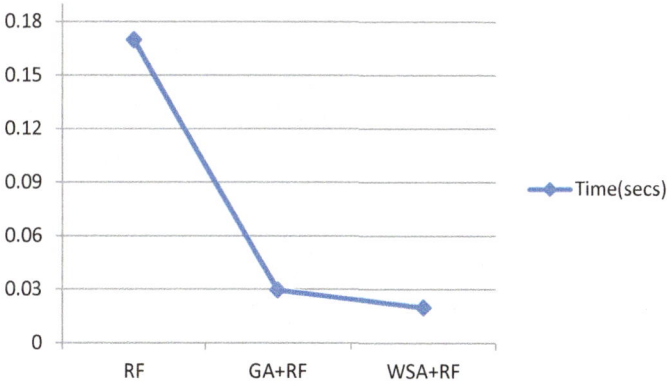

Fig. 9 Comparison of time taken for Dataset1 with RF models

Similarly, it was observed that the time taken by the WOA algorithm on the two datasets was reduced when compared with GA and RF model without feature subset selection (see Figs. 9 and 10). This finding reveals that WOA algorithm is a better candidate to produce a model with less complexity when feature dimensionality reduction is being considered to build a prototype of phishing URLs detection system. WOA and RF model took the least time, having 0.02 and 0.08 s for Dataset1 and Dataset2, respectively.

4.4 Comparison with Existing Models

This section shows the comparison of the results of the proposed models in this study with the existing model that have been developed in the literature. Table 8

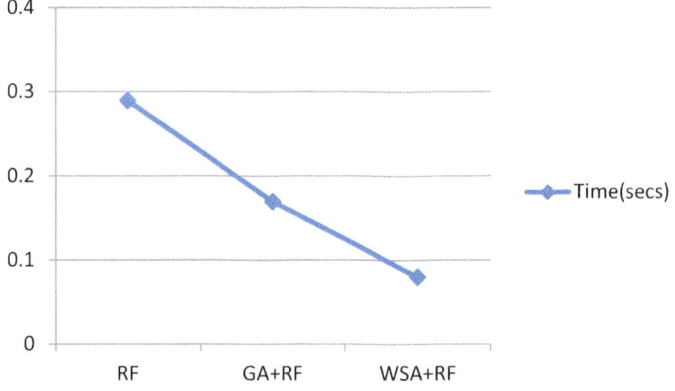

Fig. 10 Comparison of time taken for Dataset2 with RF models

Table 8 Baseline comparison with the existing model based on Dataset2

Approaches	Models	TPR	FPR	Accuracy	Precision	Recall	F-Measure	ROC
Proposed	**RF**	**0.972**	**0.031**	**0.972**	**0.972**	**0.972**	**0.971**	**0.993**
Proposed	**WOA + RF**	**0.944**	**0.059**	**0.944**	**0.944**	**0.944**	**0.944**	**0.987**
Aydin and Baykal [6]	SMO	0.938	0.066	0.938	0.938	0.938	0.938	0.936

summarizes the results of the various models under consideration. As discussed in the previous section, Naïve Bayes and SMO approach were studied in the work of [6]. From this table, the proposed models were able to outperform the state-of-the-art producing promising results across the different evaluation metrics considered in this study. In order to ensure objectivity in the models' comparison in this section, the SMO experiment in [6] was conducted and the results reported in Table 8 for comparison with the proposed models in this study.

5   Conclusion

In this study, the performance of two meta-heuristic algorithms, GA and WOA, was examined for feature subset selection towards identifying malicious web pages. Due to the high exploration capability of features selection of the two meta-heuristic algorithms, they demonstrated promising results when compared with models without feature subset selection. SVM and RF classifiers were considered to develop effective classification models for the proposed framework. From the results obtained, the models based on RF outperformed SVM model using performance metrics such as accuracy, sensitivity, specificity, recall, precision, F-measure and ROC. The results of RF and WOA with RF models were compared with existing state-of-the-art model, and the outcome revealed that the proposed models in this study gave better performance when compared with the state-of-the-art. Furthermore, this study observed that WOA meta-heuristic optimization algorithm gave the least running time and reduced model complexity when compared with other models developed in this study. Although the performance of the models without feature subset selection is better, however, models based on meta-heuristic optimization were able to reduce complexity with a slight reduction in the accuracy. In future, the authors intend to investigate other feature categories that may improve the performance of the models for phishing detection. In addition, other feature selection methods need to be investigated.

References

1. Priya, R.: An ideal approach for detection of phishing attacks using naïve bayes classifier. Int. J. Comput. Trends Technol. (Technology) **40**, 84–87 (2016)
2. Sahoo, D., Liu, C., Hoi, S.C.: Malicious URL detection using machine learning: a survey, arXiv preprint arXiv:1701.07179 (2017)
3. Lin, M.-S., Chiu, C.-Y., Lee, Y.-J., Pao, H.-K.: Malicious URL filtering—A big data application. In: 2013 IEEE International Conference on Big Data, pp. 589–596 (2013)
4. Vanhoenshoven, F., Nápoles, G., Falcon, R., Vanhoof, K., Köppen, M.: Detecting malicious URLs using machine learning techniques. In: *2016* IEEE Symposium Series on Computational Intelligence (SSCI), pp. 1–8 (2016)
5. Gupta, S.: Efficient malicious domain detection using word segmentation and BM pattern matching. In: 2016 International Conference on Recent Advances and Innovations in Engineering (ICRAIE), pp. 1–6 (2016)
6. Aydin, M., Baykal, N.: Feature extraction and classification phishing websites based on URL. In: 2015 IEEE Conference on Communications and Network Security (CNS), pp. 769–770 (2015)
7. Li, Y., Yang, Z., Chen, X., Yuan, H., Liu, W.: A stacking model using URL and HTML features for phishing webpage detection. Future Gener. Comput. Syst. **94**, 27–39 (2019)
8. Wang, R., Zhu, Y., Tan, J., Zhou, B.: Detection of malicious web pages based on hybrid analysis. J. Inform. Secur. Appl. **35**, 68–74 (2017)
9. Adewole, K.S., Akintola, A.G., Salihu, S.A., Faruk, N., Jimoh, R.G.: Hybrid rule-based model for phishing URLs detection. In: International Conference for Emerging Technologies in Computing, pp. 119–135 (2019)
10. Xuan, J., Yongzhen, L.: The detection method for two-dimensional barcode malicious URLs based on the hash function. In: 2016 3rd International Conference on Information Science and Control Engineering (ICISCE), pp. 702–705 (2016)
11. Darling, M., Heileman, G., Gressel, G., Ashok, A., Poornachandran, P.: A lexical approach for classifying malicious URLs. In: 2015 international conference on high performance computing & simulation (HPCS), pp. 195–202 (2015)
12. Lee, S., Kim, J.: Warningbird: A near real-time detection system for suspicious urls in twitter stream. IEEE transactions on dependable and secure Comput. **10**, 183–195 (2013)
13. Adewole, K.S., Han, T., Wu, W., Song, H., Sangaiah, A.K.: Twitter spam account detection based on clustering and classification methods. J. Supercomput., 1–36 (2018)
14. Bhardwaj, T., Sharma, T.K., Pandit, M.R.: Social engineering prevention by detecting malicious URLs using artificial bee colony algorithm. In: Proceedings of the Third International Conference on Soft Computing for Problem Solving, pp. 355–363 (2014)

15. Tang, R., Fong, R., Yang, X.-S., Deb, S.: Wolf search algorithm with ephemeral memory. In: Seventh International Conference on Digital Information Management (ICDIM 2012), pp. 165–172 (2012)
16. Chu, Z., Gianvecchio, S., Wang, H., Jajodia, S.: Detecting automation of twitter accounts: are you a human, bot, or cyborg? IEEE Trans. Dependable Secure Comput. 9(6), 811–824 (2012)
17. Refaeilzadeh, P., Tang, L., Liu, H.: Cross-validation. Encyclopedia of database systems, pp. 532–5382 (2009)
18. Babagoli, M., Aghababa, M.P., Solouk, V.: Heuristic nonlinear regression strategy for detecting phishing websites. Soft. Comput. 23, 4315–4327 (2019)
19. Sohrabi, M.K., Karimi, F.: A feature selection approach to detect spam in the Facebook social network. Arabian J. Sci. Eng. 43, 949–958 (2018)
20. Sahingoz, O.K., Buber, E., Demir, O., Diri, B.: Machine learning based phishing detection from URLs. Expert Syst. Appl. 117, 345–357 (2019)

Improved Cloud-Based N-Primes Model for Symmetric-Based Fully Homomorphic Encryption Using Residue Number System

K. J. Muhammed⊙**, R. M. Isiaka, A. W. Asaju-Gbolagade, K. S. Adewole, and K. A. Gbolagade**

Abstract Encryption schemes that allow computation to be performed on an encrypted data are required in modern real-world applications. This technology is a necessity for cloud computing, processing resources and share storage, in order to preserve integrity and privacy of data. The existing partial and fully homomorphic encryption (PHE and FHE) for both asymmetric and symmetric approaches are still suffering from efficiency in terms of encryption execution time and very large ciphertext file produced at the end of encryption. In this paper, we consider symmetric approaches and focus on overcoming the drawbacks of N-prime model using residue number system (RNS). The experimental results obtained show that the proposed RNS-based N-prime model for symmetric-based FHE improves the system latency and reduces the ciphertext file expansion by approximately 72% as compared to the existing N-prime model. The proposed improved N-prime model is guaranteed to provide optimum performance and reliable solution for securing integrity and privacy of user's data in the cloud.

Keywords Homomorphic encryption · Fully homomorphic encryption · Residue number system · Cloud-based · N-prime model · RNS-based N-prime model · Symmetric encryption

K. J. Muhammed (✉)
Department of Educational Technology, University of Ilorin, Ilorin, Nigeria
e-mail: jimoh.km@unilorin.edu.ng

R. M. Isiaka · K. A. Gbolagade
Department of Computer Science, Kwara State University, Malete, Nigeria
e-mail: abdulrafiu.isiaka@kwasu.edu.ng

K. A. Gbolagade
e-mail: kazeem.gbolagade@kwasu.edu.ng

A. W. Asaju-Gbolagade · K. S. Adewole
Department of Computer Science, University of Ilorin, Ilorin, Nigeria
e-mail: ayisatwuraola@gmail.com

K. S. Adewole
e-mail: adewole.ks@unilorin.edu.ng

© The Author(s), under exclusive license to Springer Nature Switzerland AG 2021
H. Chiroma et al. (eds.), *Machine Learning and Data Mining for Emerging Trend in Cyber Dynamics*, https://doi.org/10.1007/978-3-030-66288-2_8

1   Introduction

Nowadays, there is a growing interest on cloud technology globally, which informs the efforts by many companies toward implementation of cloud technologies in their business processes. Cloud technology with its own unique feature comprises cloud services, storage, computing, grid computing and distributed computing, among others. Since the idea of cloud technology was projected, there is perpetual attention for research across the world. It has become the hottest research area in the domain of information technology which has been realized as unital of the technology that postures the next-generation computing revolution [1]. Cloud computing offers many opportunities to enterprises for their customers to access and operate computing services as "pay-as-you-go". Customers saddle with responsibilities of batch processing and web applications explored cloud computing model as an efficient alternative way to acquiring and managing private data centers [2].

Cloud computing as described by US National Institute of Standards and Technology (NIST) as a model that allows and provides convenient ways of accessing unlimited pool of shared resources such as on-demand network access, servers, applications, storage and services categorized as software and hardware that can be rapidly made available to the client as soon as requested through the internet, these resources can be reviewed upward and downward automatically based on client's demand. Thus, cloud computing has five essential characteristics, three service models and four deployment models such as on-demand self-service, broad network access, resource pooling, rapid elasticity and measured service, software-as-a-service (SaaS), platform-as-a-service (PaaS), and infrastructure-as-a-service (IaaS) and private cloud, public cloud, community cloud and hybrid cloud, respectively, as shown in Fig. 1 [1, 3].

The quick move toward adoption of cloud computing has raised critical fears on an ultimate fact for the success of information systems, communication, virtualization, data availability and integrity, public auditing, scientific application and information security. Despite the demand of businesses and individual organizations to transfer their data to the cloud, security in terms of data integrity and confidentiality, trust in

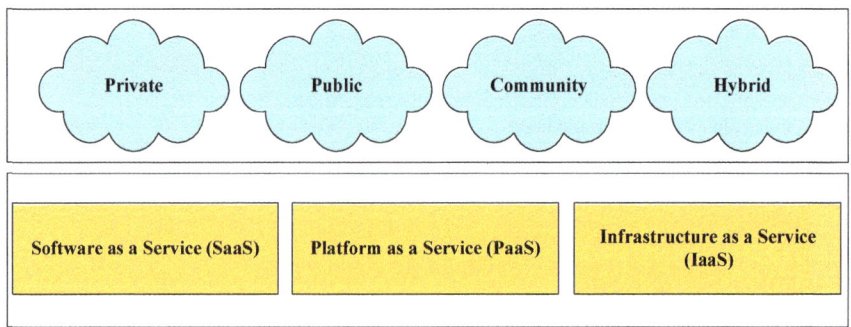

Fig. 1 Cloud solutions based on the system's deployment and service model [1]

terms of data accessibility and availability, and privacy issue have remained as the challenges for embracing cloud computing [4]. Thus, security anxieties arise when transforming data from locally owned storage to third parties cloud service provider's (CSP) storage. It is expected of CSP to encrypt user's data before storing it on their cloud server. Several cryptographic schemes such as encryption and decryption, key exchange and steganography have been used to address the security challenges facing the adoption of cloud computing. Encryption means conversion of plain text to ciphertext, and it can be used in various applications for moving data from one location to another. Encryption algorithms such as DES, AES, RSA, Blow fish and many more are meant for secured storage of data and operations on cloud computing server [5, 6]. The traditional encryption schemes seem to be weak for securing the integrity and confidentiality of stored data in the cloud computing server as it requires secret key to be provided by the users so as to decrypt the data prior to the computations, and this threatens the privacy, confidentiality and integrity of stored data.

Homomorphic encryption (HE) was introduced to overcome this concern. HE is the process of delegating processing of an encrypted data without having access to the raw data, and produces encrypted result when decrypt result matches the result of the same operations carried out on the raw data. It is considered as one of the most efficient solutions for protecting the confidentiality and user's information privacy in cloud-based services. Also, HE is grouped as partial homomorphic encryption (PHE) and fully homomorphic encryption (FHE) where PHE ropes either to additive or multiplicative operation, whereas FHE supports both operations and seems to be more secured and efficient than PHE as it poses properties of both operations [7].

The concept of FHE scheme was first introduced by Gentry and Boneh [8] and this was used to address cryptography central open issues. The scheme allows arbitrary functions to be computed over an encrypted data with no reference to decryption key [9]. The development of FHE is a revolutionary advancement, greatly extending the scope of the computations, which can be applied to process encrypted data homomorphically. Gentry [10] presented the first version of FHE scheme using ideal lattices; however, his construction was very complex and impracticable. Since then, FHE has been an interesting research area and several developments that centre on the application, implementation and improvement of FHE schemes have also been studied [11]. There are numerous practical applications of FHE in the real world such as buyer privacy in publicity, medical applications (genomics and health), national security, private prediction, private learning, education, monetary privacy and forensic image recognition, among others [12–14].

There are three major problems associated with the design and practical implementation of FHE scheme such as security problem, time efficiency problem and cipher expansion problem. However, several researchers [15, 16] have contributed to the improvement of FHE in terms of security that is based on hard mathematical problems, and the cost of breaking such scheme at a given security parameter λ should be at least 2λ. Also, in the last 11 years, progress has been made on time efficiency problem by different researchers such as [17–21], and the proposed schemes were categorized into three generations: FHE based on ideal lattices; the second generation is based on learning with error (LWE) and ring learning with error (RLWE) problems and the third generation is GSW which is also based on RLWE. However, each of these proposed schemes still needs further improvement for it to be efficiently useful in real-world applications.

It is highlighted in [7, 17] that cipher expansion for both asymmetric and symmetric FHE schemes are producing very large ciphertext size between hundreds and tens of thousands as compared to original plaintext size. For instance, ciphertext produced in [17] is till 400,000 times that of the original plaintext and likewise in [7]. These show the growth rate of ciphertext size in an existing FHE schemes which call for serious concern and constitute a major problem of the schemes because it will require large bandwidth to transfer such ciphertext through the network. We proposed scheme conversion concept for encryption scheme originally mentioned in [7] as symmetric-based FHE in order to address the issue. In addition, we provide efficient parallelization in encryption process which produces small ciphertext sizes. However, this study focuses on tackling cipher expansion problem on symmetric-based FHE scheme as proposed by [7] using residue number system.

The organization of this paper is as follows. Section 2 describes the related studies on homomorphic encryption; Sect. 3 discusses the research methodology of the proposed framework. Section 4 presents results findings and discussion. Section 5 is the conclusion to the study.

2   Related Works

HE schemes are classified into three major categories, as shown in Fig. 2 based on supported homomorphism operations and properties which are partial homomorphic encryption (PHE), somewhat homomorphic encryption (SWHE) and fully homomorphic encryption (FHE). This scheme is called PHE when it allows any number of either addition as in [22, 23] or multiplication operation as described

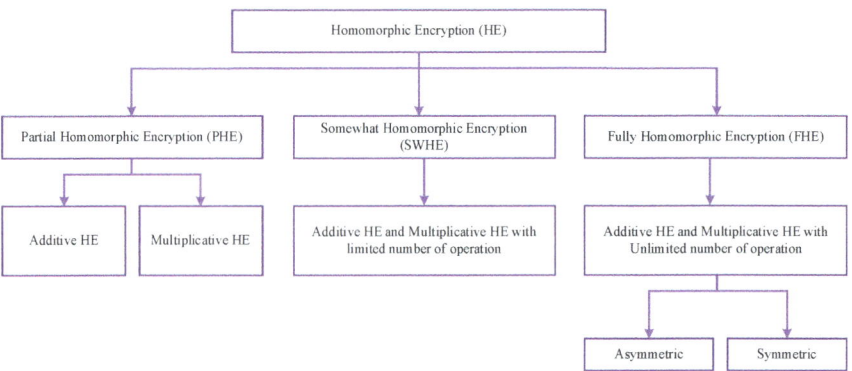

Fig. 2 Categorization of homomorphic encryption

in [24, 25]. However, SWHE supports both operations but with limited number of times allowed to evaluate the decryption function because of noise growth as illustrated in [8, 26]. Likewise in [27], the authors presented technique based on SWHE that efficiently secures pre-processing model for dishonest majority in multiparty computation over the ring $\mathbb{Z}_{2_k}$. Finally, improvement over SWHE is FHE which allows unlimited number of operations for both additive and multiplicative homomorphism to be performed on an encrypted data without increase in noise during the computation. This type of schemes is extensively discussed in [8, 10, 21, 28].

Parallel to conventional cryptography, FHE schemes are also categorized into asymmetric encryption and symmetric encryption. Asymmetric FHE schemes protect the privacy and confidentiality of data by using different pair of keys (public key and private key) to encrypt the data. Recent research on asymmetric-based FHE has been proposed by [10, 17, 19, 20, 23, 28–33]. However, these schemes suffer from very high computational overhead, cipher expansion and time efficiency which make it difficult to practice in real-world application. However, researchers such as [7, 27, 34–41] have proposed symmetric-based FHE schemes due to its strength in low computational overhead and practical consideration for real-world deployments. Moreover, some of these symmetric-based FHE schemes [7, 41–44] still suffer from large cipher expansion and low immunity against attack. This demanded that a robust security resistant and cipher expansion reduction are highly needed for these symmetric-based FHE schemes [40, 45].

3 Research Method

This section discusses the research methodology in relation to homomorphic encryption, N-prime model, residue number system and proposed RNS-based N-prime model for symmetric FHE.

3.1 Homomorphic Encryption

Just after discovery of RSA cryptosystem, HE was introduced by [24] as privacy homomorphism that protects the privacy, integrity and confidentiality of data. Generally, HE scheme allows computations on an encrypted variable without the need for decryption prior to the computation or the use of secret key [14, 46]. HE scheme contains four algorithms such as key generation (KeyGen), encryption (Enc), evaluation function (Eval(f)) and decryption algorithm (Dec) which can be described as follows:

KeyGen: (pk, sk) ← KeyGen(k), choose parameter k and the algorithm will generate *pk* as public key and *sk* as private key.

Enc: $c \leftarrow Enc(m, pk)$, where m is the plaintext-message, pk is public key and c is the ciphertext.

Dec: $m \leftarrow Dec(c, sk)$, where c is the ciphertext, sk is the private key and m is the plaintext message.

Eval(f): $c^* \leftarrow Evaluate\ (f, Enc(m1), Enc(m2), \ldots\ldots Enc(mn))$

However, for the scheme to be homomorphic in nature, it must obey the following properties:

$Dec(c^*, sk) = f(m1, m2, \ldots\ldots\ldots mn)$
$Dec(C1\ op\ C2, sk) = M1\ op\ M2$

where $C1$ and $C2$ are $Enc(M1)$ and $Enc(M2)$, respectively, op is the group operation (addition or multiplication) in the ciphertext and plaintext space, respectively.

3.2   N-Primes Model

In [7], an encryption scheme called N-primes model is investigated for constructing a symmetric-based FHE scheme which converts each plaintext character into ASCII code and passes it to the encryption algorithm $c = m + r * l * n$ where ct is the ciphertext, m is the plaintext message and $m \in [0, L-1]$, r is the noise added to the ciphertext, l is a prime big integer and $n = p1 * p2 * p3 * \ldots pi$ is the multiplication of numerous prime numbers resulting in one ciphertext for each character in the plaintext message. Thus, encryption and decryption are as follows:

Key generation

Generate n

Generate $l \in Z_n$

where n is a multiplication of various prime numbers $(n = P_1 * P_2 * P_3 * \ldots P_i)$
and l is a big prime integer value

Encryption algorithm

$c = m + r * l * n$ where c is the ciphertext, $m \in [0, L-1]$ and r is random number

Decryption algorithm

$$m = c \bmod l$$

Proof of homomorphism

Additive property

$$m_3 = c_1 + c_2 = (m_1 + r_1 * l * n) + (m_2 + r_2 * l * n)$$
$$= (m_1 + m_2) + l * n(r_1 + r_2)$$
$$since \ l * n(r_1 + r_2) mod \ l = 0 \ then$$
$$m_3 = (m_1 + m_2)$$

Multiplicative Property

$$m_3 = c_1 * c_2 = (m_1 + r_1 * l * n) * (m_2 + r_2 * l * n)$$

$$m_3 = [(m_1 * m_2 + m_1 * r_2 * l * n) + \{(m_2 * r_1 * l * n) + (r_1 * l * n)(r_2 * l * n)\}]$$

$$m_3 = \left[(m_1 * m_2 + m_1 * r_2 * l * n) + \left\{(m_2 * r_1 * l * n) + (r_1 * r_2 * l^2 * n^2)\right\}\right]$$

$$m_3 = \left(m_1 * m_2 + m_1 * l * \left\{r_2 * n + r_1 * n * m_2 + r_1 * r_2 * l * n^2\right\}\right)$$

$$Since\left[l * \left\{r_2 * n + r_1 * n * m_2 + r_1 * r_2 * l * n^2\right\}\right] mod \ l = 0$$
$$and \ multiple \ of \ l \ mod \ l = 0 \ then$$

$$m_3 = m_1 * m_2 + m_1 * 0$$

$$m_3 = m_1 * m_2$$

3.3 Residue Number System

The residue number system (RNS) is a non-weighted and carry-free number system that speeds up arithmetic operations by dividing them into smaller parallel operations.

It is characterized based on its unique properties which include fault tolerance, modularity, parallelism and carry-free operations, among others. RNS is very efficient in terms of handling arithmetic operations (addition, subtraction and multiplication) [47–50]. RNS can as well be described as moduli set of positive pairwise relatively prime numbers $(m_1, m_2 \ldots m_n)$ and the product of all the modulo $(M = m_1 \times m_2 \times \ldots \times m_n)$ that produces the dynamic range (DR) of such moduli set. Any integer X in the range $[0, M - 1]$ can be uniquely represented by an ordered set of residues $(x_1, x_2, \ldots, x_n)$. Each residue x_i is represented by $x_i = X \bmod m_i$.

To perform residue arithmetic operations such as addition or multiplication can be achieved by adding or multiplying corresponding digits relative to the modulus for their position [50]. For a given moduli set $\{m_1, m_2, \ldots, m_N\}$,

$X = \{x_1, x_2, \ldots, x_N\}$ and $Y = \{y_1, y_2, \ldots, y_N\}$ such that $x_i = |X|m_i$ and $y_i = |Y|m_i$ where $i = 1, 2, \ldots, N$, then we may define the residue arithmetic operations as follows:

$$X \otimes Y = Z$$

$$X \otimes Y = \{x_1, x_2, \ldots, x_N\} \otimes \{y_1, y_2, \ldots, y_N\} = \{z_1, z_2, \ldots, z_N\} = Z$$

where $z_i = |x_i \otimes y_i|m_i$ and $\otimes$ is residue arithmetic operator $(+, *, \text{ and } -)$. As an example, with the moduli-set $\{3, 4, 5\}$, the representation of fifteen is $\{0, 3, 0\}$, that of twenty-three is $\{2, 3, 3\}$, and adding the two residue numbers yield $\{2, 2, 3\}$ which is the representation for thirty-eight and their product is $\{0, 1, 0\}$ which is three hundred and forty-five in that system.

3.4 Proposed RNS-Based N-Primes Model for Symmetric FHE

Based on the N-prime model, RNS-based N-primes model is built without altering its homomorphic properties. The steps are illustrated in Fig. 3 and will be explained below.

The plaintext in a ring Z_n will be checked if it is numeric and pass it to RNS forward conversion module, otherwise, generate ASCII codes for each of the text characters before passing it to RNS forward conversion module.

Key Generation: This produces very big prime integer value as the secret key for the entire scheme and only to be known by owner of data.

Noise Generation: This generates a set of random integer r in a ring Zn as the noise to the entire scheme.

N-Primes Generation: This also generates product of prime numbers $\left(np_i = \prod_{i=1}^{n} np_1 * np_2 \ldots np_n \right)$

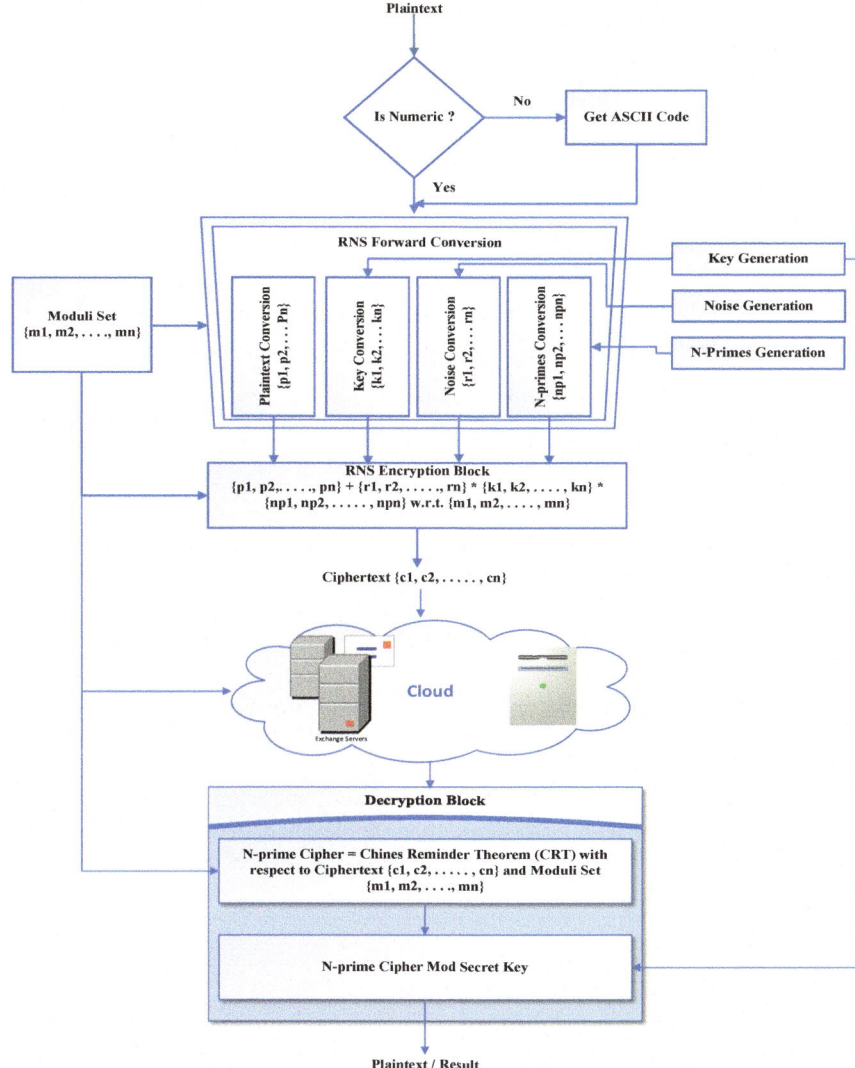

Fig. 3 Proposed RNS-based N-prime model

Moduli Set: It takes in n as the input and generates set of pairwise relatively prime integers and moduli set $\{3^n - 2, 3^n - 1, 3^n\}$ as proposed in [45] which will be used throughout this study, due to its large dynamic range.

RNS Forward Conversion: It requires five (5) parameters as input argument (plaintext (p_i), secret key (k_i), noise (r_i), n-prime numbers (np_i) and the moduli set (m_i), where $i = 1, 2,..., n$), convert p_i, k_i, r_i and np_i into residues with respect to m_i and send out four outputs into RNS encryption block.

RNS Encryption Block: Based on output received from RNS forward conversion block, the encryption block will perform RNS arithmetic operation in parallel with respect to moduli set m_i and produce the ciphertext.

$$Ciphertext\{c_1, c_2, ..., c_n\} = \{p_1, p_2, \ldots, p_n\} + \{r_1, r_2, \ldots, r_n\} * \{k_1, k_2, \ldots, k_n\}$$
$$* \{np_1, np_2, \ldots, np_n\} w.r.t.\{m_1, m_2, \ldots, m_n\}$$

However, the ciphertext can then be transmitted to the Cloud as shown in Fig. 3 and possibly, homomorphic bit operations can be carried out at the cloud server using RNS arithmetic operation based on ciphertext residues cx_i and cy_i with respect to moduli set m_i and the result will be sent to decryption block.

Decryption Block: The decryption block adopts Chinese Remainder Theorem as a backward conversion algorithm with the use of ciphertext residues (czi) and moduli set mi to obtain N-prime cipher.

$$X = \left| \sum_{i=1}^{k} \left| M_i^{-1} \right|_{mi} x_i \right|_M$$

where $M = \Pi_{i=1}^{k} m_i$, $M_i = \frac{M}{m_i}$ and M_i^{-1} is the multiplicative inverse of M_i with respect to m_i.

Finally, user will now use the generated secret key together with N-prime ciphertext to get the result of the operation.

The following experiments demonstrate the generation of the moduli set, secret key, other corresponding values and to show how these values are used for encryption and ciphertext reduction without loosen homomorphic behavior of the encryption and decryption block.

Experiment 1

Let the big prime number sk = 524287 be the secret key, random numbers r = 687529 as the noise, product of n-primes np = 23 * 73 * 101 = 69579 and moduli set $\{3^n - 2, 3^n - 1, 3^n\}$. When n = 23, then moduli set mi = [94143178825, 94143178826, 94143178827], let message m1 = 65 and message m2 = 66, the RNS-forward conversion for each of these parameters with respect to mi are:

$$m1 = [65, 65, 65]$$
$$m2 = [66, 66, 66]$$
$$sk = [524287, 524287, 524287]$$
$$r = [687529, 687529, 687529]$$
$$np = [69579, 69579, 69579]$$

To calculate the ciphertexts ct1 and ct2.

$$ct1 = |m1 + r * sk * np|_{mi}$$
$$= |[65, 65, 65] + [687529, 687529, 687529]$$
$$* [524287, 524287, 524287]$$
$$* [69579, 69579, 69579]|_{[94143178825, 94143178826, 94143178827]}$$

$$ct1 = [31330438157, 31330171748, 31329905339]$$

$$ct2 = |m2 + r * sk * np|_{mi} = |[66, 66, 66] + [687529, 687529, 687529]$$
$$* [524287, 524287, 524287]$$
$$* [69579, 69579, 69579]|_{[94143178825, 94143178826, 94143178827]}$$

$$ct2 = [31330438158, 31330171749, 31329905340]$$

Proof of Homomorphism Properties

Additive Property

$$ct3 = | ct1 + ct2 |_{mi}$$
$$= |[31330438157, 31330171748, 31329905339]$$
$$+ [31330438158, 31330171749, 31329905340]|_{[94143178825, 94143178826, 94143178827]}$$

$$ct3 = [62660876315, 62660343497, 62659810679]$$

To decrypt ct3,

$$m3 = chinese\_remainder(m_i, \ ct3) \ mod \ chinese\_remainder(m_i, sk)$$

m3 $=$ chinese_remainder([94143178825, 94143178826, 94143178827], [62660876315, 62660343497, 62659810679]) mod chinese_remainder([94143178825, 94143178826, 94143178827], [524287, 524287, 524287])

$$m3 = 131 m3 = 131$$

$$m3 = m1 + m2 = 65 + 66 = 131$$

Multiplicative Property

$ct3 = \mid ct1 * ct2 \mid_{mi}$

$\quad = \mid [31330438157, 31330171748, 31329905339]$

$\quad * [31330438158, 31330171749, 31329905340] \mid_{[94143178825, 94143178826, 94143178827]}$

$$ct3 = [37639680706, 51114847858, 18251522556]$$

To decrypt ct3,

$$m3 = \text{chinese\_remainder}(m_i, \ ct3) \bmod \text{chinese\_remainder}(m_i, sk)$$

m3=chinese_remainder([94143178825, 94143178826, 94143178827], [37639680706, 51114847858, 18251522556]) mod chinese_remainder([94143178825, 94143178826, 94143178827], [524287, 524287, 524287])

$$m3 = 4290$$

$$m3 = m1 * m2 = 65 * 66 = 4290$$

Therefore, the proof shows that the proposed model supports both homomorphic addition and multiplication.

Experiment 2

The experiment is also carried out on a string plaintext that contains message "RNS based techniques for encryption of CAP in 2020" with secret key k = 6700417, random number r = 345957466 and product of n-prime np = 31 * 53 * 71 = 116653 and moduli set $\{3^n - 2, 3^n - 1, 3^n\}$. When n = 15, then moduli set $m_i = [14348905, 14348906, 14348907]$. Thus, the encryption is as follows:

[[13889688, 9226958, 7190942], [13889684, 9226954, 7190938], [13889689, 9226959, 7190943], [13889638, 9226908, 7190892], [13889704, 9226974, 7190958], [13889703, 9226973, 7190957], [13889721, 9226991, 7190975], [13889707, 9226977, 7190961], [13889706, 9226976, 7190960], [13889638, 9226908, 7190892], [13889722, 9226992, 7190976], [13889707, 9226977, 7190961], [13889705, 9226975, 7190959], [13889710, 9226980, 7190964], [13889716, 9226986, 7190970], [13889711, 9226981, 7190965], [13889719, 9226989, 7190973], [13889723, 9226993, 7190977], [13889707, 9226977, 7190961], [13889721, 9226991, 7190975], [13889638, 9226908, 7190892], [13889708, 9226978, 7190962], [13889717, 9226987, 7190971], [13889720,

Table 1 RNS-based N-prime model vs N-prime model

File size	Proposed RNS based N-prime model		N-prime model [7]	
	Encryption time (MS)	Ciphertext size (Bytes)	Encryption time (MS)	Ciphertext size (Bytes)
2 KB	0.99	5341	1.01	18789
4 KB	1.99	10633	2.99	37529
7 KB	4.98	21217	6.97	75009
14 KB	9.97	42385	10.01	149969
28 KB	18.94	84721	39.89	299889
55 KB	30.43	169393	34.90	599729
110 KB	75.76	338737	78.79	1199409
220 KB	161.54	677425	186.53	2398769
440 KB	319.18	1354801	348.02	4797489
880 KB	741.02	2709553	838.71	9594929
2 MB	3390.83	5419057	5289.16	19189809

9226990, 7190974], [13889638, 9226908, 7190892], [13889707, 9226977, 7190961], [13889716, 9226986, 7190970], [13889705, 9226975, 7190959], [13889720, 9226990, 7190974], [13889727, 9226997, 7190981], [13889718, 9226988, 7190972], [13889722, 9226992, 7190976], [13889711, 9226981, 7190965], [13889717, 9226987, 7190971], [13889716, 9226986, 7190970], [13889638, 9226908, 7190892], [13889717, 9226987, 7190971], [13889708, 9226978, 7190962], [13889638, 9226908, 7190892], [13889673, 9226943, 7190927], [13889671, 9226941, 7190925], [13889686, 9226956, 7190940], [13889638, 9226908, 7190892], [13889711, 9226981, 7190965], [13889716, 9226986, 7190970], [13889638, 9226908, 7190892], [13889656, 9226926, 7190910], [13889654, 9226924, 7190908], [13889656, 9226926, 7190910], [13889654, 9226924, 7190908], [13889652, 9226922, 7190906]].

Experiment 3
In experiment 3, small numbers were used such as 3 for secret key, 5 for random number, $2 * 3 * 5 = 30$ for product of n-primes and n = 3 to obtain moduli set = [25–27]. These parameters were used to encrypt message of file size 931 Bytes in 0.47 MS and the ciphertext file is shown in Appendix A. The proposed model has also been tested for 11 different file sizes; encryption execution time and ciphertext file size were recorded and compared with original N-prime model as shown in Table 1.

4 Results and Discussion

The proposed RNS-based N-prime model and N-prime model implementations are done under Python using Dell Laptop having the following specifications: 64 bits

Fig. 4 Encryption time

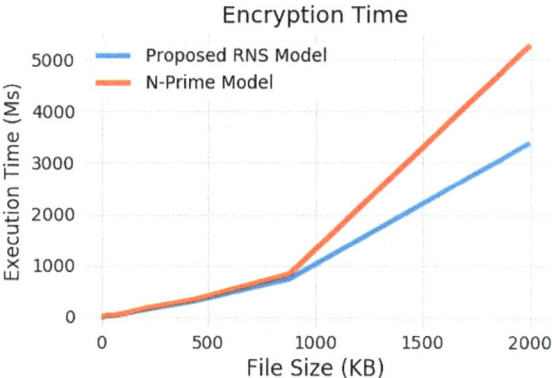

Fig. 5 Ciphertext expansion

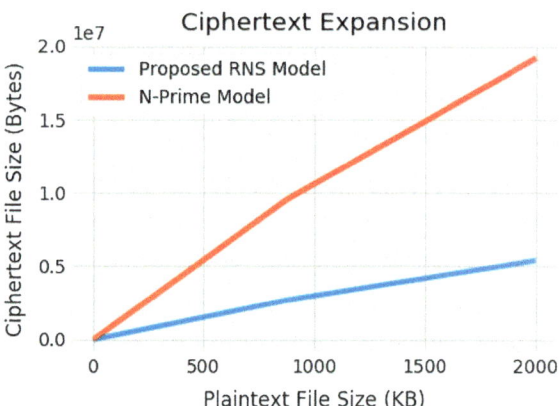

Window 10 operating system, Intel® Core™ i9 processor with 2.90 GHz and 32 GB RAM. The execution time and ciphertext expansion are studied for different plaintext size, and the results are shown in Figs. 4 and 5, respectively. Based on the execution time, the proposed RNS-based N-prime model is taking the lowest execution time, while the N-prime model is taking the highest, and this can be interpreted that during proposed RNS-based N-prime model the encryption is done based on residues number system and also all the computations are performed in parallel without carry propagation which makes the proposed model more efficient and practical. Table 2 shows the improvement of proposed RNS-based N-prime model on the N-prime model in terms of execution time.

Table 2 Execution time RATIO (RNS-based N-prime model/N-prime model)

Plaintext size in kilobytes	2	4	7	14	28	55	110	220	440	880	2000
Execution time ratio	0.98	0.67	0.72	1.00	0.48	0.87	0.96	0.87	0.92	0.88	0.64

Table 3 Ciphertext percentage reduction in N-prime model and RNS-based N-prime model

Plaintext size in kilobytes	2	4	7	14	28	55	110	220	440	880	2000
Percentage reduction (%)	71.57	71.67	71.71	71.74	71.75	71.76	71.76	71.76	71.76	71.76	71.76

Likewise, in Fig. 5, shows the ciphertext expansion between proposed RNS-based N-prime model and N-prime model where the proposed model is able to reduce the growth of ciphertext file size by approximately 72% in comparison to N-prime model as shown in Table 3. This was achieved due to encryption process that was done based on residue number system with respect to moduli set.

Finally, the security of the proposed model still depends on problem of factorization of integers to their primary numbers as mentioned in [7], such integers are product of multiple N-prime numbers and big prime number that serve as secret key that is unknown to attacker.

5 Conclusion

In this paper, we proposed RNS-based N-prime model as an improvement over N-prime model for real-world applications of symmetric-based FHE scheme that requires lower latency and reduced ciphertext file size. The proposed model is based on residue number system approach and ensures parallel computation at encryption stage which brings about efficiency and robustness in addition to the homomorphic properties. The proposed model also deals explicitly with text after converting it to ASCII value instead of binary system, pass it to RNS forward conversion module in conjunction with moduli set and then pass it to the encryption block. The results show that the proposed model outperform N-prime model in terms of encryption execution time and ciphertext expansion at 72% reduction in ciphertext file size.

Further research can be investigated to gain better decryption execution time, as it is taking larger execution time to decrypting ciphertext file back to the original plaintext or result of the computation than encrypting its corresponding plaintext.

Appendix A: Ciphertext

[[0, 6, 14], [23, 4, 12], [24, 5, 13], [24, 5, 13], [0, 6, 14], [0, 6, 14], [24, 5, 13], [23, 4, 12], [24, 5, 13], [0, 6, 14], [0, 6, 14], [0, 6, 14], [24, 5, 13], [0, 6, 14], [24, 5, 13], [0, 6, 14], [24, 5, 13], [0, 6, 14], [24, 5, 13], [24, 5, 13], [0, 6, 14], [23, 4, 12], [0, 6, 14], [23, 4, 12], [0, 6, 14], [23, 4, 12], [23, 4, 12], [24, 5, 13], [24, 5, 13], [0, 6, 14], [23, 4, 12], [0, 6, 14], [23, 4, 12], [24, 5, 13], [24, 5, 13], [0, 6, 14], [24, 5, 13], [24,

5, 13], [0, 6, 14], [24, 5, 13], [23, 4, 12], [24, 5, 13], [0, 6, 14], [24, 5, 13], [24, 5, 13], [24, 5, 13], [0, 6, 14], [24, 5, 13], [0, 6, 14], [24, 5, 13], [23, 4, 12], [0, 6, 14], [0, 6, 14], [24, 5, 13], [23, 4, 12], [24, 5, 13], [23, 4, 12], [23, 4, 12], [24, 5, 13], [23, 4, 12], [0, 6, 14], [0, 6, 14], [23, 4, 12], [23, 4, 12], [23, 4, 12], [24, 5, 13], [24, 5, 13], [23, 4, 12], [23, 4, 12], [23, 4, 12], [24, 5, 13], [0, 6, 14], [23, 4, 12], [0, 6, 14], [0, 6, 14], [0, 6, 14], [23, 4, 12], [24, 5, 13], [23, 4, 12], [24, 5, 13], [0, 6, 14], [0, 6, 14], [23, 4, 12], [23, 4, 12], [24, 5, 13], [24, 5, 13], [23, 4, 12], [24, 5, 13], [0, 6, 14], [24, 5, 13], [0, 6, 14], [24, 5, 13], [24, 5, 13], [0, 6, 14], [0, 6, 14], [0, 6, 14], [0, 6, 14], [0, 6, 14], [24, 5, 13], [0, 6, 14], [0, 6, 14], [23, 4, 12], [23, 4, 12], [0, 6, 14], [24, 5, 13], [23, 4, 12], [0, 6, 14], [0, 6, 14], [24, 5, 13], [23, 4, 12], [0, 6, 14], [23, 4, 12], [0, 6, 14], [23, 4, 12], [23, 4, 12], [24, 5, 13], [0, 6, 14], [23, 4, 12], [0, 6, 14], [0, 6, 14], [23, 4, 12], [23, 4, 12], [24, 5, 13], [23, 4, 12], [23, 4, 12], [24, 5, 13], [0, 6, 14], [23, 4, 12], [24, 5, 13], [0, 6, 14], [24, 5, 13], [23, 4, 12], [24, 5, 13], [24, 5, 13], [23, 4, 12], [0, 6, 14], [24, 5, 13], [24, 5, 13], [0, 6, 14], [24, 5, 13], [23, 4, 12], [24, 5, 13], [24, 5, 13], [24, 5, 13], [0, 6, 14], [24, 5, 13], [24, 5, 13], [0, 6, 14], [0, 6, 14], [0, 6, 14], [0, 6, 14], [0, 6, 14], [0, 6, 14], [24, 5, 13], [23, 4, 12], [23, 4, 12], [0, 6, 14], [23, 4, 12], [0, 6, 14], [24, 5, 13], [23, 4, 12], [23, 4, 12], [23, 4, 12], [23, 4, 12], [24, 5, 13], [0, 6, 14], [23, 4, 12], [0, 6, 14], [24, 5, 13], [24, 5, 13], [0, 6, 14], [23, 4, 12], [23, 4, 12], [0, 6, 14], [23, 4, 12], [23, 4, 12], [24, 5, 13], [0, 6, 14], [0, 6, 14], [23, 4, 12], [23, 4, 12], [0, 6, 14], [23, 4, 12], [0, 6, 14], [0, 6, 14], [0, 6, 14], [0, 6, 14], [24, 5, 13], [0, 6, 14], [23, 4, 12], [0, 6, 14], [23, 4, 12], [23, 4, 12], [0, 6, 14], [23, 4, 12], [0, 6, 14], [23, 4, 12], [24, 5, 13], [0, 6, 14], [0, 6, 14], [23, 4, 12], [24, 5, 13], [0, 6, 14], [23, 4, 12], [24, 5, 13], [0, 6, 14], [0, 6, 14], [24, 5, 13], [0, 6, 14], [23, 4, 12], [0, 6, 14], [23, 4, 12], [0, 6, 14], [23, 4, 12], [0, 6, 14], [24, 5, 13], [0, 6, 14], [23, 4, 12], [24, 5, 13], [0, 6, 14], [0, 6, 14], [23, 4, 12], [24, 5, 13], [0, 6, 14], [0, 6, 14], [24, 5, 13], [0, 6, 14], [23, 4, 12], [24, 5, 13], [24, 5, 13], [0, 6, 14], [23, 4, 12], [23, 4, 12], [0, 6, 14], [23, 4, 12], [0, 6, 14], [24, 5, 13], [0, 6, 14], [24, 5, 13], [0, 6, 14], [23, 4, 12], [0, 6, 14], [23, 4, 12], [0, 6, 14], [23, 4, 12], [24, 5, 13], [0, 6, 14], [0, 6, 14], [24, 5, 13], [23, 4, 12], [24, 5, 13], [24, 5, 13], [23, 4, 12], [24, 5, 13], [24, 5, 13], [0, 6, 14], [24, 5, 13], [23, 4, 12], [0, 6, 14], [24, 5, 13], [23, 4, 12], [23, 4, 12], [0, 6, 14], [0, 6, 14], [23, 4, 12], [0, 6, 14], [0, 6, 14], [0, 6, 14], [24, 5, 13], [0, 6, 14], [24, 5, 13], [23, 4, 12], [0, 6, 14], [0, 6, 14], [23, 4, 12], [24, 5, 13], [0, 6, 14], [24, 5, 13], [23, 4, 12], [24, 5, 13], [23, 4, 12], [0, 6, 14], [23, 4, 12], [0, 6, 14], [0, 6, 14], [23, 4, 12], [0, 6, 14], [23, 4, 12], [24, 5, 13], [0, 6, 14], [0, 6, 14], [24, 5, 13], [0, 6, 14], [23, 4, 12], [0, 6, 14], [0, 6, 14], [24, 5, 13], [24, 5, 13], [0, 6, 14], [23, 4, 12], [24, 5, 13], [23, 4, 12], [23, 4, 12], [24, 5, 13], [0, 6, 14], [23, 4, 12], [23, 4, 12], [0, 6, 14], [0, 6, 14], [23, 4, 12], [23, 4, 12], [0, 6, 14], [0, 6, 14], [23, 4, 12], [24, 5, 13], [23, 4, 12], [23, 4, 12], [23, 4, 12], [23, 4, 12], [24, 5, 13], [23, 4, 12], [24, 5, 13], [0, 6, 14], [0, 6, 14], [0, 6, 14], [0, 6, 14], [0, 6, 14], [23, 4, 12], [23, 4, 12], [0, 6, 14], [0, 6, 14], [24, 5, 13], [23, 4, 12], [0, 6, 14], [24, 5, 13], [0, 6, 14], [0, 6, 14], [0, 6, 14], [23, 4, 12], [23, 4, 12], [24, 5, 13], [23, 4, 12], [23, 4, 12], [24, 5, 13], [23, 4, 12], [0, 6, 14], [23, 4, 12], [24, 5, 13], [23, 4, 12], [23, 4, 12], [23, 4, 12], [23, 4, 12], [24, 5, 13], [23, 4, 12], [24, 5, 13], [0, 6, 14], [24, 5, 13], [0, 6, 14], [24, 5, 13], [24, 5, 13], [0, 6, 14], [0, 6, 14], [23, 4, 12], [24, 5, 13], [24, 5, 13], [24, 5, 13], [23, 4, 12], [24, 5, 13], [0, 6, 14], [23, 4, 12], [24, 5, 13], [0, 6, 14], [0, 6, 14], [23, 4, 12], [24, 5, 13],

[23, 4, 12], [0, 6, 14], [0, 6, 14], [23, 4, 12], [23, 4, 12], [23, 4, 12], [24, 5, 13], [24, 5, 13], [0, 6, 14], [24, 5, 13], [0, 6, 14], [0, 6, 14], [0, 6, 14], [24, 5, 13], [0, 6, 14], [23, 4, 12], [0, 6, 14], [24, 5, 13], [0, 6, 14], [0, 6, 14], [24, 5, 13], [24, 5, 13], [23, 4, 12], [23, 4, 12], [24, 5, 13], [23, 4, 12], [24, 5, 13], [0, 6, 14], [24, 5, 13], [23, 4, 12], [23, 4, 12], [24, 5, 13], [0, 6, 14], [24, 5, 13], [0, 6, 14], [0, 6, 14], [24, 5, 13], [0, 6, 14], [0, 6, 14], [23, 4, 12], [0, 6, 14], [23, 4, 12], [23, 4, 12], [24, 5, 13], [23, 4, 12], [0, 6, 14], [0, 6, 14], [0, 6, 14], [23, 4, 12], [23, 4, 12], [23, 4, 12], [0, 6, 14], [24, 5, 13], [24, 5, 13], [0, 6, 14], [0, 6, 14], [23, 4, 12], [24, 5, 13], [24, 5, 13], [24, 5, 13], [23, 4, 12], [24, 5, 13], [0, 6, 14], [24, 5, 13], [23, 4, 12], [0, 6, 14], [24, 5, 13], [0, 6, 14], [0, 6, 14], [23, 4, 12], [23, 4, 12], [24, 5, 13], [23, 4, 12], [23, 4, 12], [24, 5, 13], [0, 6, 14], [24, 5, 13], [0, 6, 14], [23, 4, 12], [0, 6, 14], [23, 4, 12], [24, 5, 13], [23, 4, 12], [23, 4, 12], [24, 5, 13], [0, 6, 14], [23, 4, 12], [23, 4, 12], [23, 4, 12], [24, 5, 13], [23, 4, 12], [24, 5, 13], [0, 6, 14], [23, 4, 12], [23, 4, 12], [0, 6, 14], [24, 5, 13], [23, 4, 12], [24, 5, 13], [0, 6, 14], [0, 6, 14], [0, 6, 14], [23, 4, 12], [24, 5, 13], [24, 5, 13], [0, 6, 14], [0, 6, 14], [0, 6, 14], [24, 5, 13], [24, 5, 13], [0, 6, 14], [23, 4, 12], [0, 6, 14], [23, 4, 12], [24, 5, 13], [0, 6, 14], [0, 6, 14], [23, 4, 12], [23, 4, 12], [24, 5, 13], [0, 6, 14], [24, 5, 13], [24, 5, 13], [23, 4, 12], [23, 4, 12], [23, 4, 12], [24, 5, 13], [0, 6, 14], [23, 4, 12], [23, 4, 12], [0, 6, 14], [23, 4, 12], [23, 4, 12], [0, 6, 14], [23, 4, 12], [24, 5, 13], [0, 6, 14], [0, 6, 14], [23, 4, 12], [23, 4, 12], [0, 6, 14], [23, 4, 12], [0, 6, 14], [23, 4, 12], [24, 5, 13], [0, 6, 14], [23, 4, 12], [23, 4, 12], [23, 4, 12], [0, 6, 14], [24, 5, 13], [0, 6, 14], [23, 4, 12], [0, 6, 14], [0, 6, 14], [0, 6, 14], [24, 5, 13], [23, 4, 12], [23, 4, 12], [24, 5, 13], [23, 4, 12], [0, 6, 14], [24, 5, 13], [0, 6, 14], [0, 6, 14], [0, 6, 14], [0, 6, 14], [23, 4, 12], [24, 5, 13], [0, 6, 14], [24, 5, 13], [23, 4, 12], [23, 4, 12], [0, 6, 14], [24, 5, 13], [0, 6, 14], [23, 4, 12], [24, 5, 13], [24, 5, 13], [23, 4, 12], [24, 5, 13], [0, 6, 14], [24, 5, 13], [23, 4, 12], [23, 4, 12], [23, 4, 12], [23, 4, 12], [0, 6, 14], [24, 5, 13], [23, 4, 12], [0, 6, 14], [0, 6, 14], [24, 5, 13], [24, 5, 13], [0, 6, 14], [0, 6, 14], [0, 6, 14], [0, 6, 14], [23, 4, 12], [23, 4, 12], [0, 6, 14], [24, 5, 13], [0, 6, 14], [23, 4, 12], [0, 6, 14], [0, 6, 14], [0, 6, 14], [23, 4, 12], [23, 4, 12], [0, 6, 14], [24, 5, 13], [24, 5, 13], [23, 4, 12], [24, 5, 13], [23, 4, 12], [24, 5, 13], [23, 4, 12], [23, 4, 12], [0, 6, 14], [24, 5, 13], [0, 6, 14], [0, 6, 14], [23, 4, 12], [23, 4, 12], [0, 6, 14], [24, 5, 13], [0, 6, 14], [0, 6, 14], [0, 6, 14], [24, 5, 13], [0, 6, 14], [23, 4, 12], [24, 5, 13], [23, 4, 12], [23, 4, 12], [23, 4, 12], [0, 6, 14], [23, 4, 12], [24, 5, 13], [0, 6, 14], [24, 5, 13], [23, 4, 12], [23, 4, 12], [0, 6, 14], [23, 4, 12], [24, 5, 13], [24, 5, 13], [0, 6, 14], [24, 5, 13], [0, 6, 14], [0, 6, 14], [23, 4, 12], [24, 5, 13], [0, 6, 14], [0, 6, 14], [0, 6, 14], [23, 4, 12], [23, 4, 12], [24, 5, 13], [0, 6, 14], [24, 5, 13], [24, 5, 13], [23, 4, 12], [23, 4, 12], [23, 4, 12], [24, 5, 13], [0, 6, 14], [0, 6, 14], [23, 4, 12], [23, 4, 12], [0, 6, 14], [0, 6, 14], [24, 5, 13], [23, 4, 12], [23, 4, 12], [0, 6, 14], [24, 5, 13], [23, 4, 12], [0, 6, 14], [23, 4, 12], [23, 4, 12], [0, 6, 14], [23, 4, 12], [24, 5, 13], [24, 5, 13], [0, 6, 14], [0, 6, 14], [23, 4, 12], [24, 5, 13], [23, 4, 12], [24, 5, 13], [0, 6, 14], [0, 6, 14], [24, 5, 13], [23, 4, 12], [23, 4, 12], [0, 6, 14], [23, 4, 12], [23, 4, 12], [23, 4, 12], [0, 6, 14], [23, 4, 12], [24, 5, 13], [23, 4, 12], [0, 6, 14], [0, 6, 14], [24, 5, 13], [23, 4, 12], [0, 6, 14], [23, 4, 12], [0, 6, 14], [23, 4, 12], [23, 4, 12], [24, 5, 13], [0, 6, 14], [23, 4, 12], [0, 6, 14], [24, 5, 13], [0, 6, 14], [0, 6, 14], [23, 4, 12], [23, 4, 12], [0, 6, 14], [23, 4, 12], [0, 6,

14], [0, 6, 14], [23, 4, 12], [0, 6, 14], [23, 4, 12], [23, 4, 12], [24, 5, 13], [23, 4, 12], [0, 6, 14], [0, 6, 14], [0, 6, 14], [0, 6, 14], [24, 5, 13], [23, 4, 12], [23, 4, 12], [23, 4, 12], [23, 4, 12], [0, 6, 14], [0, 6, 14], [23, 4, 12], [23, 4, 12], [24, 5, 13], [0, 6, 14], [24, 5, 13], [0, 6, 14], [0, 6, 14], [23, 4, 12], [23, 4, 12], [23, 4, 12], [23, 4, 12], [24, 5, 13], [0, 6, 14], [23, 4, 12], [23, 4, 12], [0, 6, 14], [24, 5, 13], [0, 6, 14], [0, 6, 14], [23, 4, 12], [24, 5, 13], [0, 6, 14], [0, 6, 14], [0, 6, 14], [24, 5, 13], [23, 4, 12], [23, 4, 12], [0, 6, 14], [24, 5, 13], [23, 4, 12], [24, 5, 13], [24, 5, 13], [0, 6, 14], [23, 4, 12], [23, 4, 12], [0, 6, 14], [24, 5, 13], [23, 4, 12], [24, 5, 13], [0, 6, 14], [0, 6, 14], [0, 6, 14], [23, 4, 12], [24, 5, 13], [0, 6, 14], [0, 6, 14], [23, 4, 12], [0, 6, 14], [23, 4, 12], [0, 6, 14], [0, 6, 14], [24, 5, 13], [23, 4, 12], [0, 6, 14], [24, 5, 13], [0, 6, 14], [0, 6, 14], [24, 5, 13], [23, 4, 12], [0, 6, 14], [23, 4, 12], [0, 6, 14], [23, 4, 12], [23, 4, 12], [24, 5, 13], [0, 6, 14], [23, 4, 12], [0, 6, 14], [0, 6, 14], [23, 4, 12], [24, 5, 13], [23, 4, 12], [23, 4, 12], [0, 6, 14], [23, 4, 12], [23, 4, 12], [0, 6, 14], [0, 6, 14], [23, 4, 12], [23, 4, 12], [0, 6, 14], [0, 6, 14], [0, 6, 14], [23, 4, 12], [0, 6, 14], [23, 4, 12], [23, 4, 12], [24, 5, 13], [23, 4, 12], [23, 4, 12], [24, 5, 13], [24, 5, 13], [0, 6, 14], [0, 6, 14], [0, 6, 14], [24, 5, 13], [0, 6, 14], [23, 4, 12], [0, 6, 14], [23, 4, 12], [23, 4, 12], [23, 4, 12], [24, 5, 13], [0, 6, 14], [24, 5, 13], [0, 6, 14], [0, 6, 14], [0, 6, 14], [0, 6, 14], [23, 4, 12], [24, 5, 13], [24, 5, 13], [23, 4, 12], [24, 5, 13], [0, 6, 14], [24, 5, 13], [23, 4, 12], [23, 4, 12], [24, 5, 13], [23, 4, 12], [24, 5, 13], [0, 6, 14], [23, 4, 12], [0, 6, 14], [0, 6, 14], [23, 4, 12], [24, 5, 13], [23, 4, 12], [23, 4, 12], [23, 4, 12], [0, 6, 14], [23, 4, 12], [24, 5, 13], [0, 6, 14], [23, 4, 12], [23, 4, 12], [23, 4, 12], [23, 4, 12], [0, 6, 14], [23, 4, 12], [23, 4, 12], [23, 4, 12], [0, 6, 14], [23, 4, 12], [24, 5, 13], [0, 6, 14], [0, 6, 14], [0, 6, 14], [0, 6, 14], [23, 4, 12], [23, 4, 12], [0, 6, 14], [23, 4, 12]].

References

1. Puthal, D., Sahoo, B., Mishra, S., Swain, S.: Cloud computing features, issues, and challenges: a big picture. In: 2015 International Conference on Computational Intelligence and Networks, pp. 116–123 (2015)
2. Stojmenovic, I., Wen, S.: The fog computing paradigm: Scenarios and security issues. In: 2014 Federated Conference on Computer Science and Information Systems, pp. 1–8 (2014)
3. Bokhari, M.U., Makki, Q., Tamandani, Y.K.: A survey on cloud computing. In: Big Data Analytics, pp. 149–164. Springer (2018)
4. Chang, V., Kuo, Y.-H., Ramachandran, M.: Cloud computing adoption framework: a security framework for business clouds. Future Gener. Comput. Syst. **57**, 24–41 (2016)
5. Patel, N., Oza, P., Agrawal, P.: Homomorphic cryptography and its applications in various domains. In: International Conference on Innovative Computing and Communications, pp. 269–278 (2019)
6. Prasad, J.P., Ramulu, B.S.: Homomorphic encryption security for cloud computing. Global J. Comput. Sci. Technol. (2019)
7. Mohammed, M.A., Abed, F.S.: An improved Fully Homomorphic Encryption model based on N-Primes. Kurdistan J. Appl. Res. **4**, 40–49 (2019)
8. Gentry, C., Boneh, D.: A fully homomorphic encryption scheme vol. 20: Stanford University Stanford (2009)
9. Ogburn, M., Turner, C., Dahal, P.: Homomorphic encryption. Procedia Compu. Sci. **20**, 502–509 (2013)

10. Gentry, C.: Fully homomorphic encryption using ideal lattices. In: Stoc, pp. 169–178 (2009)
11. Archer, D., Chen, L., Cheon, J.H., Gilad-Bachrach, R., Hallman, R.A., Huang, Z. et al.: Applications of homomorphic encryption. Technical report, HomomorphicEncryption. org, Redmond, WA (2017)
12. Brenner, M., Perl, H., Smith, M.: Practical applications of homomorphic encryption. In: SECRYPT, pp. 5–14 (2012)
13. Armknecht, F., Boyd, C., Carr, C., Gjøsteen, K., Jäschke, A., Reuter, C.A., et al.: A guide to fully homomorphic encryption. IACR Cryptology ePrint Archive **2015**, 1192 (2015)
14. Laine, K.: Homomorphic encryption. In: Responsible Genomic Data Sharing, pp. 97–122. Elsevier (2020)
15. Brakerski, Z., Langlois, A., Peikert, C., Regev, O., Stehlé, D.: Classical hardness of learning with errors. In: Proceedings of the Forty-Fifth Annual ACM Symposium on Theory of Computing, pp. 575–584 (2013)
16. Micciancio, D., Peikert, C.: Hardness of SIS and LWE with small parameters. In: Annual Cryptology Conference, pp. 21–39 (2013)
17. Chillotti, I., Gama, N., Georgieva, M., Izabachene, M.: Faster fully homomorphic encryption: Bootstrapping in less than 0.1 seconds. In: International Conference on the Theory and Application of Cryptology and Information Security, pp. 3–33 (2016)
18. Chillotti, I., Gama, N., Georgieva, M., Izabachene, M.: Improving TFHE: faster packed homomorphic operations and efficient circuit bootstrapping (2017)
19. Ducas, L., Micciancio, D.: FHEW: bootstrapping homomorphic encryption in less than a second. In: Annual International Conference on the Theory and Applications of Cryptographic Techniques, pp. 617–640 (2015)
20. Gentry, C., Sahai, A., Waters, B.: Homomorphic encryption from learning with errors: conceptually-simpler, asymptotically-faster, attribute-based. In: Annual Cryptology Conference, pp. 75–92 (2013)
21. Gao, S., MathCrypt, S.B.: Efficient fully homomorphic encryption scheme. IACR Cryptology ePrint Archive **2018**, 637 (2018)
22. Dhakar, R.S.,Gupta, A.K., Sharma, P.: Modified RSA encryption algorithm (MREA). In: 2012 Second International Conference on Advanced Computing & Communication Technologies, pp. 426–429 (2012)
23. Hu, Y.: Improving the efficiency of homomorphic encryption schemes (2013)
24. Rivest, R.L., Adleman, L., Dertouzos, M.L.: On data banks and privacy homomorphisms. Foundations Secure Comput. **4**, 169–180 (1978)
25. ElGamal, T.: A public key cryptosystem and a signature scheme based on discrete logarithms. IEEE Trans. Inf. Theory **31**, 469–472 (1985)
26. Smart, N.P., Vercauteren, F.: Fully homomorphic encryption with relatively small key and ciphertext sizes. In: International Workshop on Public Key Cryptography, pp. 420–443 (2010)
27. Orsini, E., Smart, N.P., Vercauteren, F.: Overdrive2k: efficient secure MPC over Z_2k from Somewhat Homomorphic Encryption. In: Cryptographers' Track at the RSA Conference, pp. 254–283 (2020)
28. Gentry, C., Halevi, S.: Fully homomorphic encryption without squashing using depth-3 arithmetic circuits. In: 2011 IEEE 52nd Annual Symposium on Foundations of Computer Science, pp. 107–109 (2011)
29. Brakerski, Z., Vaikuntanathan, V.: Fully homomorphic encryption from ring-LWE and security for key dependent messages. In: Annual Cryptology Conference, pp. 505–524 (2011)
30. Brakerski, Z., Vaikuntanathan, V.: Efficient fully homomorphic encryption from (standard) LWE. SIAM J. Comput. **43**, 831–871 (2014)
31. Chen, Y., Gu, B., Zhang, C., Shu, H.: Reliable fully homomorphic disguising matrix computation outsourcing scheme. In: 2018 14th International Wireless Communications & Mobile Computing Conference (IWCMC), pp. 482–487 (2018)
32. Van Dijk, M., Gentry, C., Halevi, S., Vaikuntanathan, V.: Fully homomorphic encryption over the integers. In: Annual International Conference on the Theory and Applications of Cryptographic Techniques, pp. 24–43 (2010)

33. Brakerski, Z.: Fundamentals of fully homomorphic encryption. In: Providing Sound Foundations for Cryptography, pp. 543–563 (2019)
34. Burtyka, P., Makarevich, O.: Symmetric fully homomorphic encryption using decidable matrix equations. In: Proceedings of the 7th International Conference on Security of Information and Networks, p. 186 (2014)
35. Gupta, C., Sharma, I.: A fully homomorphic encryption scheme with symmetric keys with application to private data processing in clouds. In: 2013 Fourth International Conference on the Network of the Future (NoF), pp. 1–4 (2013)
36. Li, J., Wang, L.: Noise-free Symmetric Fully Homomorphic Encryption based on noncommutative rings. IACR Cryptology ePrint Arch. **2015**, 641 (2015)
37. Liang, M.: Symmetric quantum fully homomorphic encryption with perfect security. Quantum Inf. Process. **12**, 3675–3687 (2013)
38. Umadevi, C., Gopalan, N.: Outsourcing private cloud using symmetric fully homomorphic encryption using Q^n_p matrices with enhanced access control. In: 2018 International Conference on Inventive Research in Computing Applications (ICIRCA), pp. 328–332 (2018)
39. Qu, Q., Wang, B., Ping, Y., Zhang, Z.: Improved cryptanalysis of a fully homomorphic symmetric encryption scheme. Secur. Commun. Netw. **2019** (2019)
40. Vizár, D., Vaudenay, S.: Cryptanalysis of Enhanced MORE (2018)
41. Hariss, K., Noura, H., Samhat, A.E.: Fully Enhanced Homomorphic Encryption algorithm of MORE approach for real world applications. J. Inform. Secur. Appl. **34**, 233–242 (2017)
42. Hariss, K., Noura, H., Samhat, A.E., Chamoun, M.: An efficient solution towards secure homomorphic symmetric encryption algorithms. In: ITM Web of Conferences, p. 05002 (2019)
43. Hariss, K., Noura, H., Samhat, A.E., Chamoun, M.: Design and realization of a fully homomorphic encryption algorithm for cloud applications. In: International Conference on Risks and Security of Internet and Systems, pp. 127–139 (2018)
44. Kipnis, A., Hibshoosh, E.: Efficient methods for practical fully homomorphic symmetric-key encrypton, randomization and verification. IACR Cryptology ePrint Arch. **2012**, 637 (2012)
45. Vizár, D., Vaudenay, S.: Cryptanalysis of chosen symmetric homomorphic schemes. Studia Scientiarum Mathematicarum Hungarica **52**, 288–306 (2015)
46. Kim, J., Shim, H., Han, K.: Comprehensive Introduction to fully homomorphic encryption for dynamic feedback controller via LWE-based cryptosystem. In: Privacy in Dynamical Systems, pp. 209–230. Springer (2020)
47. Hosseinzadeh, M., Navi, K.: A new moduli set for residue number system in ternary valued logic. J. Appl. Sci. **7**, 3729–3735 (2007)
48. Bankas, E.K., Gbolagade, K.A.: A New efficient RNS reverse converter for the 4-moduli set world academy of science, engineering and technology. Int. J. Comput. Electr. Autom. Control Inform. Eng. **8**, 328–332 (2014)
49. Younes, D., Steffan, P.: A comparative study on different moduli sets in residue number system. In: 2012 International Conference on Computer Systems and Industrial Informatics, pp. 1–6 (2012)
50. Omondi, A.R., Premkumar, B.: Residue number systems: theory and implementation vol. 2: World Scientific (2007)

Big Data Analytics: Partitioned B+-Tree-Based Indexing in MapReduce

Ali Usman Abdullahi⃝, Rohiza Ahmad, and Nordin M. Zakaria

Abstract Big data analytics platforms are designed to improve performance by avoiding the extract transfer load approach. Also, there are techniques which have worked very well in performance optimization for relational databases. Yet these techniques are in the process of integration into big data analytics. Indexing and its data structure is an example of such techniques. Despite its popularity in query optimization for efficient data mining, the indexing was not integrated into the MapReduce platform. By design the MapReduce was made to perform a full scan of the input data. However, there were attempts made to incorporate the indexing for performance improvement in MapReduce in recent years. However, these attempts have not exhausted the potentials of indexing in the MapReduce query processing. Consequently, this chapter presents an indexing approach that uses the partitioned B+-Tree as its data structure to index the InputSplit component of the Hadoop distributed file system. This was done to achieve efficient data mining query processing when used with the Hadoop MapReduce. The results of this study showed that the proposed index method has significantly reduced the index size as well as the execution runtime of all search queries by at least 50% for all the used data sizes when compared with the Normal MapReduce processing and another clustered index approach. Thus, the use of the proposed index approach has the potential to significantly reduce the time taken in mining data within a dataset by half.

A. U. Abdullahi (✉)
Computer Science Education Department, Federal College of Education (Tech), Gombe, Nigeria
e-mail: usmanali@fcetgombe.edu.ng
URL: http://www.fcetgombe.edu.ng

R. Ahmad · N. M. Zakaria
Computer and Information Sciences Department, Universiti Teknologi PETRONAS,
Seri Iskandar, Perak, Malaysia
e-mail: rohiza_ahmad@utp.edu.my

N. M. Zakaria
e-mail: nordinzakaria@utp.edu.my

217

1 Introduction

Indexing, as an information retrieval technique, is the process of generating all the suitable data structures that allow for efficient retrieval of stored information [9, 21]. The term index referred to the suitable data structure needed to allow for the efficient information retrieval [21]. The data structures used in index in most cases do not store the information itself rather it uses other data structures and pointers to locate where the data has been stored. Also, the indexing have played a very important role in performance improvement of relational databases. The indexing techniques in relational database management systems (RDBMS) are well developed and matured [9, 18].

However, in big data analytics the indexing concept is still at the developmental stage and needs more attention from researchers as pointed out by Yang and Parker [34], An [2], Richter et al. [25] and Sevugan et al. [28]. There are two indexing concepts when it comes to indexing in big data analytics. The first one is the document indexing, which is the use of the MapReduce process to index the contents of any stored document for easy retrieval. Usually, the most used indexing technique for this task is the inverted index [21]. In this concept the main goal of the MapReduce job is the indexing of content of the document itself [20]. Early works of indexing in MapReduce were concentrated on this aspect. This was supported by the fact that the developers of MapReduce (Google) dealt mainly with searching through large document datasets [6, 7, 31]. Basically the search engine companies are interested in fast retrieval of their queried stored data [15, 18, 19, 30, 35].

The second concept is indexing for speeding up the process of the MapReduce's job execution. The indexing here performs the same function it does with the traditional databases of restricting the query processing to only the input data to be affected by any given query [2, 5, 10, 17, 34]. This indexing is similar to the one found in most structured query language (SQL)-on-Hadoop, and even in other variants of distributed files systems (DFS), like BigTable. However, the way and manner in which the index approach works in the traditional database, SQL-on-Hadoop and even on table-based DFS, is that it is usually based on the storage structure, i.e., table for those platforms that use tables as storage structure [14, 16, 19]. This fact means that indexes are usually provided for each table making the amount of data to be scanned before query processing to increase. On the contrary, the user-defined indexing approach is quite different when it is used directly on Hadoop, where Hadoop distributed files systems (HDFS) components that handle the stored data are indexed [22, 33].

Particularly in the case of MapReduce, this indexing concept ensures the scanning of only those blocks/chunks/splits that are targeted by the query to be processed. The study presented in this chapter is focused on this second indexing concept as a way of improving big data analytics using the MapReduce programming paradigm for data mining [14].

This concept of indexing, if integrated into the MapReduce execution process can help solve the following problems: (1) The scanning of the whole input data, (2)

The same ways of handling both low and high selectivity tasks and (3) The lack of robust access method for information retrieval in online access processing (OLAP) situations.

In an effort to overcome these problems some researchers have attempted to incorporate the indexing mechanism into the MapReduce framework. For instance, Yang and Parker [34] have employed HDFS files as B-Tree nodes to achieve indexing. Also, An [2] used blockIds from the HDFS as B+Tree search keys for determining the start and the end of the contiguous blocks to scan, which is used as index. In addition, Richter et al. [25] used the copies of the replica stored by HDFS, to form a clustered index of different data attributes that may likely be used as incoming query predicates. Clustered indexes are built on those fields, which is subsequently used for query processing.

In all of these studies, the authors used an indexing data structure that scale logarithmically, thereby preventing the MapReduce from performing a full scan of the input data. Thus, guiding the process is to just scan and process the data that corresponds to the output of the indexes. Notwithstanding the prevention of full scanning of the input data by these approaches, their indexes are built using all of the keys from the stored data. So, the full scanning of the indexing structure itself is done due to the nature of the components of the HDFS used. Furthermore, when the B-Tree and B+-Tree as indexing data structures are used in MapReduce it can facilitate the scanning of relevant portions of the stored data to answer range queries covering small sections (low selectivity) or wider sections of the data (high selectivity).

However, the search algorithm in the B-Tree takes a non-uniform time for different types of queries, due to the storing of the search key's corresponding values/records at all levels of the tree [29], while the search algorithm in the B+-Tree involves searching the whole of the tree all the time [3]. This is relevant because the search in the tree structures is the basic process that all other processes are based on. The search algorithm of B-Tree has been improved in the B+-Tree variants and those variants can be used to further improve the performance in the MapReduce query processing.

Lastly, any indexing approach to be used in MapReduce should improve the two problems mentioned for the effectiveness of MapReduce in mining data stored in the warehouse for OLAP. Thus, the indexing approach proposed in the chapter chooses an improved variant of the B+-Tree called partitioned B+Tree introduced by Abdullahi et al. [1]. The partitioned B+-Tree provides an improved search algorithm, which ensures reduced execution time for query retrieval. This fact is what influences a quick record retrieval during mining of big data using MapReduce platforms [16, 26].

1.1 Objective of the Chapter

The main objective of this chapter was to present an indexing approach that puts together partitioned B+-Tree as data structure and its algorithms for inserting and searching of data. The strategy of the proposed index approach is to restrict the amount of the input data to be processed by the MapReduce platform to only the part of the input data that is required for the query processing.

The remaining sections of the chapter are as follows: Sect. 2 presents the review of the literatures, Sect. 3 discusses the methodology used in the study and Sect. 4 presents the experimental setup and the results of the experiments conducted. The conclusion and future work are discussed in Sect. 6.

2   Literature Review

In this section the review of the two concepts of indexing that were used in MapReduce is presented.

2.1   Inverted Index in MapReduce

The first concept of the indexing involves a simple inverted index which was said to be implemented trivially in MapReduce as one of the effective tasks for textual data retrieval. This was highlighted in the original paper on MapReduce [8] cited by Graefe and Kuno [12] and discussed in Cao et al. [4], Stewart et al. [32] and Mofidpoor et al. [22]. The inverted index works by parsing the split of the input data to the map function, which performs the filtering of the data as defined in it and emits a sequence of <*key,value*> pairs. Then, the reduce function accepts all of the pairs of the same key, sorts the corresponding values and emits <*key,list(value)*> pair. The set of all the output pairs form a simple inverted index. McCreadie, Macdonald and Ounis [20] concluded that two interpretations of the above scenario can be a per-token indexing or a per-term indexing in relation to the indexing of corpus datasets: The per-token indexing strategy involves emitting <*term,doc−ID*> pairs for each token in a document by the map function, while the reduce function does the aggregation of each unique term with its corresponding doc-ID to obtain the term frequencies (tf), after which the completed posting list for that term is written to a disk.

On the other hand, the per-term indexing is the one in which the map function emits tuples in the form <*term,(doc−ID,tf)*> and this reduces the number of emitted operations as only unique terms per document are emitted. The reduce function only sorts the instances by document to obtain the final posting list sorted by ascending doc-ID.

Apart from these two methods the inverted index is also used by MapReduce to index a whole document in what is called per-document indexing. In the per-document indexing, the map function emits tuples in the form of <*document,doc−ID*>, while the reduce phase writes all the index structures. Though this strategy emits fewer key/value pairs, the value of each pair emitted used to have more data and has a reduced intermediate result. Thus, this approach achieves higher levels of compression than the single-term approaches. Also, the documents are easily indexed on the same reduce task due to the sorting of the document names [20].

In addition, there is the per-posting list indexing. This approach is based on a single-pass indexing, which splits data onto multiple map tasks with each operating on its own data subset. The map task serves as the scanning phase of the single-pass indexing. As this process runs on the document, a compressed posting list is built in the memory for each term. This partial index is flushed by the map task when the memory runs low or when all the documents are processed. The flushing is done by emitting a set in the form of *<term,postinglist>* pairs for all the term present in the memory. In all of these four strategies of document indexing in MapReduce the main task focus is on and carried out by the MapReduce job, which is the indexing of the document's contents [20].

2.2 User-Defined Indexing in MapReduce

The indexing here focuses on restricting the query processing to include only those parts of the input data to be effected by the query [2, 34]. For the MapReduce framework, this concept achieves its purpose by only scanning of the blocks or splits or files that contain the input data. This study has focused on the second indexing concept as a way of improving big data analytics using the MapReduce programming paradigm.

In big data analytics, the second indexing concept has been initially seen as being less important or a costly one due to the fact that infrequent updating is not a common characteristic of big data. However, later works in the area indicated that indexing can play a major role in the information retrieval aspect of the big data analytics. This makes its benefit to surpass its cost. Yang and Parker [34], for instance, implemented indexing by employing a HDFS file as B-Tree node. The index used by the authors is considered non-clustered index [27]. This in turn was due to the fact that the reading of the index is not done sequentially. Their approach adds a primitive function to traverse the tree in order to locate the required segment of the data to be processed by their improved Map-reduce-merge-traverse version of MapReduce. This introduction of the B-Tree index has improved the performance of the MapReduce (changed to Map-reduce-merge-traverse). The improvement is up to 50% and above for a three-node cluster when compared to any number of nodes less than 10 with the normal MapReduce.

On the other hand, An [2] used blockId as the B+-Tree's search keys and the offset of the keys as a value to build an index. The built index is stored in a HDFS file as contiguous blocks rising from the node at the bottom to the root at the top. The index-based execution of the MapReduce is done by first searching for the keys that are contained in the queries' predicate from the already built index. This is done to determine the starting and the end of the contiguous blocks that represent the search keys in the index. The blockIds that fall between the two points are those that are scanned for query processing by the MapReduce. With these two mentioned techniques, the amount of data to be scanned by the MapReduce is restricted. The B+-Tree with blockIds as the keys index scan for MapReduce, outperformed the normal MapReduce by at most 50%, and below for varying percentage of I/O volume used by in the study.

Furthermore, Richter et al. [25] used the copy of the replicas stored by the HDFS to index different data attributes that may likely be used as predicates of incoming queries. Their approach used a clustered index [24] based on each of these replicas. The clustered index is first scanned to get the blockId of the data to be processed from the index. Then, it uses the returned blockIds to locate the input data to be scanned and used by the MapReduce for processing the query in the question. Then, it used its customized input-split policy that helped their approach to determine the part of the input data that could take part in the query processing. After the index result is returned, the results of using their indexing with other Hadoop settings indicated a 20% reduction of the execution runtime.

3   Methodology

This section of the chapter discusses the methodology used in the study. It gives the details of how various component of the study come together to achieve its object. Figure 1 shows the overall picture of the proposed indexing scheme.

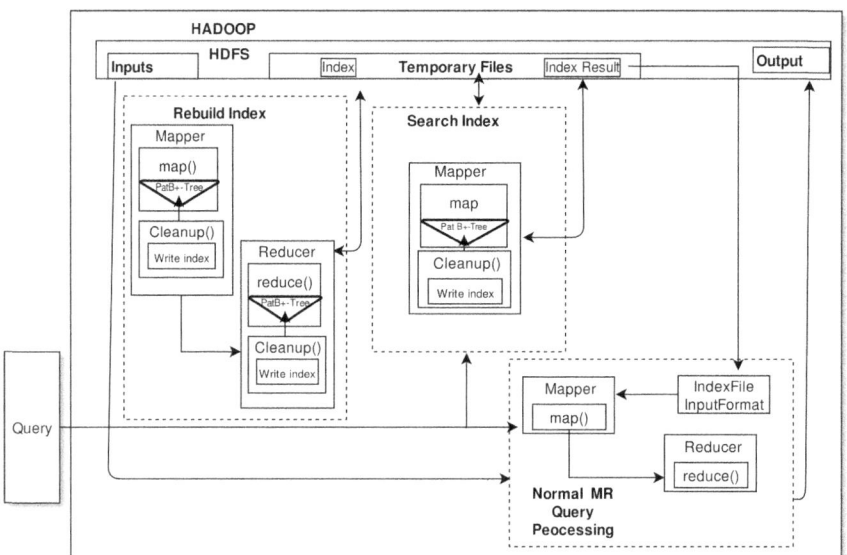

Fig. 1 The block diagram of the scheme indexing for the MapReduce query processing

The proposed approach basically consists of several components. They include the query, which is the statement that specifies how the process should be done; the rebuild index component, which is used for building and subsequent re-building of the index when needed. Then, there is the search index component, which is used for searching the index using the values provided in the query's predicate. This component searches and returns the location of the data required by the query. Lastly, the scheme also has another component which is the query processing component. This component does the actual processing of the query been issued.

The approach works first by building an index using partitioned B+-Tree (the data structure) with search keys from the input data and their corresponding values (input-splits) using the rebuild index component. The built index is stored in a temporary file on the HDFS. When a query for certain processing is to be executed, its predicates will first be used to search the index to get the input-splits using the search index component. The search index component also uses the Pat B+-Tree to re-construct the index and use it to search and return the targeted results.

By doing that, only the input-splits that cover the part of the input data that is required by the query are returned and stored in yet another temporary file on the HDFS. The returned input-splits are then read and used by the query processing component to process the given query. The framework view of the approach's working flow, as mentioned, is captured in Fig. 1.

3.1 Partitioned B+-Tree

The partitioned B+-Tree mentioned is an adaptation of the partitioned B-Tree [11], in which the author proposed the addition of an artificial lead key to the BTree. This addition resulted in a logical partitioning of the B-Tree based on run generations that feed the B-Tree during the initial index creation. Graefe [11] implemented his tree based on the traditional B-Tree. However, in this study the same idea was implemented with some improvements on B+Tree. The choice of partitioned B+-Tree for this study was due to its performance in a preliminary experiment in comparison with other variants of B+-Tree as reported in [1].

3.2 InputSplit as Component of Choice in the HDFS

The corresponding values for the search keys used by the proposed indexing approach are the input-split components of the Hadoop distributed file system (HDFS). The input-split is the HDFS's feature that uniquely identifies the chunks of data that the Hadoop made use of for the proper managing of the blocks of the stored data on its HDFS [13, 23, 36].

The reason for choosing the input-split is because it provides a better balance between the index size and searching cost on one side and the size of the input data to be scanned on the other. So, even if the blockId is to be used for indexing, the input-splits that cover those blocks have to be generated first, before extracting the blockIds from them.

Index Rebuild in MapReduce First, the index needs to be built over the complete input data before it is used as mentioned earlier. A complete MapReduce program is used to extract the HDFS components alongside the search key values from the input data in order to build the index.

Figure 2 has shown the expanded version of the rebuild index block of Fig. 1; here, the internal workings of the component are depicted. The rebuild index reads the inputs from the stored data on the HDFS using its Mapper class called the RebuildIndexMapper. This Mapper class has three methods, namely, setup(), map() and the cleanup(). The setup() method takes a context variable of the context class type as the argument. The context class is an optimized record reader/writer handler used by the Hadoop. The setup() method used the variable to extract the Hadoop internal component's settings and properties. One such component is the input-split, which is the targeted component for the proposed index approach as mentioned earlier.

Also, the context variable in the setup() method is used alongside the extracted input-split to further get the properties of other components, such as the filename of the input files. In some cases, the extracted filenames are concatenated with the extracted values from the records in order to form unique search key values. The use of the filename as part of the unique search key values is to avoid the treating of the keys from files with the same values as one and the same entry.

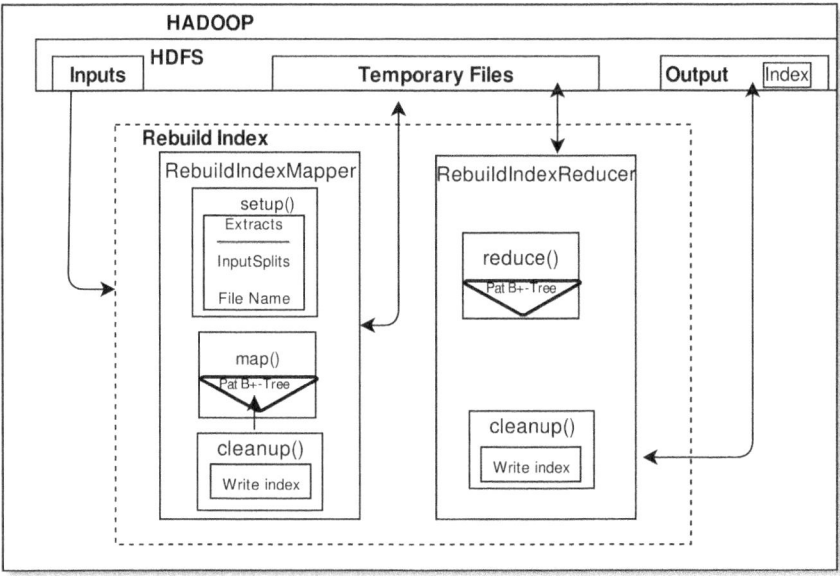

Fig. 2 Block diagram for index building process

However, if the input data is contained in a larger file, then only the unique key values will be extracted and used as the index search keys.

Then, the map() method of the RebuildIndexMapper class reads each record from the input file(s) using its defined key/value arguments. The method passes the values as arguments to an external method. The function of this external method is just to extract and return the values designated for use as the index search keys. After that, the input data is grouped in small files with similar values in the field intended for use in the index.

Then, the map() method concatenates the returned values with the file names extracted in the setup() method to form the unique search key values; else, the returned values from the extra method are used as the search keys. The search key extracted formed together with the input-split for each record are passed to the Pat B+-Tree for index building.

Furthermore, the index building is done by the build() method of the Pat B+-Tree. The build() builds the tree as the Map() method iterates over the input records and passes the key/value pairs to it. A modification was made to the insertion condition in the insertRecord() of the Pat B+-Tree to only allow the insertion of one search key value for a range of keys having the same input-split value. This was done to prevent multiple insertion of input-split values for all of the keys that fall under it, hence, compressing and compacting the index to a smaller size.

In addition, the cleanup() method was used to capture the final state of the tree and to write the subtree as a single object from a chunk of the stored data. The cleanup() method calls the toString() method of the Pat B+-Tree, which in turn, prints the subtree onto temporary file(s) within the HDFS. The output at the mapper stage produces subtrees from the chunks of data processed by different map jobs.

Searching the Index using MapReduce The second step for the integration, which was represented by the second block in Fig. 1, is the searching of the index. Figure 3 presents the details of the internal workings for the processes involved in the searching of the index. Here also, the mapper class's only program is used to read from the index file generated by the previous scheme's component, the rebuild index. This simply means that the program only has a mapper called the SearchIndexMapper class.

Also, this mapper class has the same three methods as the previous one, which are: the setup(), the map() and cleanup() methods. The setup() uses its context variable as an argument to read the range of keys, to be searched for within the index, through the context variable's configuration instance, which helps in accepting arguments at runtime.

Then, the map() method of the SearchIndexMapper class reads each record from the index file(s) using its defined key/value arguments to rebuild the tree in the same way as described earlier. This is done to ensure that the tree is rebuilt and retained/cached by the Hadoop for faster searching. Algorithm 1 shows the pseudo-code of the map() method highlighting the functions explained

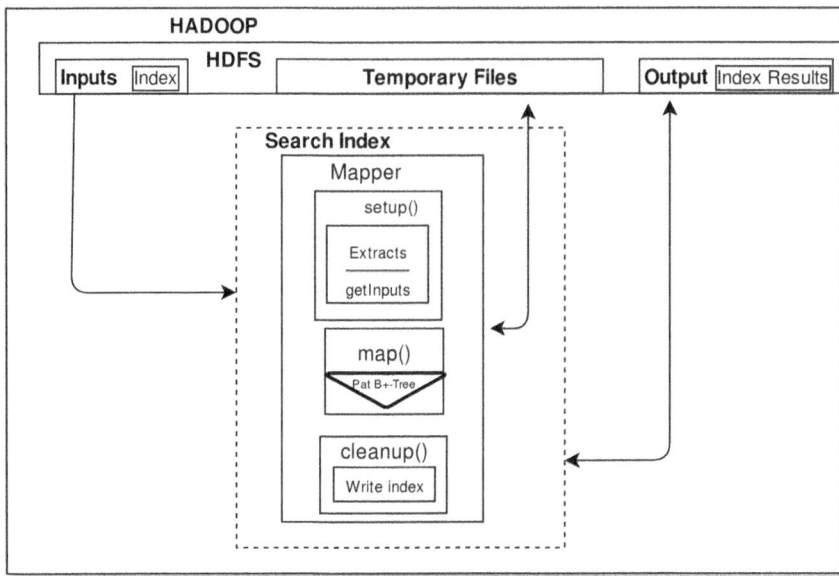

Fig. 3 Block diagram for the index searching process

Algorithm 1: Algorithm for SearchIndexMapper Class

Input: Range of Keys
Result: Input-Split values
SearchIndexMapper extends Mapper<inkeyType ,
invalue,outkeyType,outvalueType >{
Setup(Context context){ get configuration args at runtime }
map (inkey,invalue, contetx){
if *indexValue* **then**
   | split the indexValue
   | **if** *indexValuelength* >= *3* **then**
   | | build Pat B+-Tree
   | **end**
end
} **cleanup** (Context context){
iterate through the values returned by Pat B+-Tree search
print with context(null, value) }
}

After that, the cleanup() method of this class captures the final state of the tree and calls the searchRange() method of the Pat B+-Tree by passing the range of keys from the query's predicates to it. The searchRange() method of the Pat B+-Tree has been modified to search and return the appropriate value that corresponds to any given search key. The search is done by comparing the given search key with a range of separator keys in the tree. This is possible due to the logic used in building and compressing the Pat B+-Tree explained earlier.

Then, the cleanup() method uses its context's write() method to iterate and print the returned values onto the output file(s) within the HDFS as the index's result.

Using the Index in MapReduce Execution Process: In order to make use of the index in the main MapReduce query execution process, i.e., query processing of Fig. 1, another complete MapReduce program is required. The program operates in a normal way as an ordinary MapReduce function, with the exception of using a customized FileInputformat. Figure 4 shows the processes, especially the IndexFileInputFormat component.

The query processing uses three classes, namely, The Mapper, the Reducer and the IndexFileInputFormat classes. The Mapper class consists of two methods: the setup() and the map() methods.

The setup() method of the Mapper class uses its context variables' configuration property to get runtime arguments in the same way as explained in the details earlier.

The map() method, on the other hand, iterates through the input data records comparing the keys/values from the records with the lower and upper bounds from the query's predicates. The map() method emits only those records that have met the condition set by the query's predicates.

Then, the class's getSplit() method calls yet another feeder method the readInputSplitFromFile() method that reads the index result which was returned by the SearchIndex class discussed earlier. The readInputSplitFromFile() method iterates

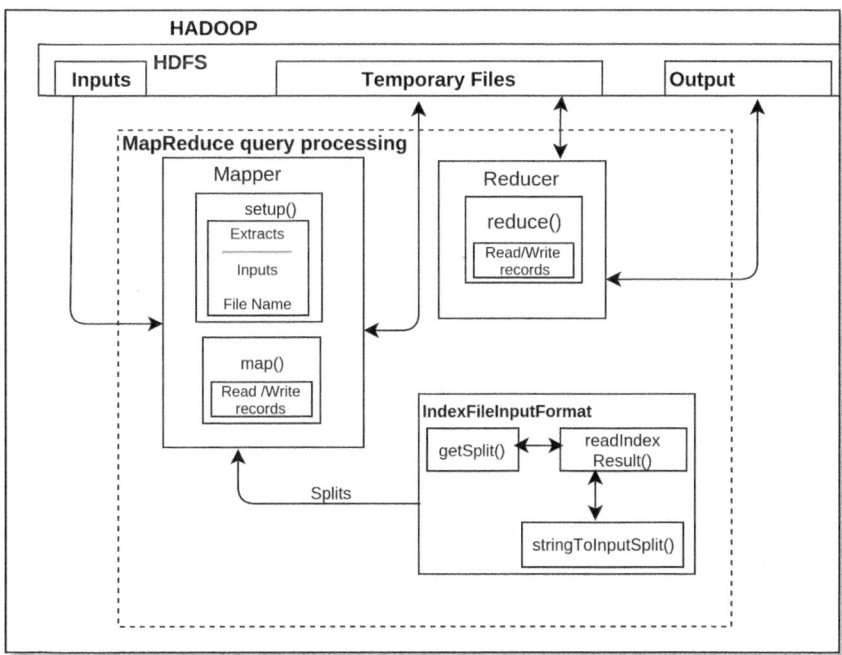

Fig. 4 Block diagram for integrating the index using IndexFileInputFormat class

through the index of the result entries, which has "null' as the value of all the keys and a string form of the input-splits as their corresponding values. The index results are the ones printed out during the searching of the index using the query's predicates.

The string form of the input-splits is not recognized by the FileInputFormat as valid input-splits, hence another method that transforms the strings into a valid input-split is needed.

4 Experimental Results and Discussion

In this section the setup of the experiment as well as the results obtained and their interpretations are discussed.

4.1 The Dataset

In this study, three different datasets were used.

The dataset, which is related to oil and gas domain. Meteorological and Oceanography data (Modata) is the data used by the Oil and Gas (O&G) industries for exploration sites analysis and forecasting. Modata is quite different from the previous datasets discussed, because it comes in multiple small files, as opposed to one large file like the other benchmark datasets. Each file contains data for a single point out of several thousand others in an exploration site.

In addition, the files have varied sizes according to the number of information/columns/parameters that they capture. For instance, files that have the minimum number of columns, i.e., 20 columns, have the size of 60 MB, while files which have maximum number of columns, i.e., 300 columns, have the size of 307 MB.

Furthermore, each tuple in the file represents data taken for a particular hour of the day. For example, in a site called Sea Fine central (SF-Central), there are 14,881 points, and each point will have several tuples of data which are recorded every hour. Since site SF-Central is nearer to the shoreline, there are 20 parameters or features recorded in each tuple such as 'wind direction', 'Wind speed', 'date' and 'time'. Besides raw data, other derived statistical data such as the 'mean' and the 'variation' of the parameters are also included in the data collection.

For sites that are more toward the ocean, the parameters collected will be even more. Moreover, each file contains the tuples described for six decades. From this description of the Modata, one can imagine the hugeness and complexity the data files. Figure 5 shows sample of data from one file of Modata.

On the other hand, for the Modata, a record volume of 10, 20, 40, 80, 160 and 320 million were queried from the data sizes mentioned above, respectively. Also, Table 1 shows the summary of the query sets with their expected returning records.

```
CCYYMM DDHHmm  WD   WS   ETOT    TP  VMD ETTSea TPSea VMDSea ETTSw  TPSw VMDSw  MO1   MO2   HS
DMDIR ANGSPR INLINE
195607 010000 107.1 9.04 0.066 4.621 288.5 0.065 4.622 288.5 0.000 6.057 165.9 0.100 0.160 1.024 289.5 0.8748 0.7926
195607 010100 107.8 9.03 0.075 4.854 289.1 0.075 4.854 289.1 0.000 6.047 153.1 0.112 0.177 1.092 290.0 0.8803 0.7996
195607 010200 108.6 9.02 0.082 4.971 289.9 0.082 4.971 289.9 0.000 5.989 132.4 0.122 0.191 1.148 290.7 0.8819 0.8013
195607 010300 109.4 9.01 0.089 5.042 290.6 0.089 5.042 290.6 0.000 5.898 79.7 0.130 0.201 1.192 291.4 0.8816 0.8006
195607 010400 110.2 9.00 0.094 5.114 291.2 0.094 5.114 291.2 0.000 5.847 53.8 0.136 0.209 1.227 292.0 0.8807 0.7990
195607 010500 111.0 8.99 0.098 5.219 291.8 0.098 5.219 291.8 0.000 5.845 44.0 0.141 0.214 1.253 292.5 0.8795 0.7972
195607 010600 111.8 8.99 0.102 5.341 292.3 0.101 5.340 292.2 0.000 5.889 36.9 0.145 0.219 1.274 292.9 0.8784 0.7955
195607 010700 109.4 8.55 0.102 5.422 292.5 0.102 5.421 292.4 0.000 5.940 33.7 0.146 0.220 1.280 293.0 0.8734 0.7885
195607 010800 106.6 8.11 0.100 5.485 292.4 0.099 5.481 292.1 0.001 5.508 334.0 0.142 0.214 1.264 292.9 0.8674 0.7804
195607 010900 103.6 7.68 0.095 5.529 292.1 0.092 5.508 291.1 0.003 5.557 322.1 0.135 0.203 1.235 292.5 0.8607 0.7718
195607 011000 100.1 7.24 0.090 5.561 291.6 0.082 5.285 289.1 0.008 5.643 316.5 0.127 0.190 1.198 292.0 0.8559 0.7659
195607 011100 96.2 6.81 0.084 5.592 291.0 0.069 5.059 286.3 0.015 5.737 312.6 0.118 0.177 1.158 291.5 0.8521 0.7616
195607 011200 91.9 6.37 0.078 5.625 290.5 0.052 4.670 282.8 0.026 5.789 305.3 0.109 0.163 1.117 291.0 0.8493 0.7586
195607 011300 95.3 6.74 0.073 5.660 289.9 0.060 5.113 285.2 0.013 5.839 310.9 0.103 0.154 1.084 290.4 0.8486 0.7583
```

Fig. 5 Sample of data from Modata files

Table 1 Data sizes and expected returning records for Modata	S/no.	Data size	Number of target records
	1	20 GB	10,000,000
	2	50 GB	20,000,000
	3	100 GB	40,000,000
	4	200 GB	80,000,000
	5	500 GB	160,000,000
	6	1 TB	320,000,000

4.2 Index Building Using the Datasets

On the other hand, the search keys for the Modata were formed by concatenating the individual filename to the timestamps. The timestamps are the field with the unique values in each of the files.

However, the same timestamps are found in all other files, which are in their thousands in the Modata dataset. Hence, the search keys in the Modata are lengthier. For the search keys' corresponding values, the proposed index method used the input-split component of the HDFS. The decision to use the input-split comes through the observation made from the reviewed literatures. This is due to the fact that, using any other HDFS components required the authors to develop algorithms to work with the index.

4.3 Test Queries

The queries used in the experiment of this research were formed into sets targeting the restrictive clauses that were relevant to indexing, namely, WHERE, JOIN and

Table 2 Sample select queries with WHERE, JOIN and GROUP BY for Modata

Query type no	MOData dataset
1	select * from Modata.Mdata20 where gpt <= Modata 'SF000150' and datadate >='1956-07-31 00:00:00' AND 'SF000320' and '2000-12-31 00:00:00' distribute by wd
2	select sc.gpt, sc.wd, sc.datadate, mk.gpt, mk.wd, Modata mk.datadate from Modata.Mdata10a sf JOIN Modata.Mdata10b sc ON sc.datadate = sf.datadate where sc.gpt <='SF000150' and sf.gpt <= 'SF000320' and sc.datadate = '2000-12-31 00:00:00' distribute by sf.wd, sc.wd
3	select year(sf.datadate), sf.ws, avg(sf.wd) from Mo- Modata data.Mdata20 sf where sf.gpt <='SF000150' and datadate >= '1956-07-31 00:00:00' AND 'SF000320' and '2000-12-31 00:00:00' group by year(sf.datadate) order by dd

Note 'wd' = Wind Direction, 'ws' = Wind Speed, 'gpt' = location 'datadate' = timestamps

GROUP BY. The reason for their choice was due to the fact that they were responsible for determining the data size to be considered in their query process. Furthermore, indexing was also used to restrict the amount of data to take part in the query processing.

The query sets are given in Table 2.

4.4 The Experiment and Its Setup

This section describes, in detail, the steps followed in conducting the evaluation of the experiment. The experiment was performed on the datasets mentioned earlier.

The research experiment was setup to evaluate the indexing scheme. The experiment was setup on a cluster of nine nodes, i.e., one Namenode (master) and eight Datanodes (slaves). Each had the following configuration: 8 core CPU, 8 GB RAM, 1 TB HDD, 1 GB bandwidth Ethernet card and running the Ubuntu 12.0.4 (LINUX) Operating System. The platform used was the Hadoop-2.7.1.

4.5 Index Creation Performance Evaluation

In order to see how the index creation time of the proposed index fares, it was compared to yet another indexing approach that is based on the same HDFS component of input-splits, but uses clustered index approach. Also, due to the nature of the queries that would use the proposed indexing when created, two different indexes are needed. One for queries, which work on single file, the SELECT … WHERE and SELECT … WHERE… GROUP BY queries, and the other for multiple files queries, that use SELECT… WHERE…JOIN query.

5 Results and Discussions

This section presents the experimental results and their discussions for both the index creation and query performance evaluation.

Modata Dataset Index Creation Performance

Table 3 shows the execution runtime for both the proposed index and the clustered index approaches to create index for queries on Modata. The same results were presented in Fig. 6 for better view.

From both the table and figure, it can be seen that the index creation for the proposed index was still faster than that of the clustered index. The index for the clustered index takes time more than that of the proposed indexing for the data sizes.

Similarly, Table 4 and Fig. 7 show that the index creation time for all data sizes using the proposed indexing fared better than that of the clustered index when used with join query. The index creation runtime for the clustered index was at least double the time of the creation of the proposed indexing for all the data sizes.

Table 3 Execution runtime for index creation on single file Modata

Data size	Index creation time (sec)	
	Clustered index	Proposed Index Method
20 GB	1167	596
50 GB	1673	1367
100 GB	3860	3619
200 GB	6731	5577
500 GB	12209	11591
1 TB	21821	20837

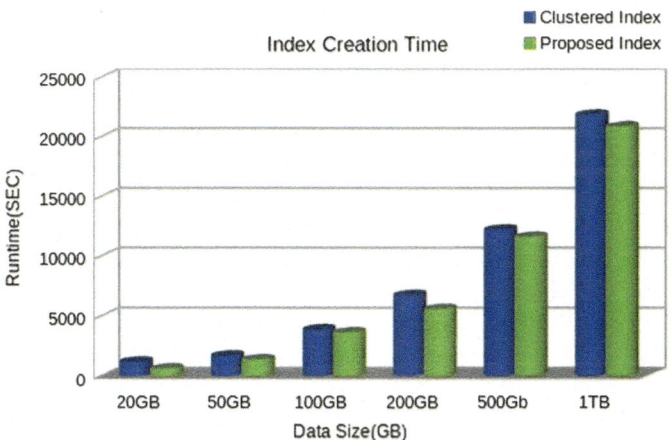

Fig. 6 Execution runtime for the index creation on single file Modata

Table 4 Execution runtime for multiple files index creation on Modata

Data size	Index creation time	
	Clustered index	Proposed index
20 GB	1021	410
50 GB	3025	879
100 GB	10689	1695
200 GB	10128	3646
500 GB	66227	8421
1 TB	75600	17014

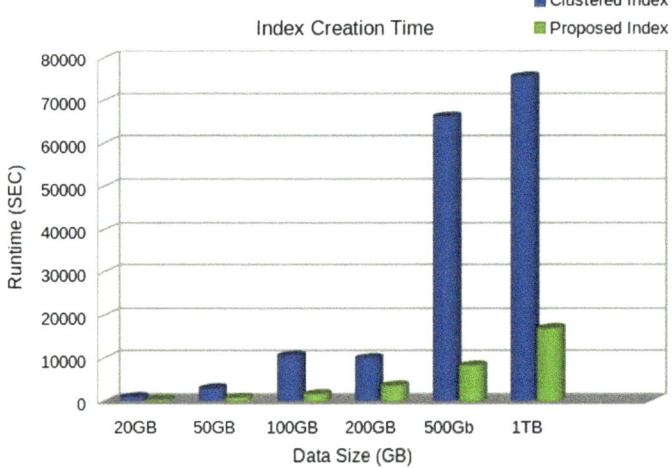

Fig. 7 Execution runtime for the multiple files index creation when applied to Modata

Thus, from the experiment conducted, we could say that our proposed indexing method performed better by taking lesser time to create its index for both single file query as well as multiple files for all different datasets and different data sizes. As mentioned earlier, this was attributed to the way the approach handled the creation of the index's data structure, the partitioned B+-Tree. The partitioned B+-Tree has an internal sorting that enabled it to compare and store only search keys that have unique corresponding values in the tree. So, for the proposed indexing the major determinant for runtime is the comparison amongst key/value entries, while disk I/O operation was less. On the other hand, the clustered index's runtime is determined by the I/O operation time as all search keys and their corresponding values must be stored.

Also, the high increase in the execution runtime for creating the index noticed, for both index approaches for 500 GB and 1 TB, may be largely attributed to the high increase in the number of map tasks (the total number of map tasks lunched): Data local map task (the total number of map tasks that reside on the same node on which they are processed) and the rack map tasks (the number of map tasks that need to be transferred to another node for processing). Increase in number of map tasks is translated into increase in time for each of the map tasks to run to its completion. While the increase in the difference between the map tasks launched and that of data local map tasks means there is going to be more of shuffling and sorting, which in turn led to transferring of data across the nodes of the Hadoop cluster, and that also increases the execution runtime.

Modata Execution Runtime Performance: Similarly, this subsection also presents the results of running the three queries in Table 2 using the normal MR, the clustered index and proposed indexing. However, the dataset used here is called the Modata dataset. First, the performance of the normal MR was compared with that of the proposed index as well as the clustered index approaches.

Query Execution Runtimes for the Normal MapReduce and Clustered Index versus Proposed indexing Methods: Query 1 **in** Table 2
Table 5 shows the results of the MapReduce job execution runtime for the Query 1, i.e., SELECT .. WHERE, on the Modata for all data sizes using the normal Hadoop MapReduce, the clustered index and the proposed indexing approaches. In similar pattern with the previous dataset, Fig. 8 presents the column chart for the results. Based on the results, the normal MR has the highest runtime for all the data since full input scan was used during the query processing. Then, the next highest runtime was that of the clustered index approach.

The runtime of the proposed indexing was 399 s lower than that of the normal MR and 78 s lower than that of the clustered index for 20 GB of data. The runtime was also lower, i.e., 807 s lower than that of normal MR and 31 s lower than that of the clustered index for 50 GB of data. For 100 GB of data size the runtime for the proposed indexing takes 2125 s less than the normal MR and 357 s lower than that of the clustered index. The same goes for 200 GB of data; the runtime of the proposed indexing takes lesser time, i.e., 2575 s to complete than that of normal MR as well as 1044 s lower than that of the clustered index. The runtime of the proposed indexing was 143 s lower than that of the normal Hadoop MR and 11 s lower than that of the clustered index for 500 GB of data. Then, for 1 TB data size, the runtime for the

Table 5 Query 1 execution runtimes for the normal MR versus clustered index versus proposed index methods (Modata)	Data size	Query 1 execution runtime (sec)		
		Normal MR	Clustered index	Proposed index method
	20 GB	724	403	325
	50 GB	1641	865	834
	100 GB	3578	1810	1453
	200 GB	7236	5705	4661
	500 GB	12924	12892	12781
	1 TB	22451	16010	15909

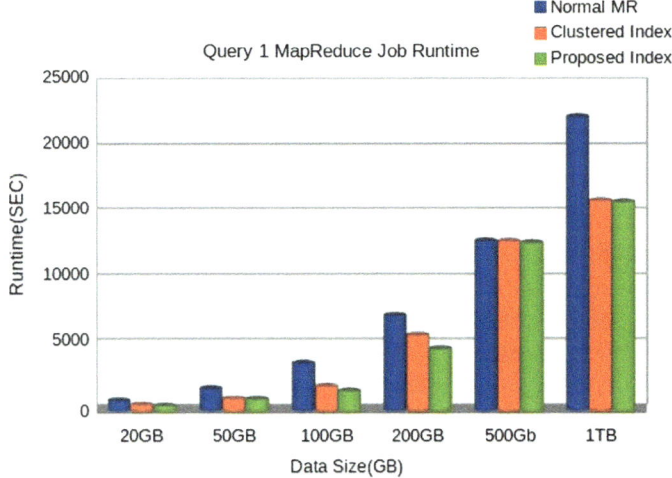

Fig. 8 Query 1 execution runtimes for the normal MR versus clustered index versus proposed indexing methods (Modata)

proposed indexing was also 6542 s lower than the normal MR and 101 s lower than that of the clustered index. The trend of the runtime increased with the increase of the data size significantly for both approaches. However, the time difference between the clustered index and proposed indexing was not so wide. This was due to the small files nature of the Modata. This may require more resources for their handling as the number of input-splits increased significantly.

Query 2 in Table 2

Table 6 presents the Query 2 type, i.e., SELECT .. WHERE .. GROUP BY query, job processing runtime for all data sizes using the normal MR, the clustered index and the proposed indexing approaches, and Fig. 9 gives the results in column chart form.

Table 6 Query 2 execution runtimes for the normal MR versus clustered index versus proposed indexing methods (Modata)

Data size	Query 2 execution runtime (sec)		
	The normal MR	Clustered index	ISPB method
20 GB	744	279	176
50 GB	1711	555	404
100 GB	3850	801	633
200 GB	5707	2213	2141
500 GB	12892	4700	4607
1 TB	19562	4841	4580

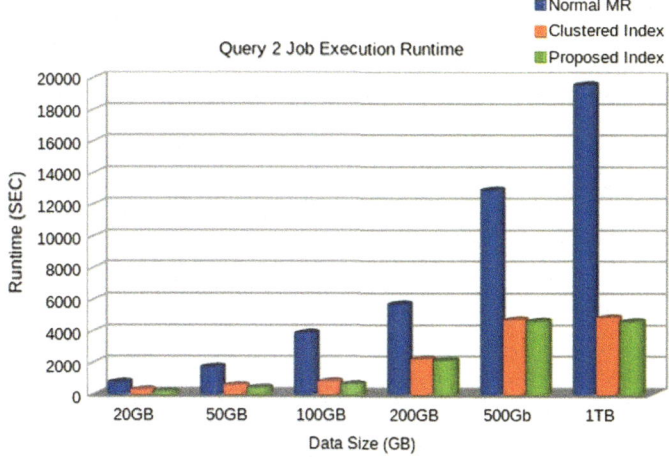

Fig. 9 Query 2 execution runtimes for the normal MR versus clustered index versus proposed indexing method (Modata)

The runtime results for individual data sizes are generally lower here than those of the Query 1. This was so, because the returned record set for the GROUP BY query was less voluminous. For 20 GB of data size the runtime for the proposed indexing method takes 568 s lesser than that of the normal MR and 103 s lower than that of the clustered index. For 50 GB and 100 GB of data sizes the runtime of the proposed indexing are lower by 1307 s and 3217 s lower than that of the normal MR and 151 s and 168 s lower than that of the clustered index, respectively. As for 200 GB of data size, the runtime was lower by 3566 s than that of the normal MR and 72 s lower than that of the clustered index. At the same time, the runtime of proposed indexing was 8825 s lower than the normal MR and 633 s lower than that of the clustered index for 500 GB of data. Lastly, for 1 TB data size, the runtime for the proposed indexing was also 14982 s lower than the normal MR and 261 s lower than that of the clustered index. Again, the proposed indexing outperformed the normal MR and the clustered index in all cases.

Query 3 in Table 2
Running Query 3, which was a query with JOIN clause using both normal MR, clustered index and the proposed indexing produced results as in Table 7. The same results are also presented in a chart form in Fig. 10.

The results show that the normal MR has the longest runtime for all the data sizes as compared to the proposed indexing. This was due to the complete scanning of the input during the query processing, followed by the clustered index. The runtime of proposed indexing method was 198 s lower than that of the normal MR and 189 s lower than that of the clustered index for 20 GB of data. The runtime was also 103 s lower than that of the normal MR and 293 s lower than that of the clustered index for 50 GB of data. For 100 GB of data size, the runtime for the proposed indexing was

Table 7 Query 3 execution runtimes for the normal MR versus clustered index versus proposed indexing method (Modata)

Data size	Query 3 execution runtime (sec)		
	Normal MR	Clustered index	ISPB method
20 GB	569	560	371
50 GB	795	985	692
100 GB	1866	1238	852
200 GB	3269	2091	1742
500 GB	10199	6061	5865
1 TB	23488	11714	9727

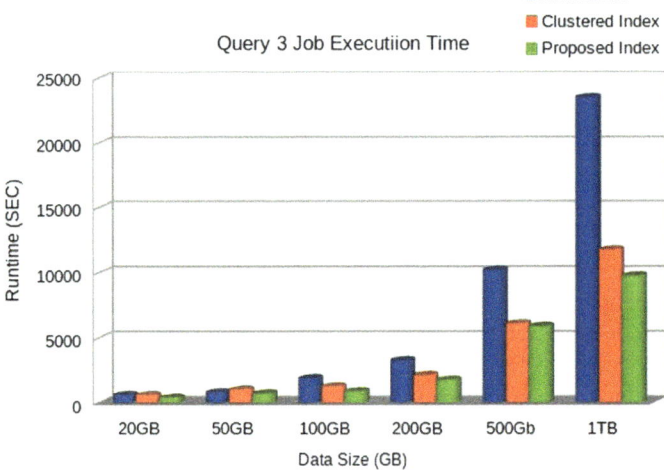

Fig. 10 Query 3 execution runtimes for the normal MR versus clustered index versus proposed indexing methods (Modata)

1014 s lower than that of the normal MR and 386 s lower than that of the clustered index. Similar trend was observed for 200 GB of data size, where the runtime of the proposed indexing was 1527 s lower than that of the normal MR and 349 s lower than that of the clustered index. Furthermore, the runtime of the proposed indexing method takes 4334 s less than that of the normal MR and 197 s lower than that of the clustered index for 500 GB of data. Then, for 1 TB data size, the runtime of the proposed indexing was 13761 s lower than that of the normal MR and 1987 s lower than that of the clustered index. The trend of the runtime increased significantly with the increase of the data size for both approaches.

In summary, after all these experiments have been conducted, we can say that query processing runtime becomes faster with the involvement of index in all cases.

6 Conclusion

As stated earlier, the index created and used by the proposed index was found to be far smaller than those created by clustered index approaches reported in this study. The index of proposed index was at least 1000 times smaller than those produced by other approaches. The reason for a smaller index size was first due to the fact that the data structure used by the proposed index. The partitioned B+-Tree enabled the building of the index using only the search keys, with unique input-splits as a representative of other search keys as the key value pair entry to the tree.

Secondly, the HDFS component chosen for the indexing in the proposed indexing, which was the input-split, helped in reducing size of both index and the input data. By default, the input-split in the Hadoop MapReduce process was what determined which part of the input data will take part in the query processing. Also, the input-split determines the number of mappers that the Hadoop needed to deploy to execute certain task.

Moreover, another reason for the better performance on the side of the proposed indexing during the query processing was due to the restriction brought about by the proposed index of processing only those input-splits that are returned by the index component of the scheme.

In summary, the proposed indexing has through its data structure the Pat B+-Tree, algorithms and approach (proposed index) has been able to significantly improve the MapReduce's query processing capability.

References

1. Abdullahi, A.U., Ahmad, R., Zakaria, M.N.: Experimental performance analysis of B+-trees with big data indexing potentials. In: International Conference of Reliable Information and Communication Technology, pp. 20–29. Springer (2017)
2. An, M., Wang, Y., Wang, W.: Using index in the mapreduce framework. In: Web Conference (APWEB), 2010 12th International Asia-Pacific, pp. 52–58. IEEE (2010)
3. B+-tree, B-tree: http://scienceblogs.com/goodmath/2008/07/06/btrees-balancedsearch-trees-f/ (2016)
4. Cao, J., Han, H., Zhao, M., Ye, S., Zhu, D., Li, L.: An optimized method oftranslating sql to more efficient map-reduce tasks. Int. J. Grid Distrib. Comput. **8**(4), 249–256 (2015)
5. Chaudhuri, S., Narasayya, V.: Self-tuning database systems: a decade of progress. In: Proceedings of the 33rd International Conference on Very Large Data Bases, pp. 3–14. VLDB Endowment (2007)
6. Chen, C.P., Zhang, C.Y.: Data-intensive applications, challenges, techniques andtechnologies: a survey on big data. Inf. Sci. **275**, 314–347 (2014)
7. Chen, M., Mao, S., Liu, Y.: Big data: a survey. Mobile Netw. Appl. **19**(2), 171–209 (2014)
8. Dean, J., Ghemawat, S.: Mapreduce: a flexible data processing tool. Commun. ACM **53**(1), 72–77 (2010)
9. Gani, A., Siddiqa, A., Shamshirband, S., Hanum, F.: A survey on indexing techniques for big data: taxonomy and performance evaluation. Knowl. Inf. Syst. **46**(2), 241–284 (2016)
10. Glombiewski, N., Seeger, B., Graefe, G.: Waves of misery after index creation. BTW 2019 (2019)

11. Graefe, G.: Sorting and indexing with partitioned b-trees. CIDR **3**, 5–8 (2003)
12. Graefe, G., Kuno, H.: Self-selecting, self-tuning, incrementally optimized indexes. In: Proceedings of the 13th International Conference on Extending Database Technology, pp. 371–381. ACM (2010)
13. Hadoop, A.: Apache hadoop. http://hadoop.apache.org/ (2017)
14. He, J., Yao, S.w., Cai, L., Zhou, W.: Slc-index: A scalable skip list-based indexfor cloud data processing. J. Central South Univ. **25**(10), 2438–2450 (2018)
15. Hong, Z., Xiao-Ming, W., Jie, C., Yan-Hong, M., Yi-Rong, G., Min, W.: A optimized model for mapreduce based on hadoop. TELKOMNIKA (Telecommunication Computing Electronics and Control) **14**(4) (2016)
16. Ibrahim, H., Sani, N.F.M., Yaakob, R., et al.: Analyses of indexing techniques onuncertain data with high dimensionality. IEEE Access **8**, 74101–74117 (2020)
17. Idreos, S., Kersten, M.L., Manegold, S.: Database cracking. In: CIDR. vol. 7, pp. 7–10 (2017)
18. Khasawneh, T.N., AL-Sahlee, M.H., Safia, A.A.: Sql, newsql, and nosql databases: a comparative survey. In: 2020 11th International Conference on Information and Communication Systems (ICICS), pp. 013–021 (2020)
19. Lee, S., Jo, J.Y., Kim, Y.: Performance improvement of mapreduce process bypromoting deep data locality. In: Data Science and Advanced Analytics (DSAA), 2016 IEEE International Conference on, pp. 292–301. IEEE (2016)
20. McCreadie, R., Macdonald, C., Ounis, I.: On single-pass indexing with mapreduce. In: Proceedings of the 32nd International ACM SIGIR Conference on Research and Development in Information Retrieval, pp. 742–743. ACM (2009)
21. McCreadie, R., Macdonald, C., Ounis, I.: Mapreduce indexing strategies: Studyingscalability and efficiency. Inf. Process. Manage. **48**(5), 873–888 (2012)
22. Mofidpoor, M., Shiri, N., Radhakrishnan, T.: Index-based join operations in hive. In: Big Data, 2013 IEEE International Conference on, pp. 26–33. IEEE (2013)
23. Philip Chen, C., Zhang, C.Y.: Data-intensive applications, challenges, techniquesand technologies: a survey on big data. Information Sciences **275**, 314–347 (2014) 24
24. Ramakrishnan, R., Gehrke, J., Gehrke, J.: Database management systems, vol. 3. McGraw-Hill New York (2010)
25. Richter, S., Quian´e-Ruiz, J.A., Schuh, S., Dittrich, J.: Towards zero-overhead staticand adaptive indexing in hadoop. VLDB J. **23**(3), 469–494 (2014)
26. Roy, S., Mitra, R.: A survey of data structures and algorithms used in the contextof compression upon biological sequence. Sustain. Humanosphere **16**(1), 1951–1963 (2020)
27. Rys, M.: Xml and relational database management systems: inside microsoft sqlserver 2005. In: Proceedings of the 2005 ACM SIGMOD international conference on Management of data, pp. 958–962. ACM (2005)
28. Sevugan, P., Shankar, K.: Spatial data indexing and query processing in geocloud. J. Testing and Eval. **47**(6) (2019)
29. Silberschatz, A., Korth, H.F., Sudarshan, S., et al.: Database system concepts, vol. 4. McGraw-Hill New York (1997)
30. Silva, Y.N., Almeida, I., Queiroz, M.: Sql: From traditional databases to big data. In: Proceedings of the 47th ACM Technical Symposium on Computing Science Education, pp. 413–418. ACM (2016)
31. Statista: Volume of data worldwide from 2010-2025. https://www.statista.com/statistics/871513/worldwide-data-created/ (2020)
32. Stewart, R.J., Trinder, P.W., Loidl, H.W.: Comparing high level mapreduce querylanguages. In: Advanced Parallel Processing Technologies, pp. 58–72. Springer (2011)
33. Suman, A.K., Gyanchandani, M.: Improved performance of hive using index-basedoperation on big data. In: 2018 Second International Conference on Intelligent Computing and Control Systems (ICICCS), pp. 1974–1978. IEEE (2018)
34. Yang, H.C., Parker, D.S.: Traverse: simplified indexing on large map-reduce-mergeclusters. In: International Conference on Database Systems for Advanced Applications, pp. 308–322. Springer (2009)

35. Zhang, Q., He, A., Liu, C., Lo, E.: Closest interval join using mapreduce. In: DataScience and Advanced Analytics (DSAA), 2016 IEEE International Conference on, pp. 302–311. IEEE (2016)
36. Zikopoulos, P., Eaton, C.: Understanding big data: analytics for enterprise classhadoop and streaming data. McGraw-Hill Osborne Media (2011)

Internet of Vehicle for Two-Vehicle Look-Ahead Convoy System Using State Feedback Control

Mu'azu Jibrin Musa⑩, **Olakunle Elijah**⑩, **Shahdan Sudin**,
Suleiman Garba, Saliu Aliu Sala, and Abubakar Abisetu Oremeyi

Abstract In recent years, the success of the Internet of Things has been extended into the field of transportation making it possible to have the internet of Vehicle (IoV). The combination of IoV and machine learning is expected to provide a robust system for high mobility vehicles that exhibit strong dynamics in the wireless channels, network topologies and traffic dynamics. We examine the use of IoV for two-vehicle look-ahead convoy in addressing the problem of safety and comfort in a two-vehicle convoy system. The two-vehicle look-ahead convoy systems have been getting much attention by researchers due to its inherent problem of string-instability, passengers' discomfort and its involvement in many areas of applications, which include the public and defence sectors. In this chapter, the dynamics model of the convoy was formed using the simplified vehicle model. The conventional control strategy was adopted to control the convoy. The best response of the controlled vehicle was achieved with the design parameters of 0.28 for kp1 and kv1, and 0.36 for kp2 and kv2,

M. J. Musa (✉) · S. Sudin
Division of Control and Mechatronics Engineering, Universiti Teknologi Malaysia, 81310 Johor
Bahru, Johor, Malaysia
e-mail: mmjibrin@utm.my

S. Sudin
e-mail: shahdan@utm.my

O. Elijah
Wireless Communication Center, School of Electrical Engineering, Universiti Teknologi
Malaysia, 81310 Johor Bahru, Malaysia
e-mail: elij_olak@yahoo.com

S. Garba
Nigerian Communications Commission, Abuja, Nigeria
e-mail: sgarba@ncc.gov.ng

S. A. Sala
Centre for Geodesy and Geodynamics, National Space Research and Development Agency,
Federal Ministry of Science and Technology, Toro, Bauchi State, Nigeria
e-mail: salisusala@yahoo.com

A. A. Oremeyi
School of Engineering, Kogi State Polytechnic Lokoja, Itakpe Campus, Nigeria
e-mail: abubakarabisetu@kogipolytechnic.edu.ng

© The Author(s), under exclusive license to Springer Nature Switzerland AG 2021   241
H. Chiroma et al. (eds.), *Machine Learning and Data Mining for Emerging Trend
in Cyber Dynamics*, https://doi.org/10.1007/978-3-030-66288-2_10

respectively. An improvement was made in the controller design through the design and implementation of a state-feedback controller (SFC) using the pole placement technique. The performance of the two controllers was evaluated with respect to the controlled vehicle's speed, acceleration and jerk. It was clearly seen in the presented result that the SFC controller gave a perfect speed tracking of the controlled vehicle without any overshoot. An acceptable acceleration of 2 ms^{-2} was achieved with zero value of jerk in the controlled vehicle. This signifies string-stability, achievement of acceptable acceleration and passengers' comfort in the controlling vehicle using the proposed SFC.

Keywords Acceleration · Connected cars · Internet of Vehicle · State-feedback controller · Vehicle-to-Vehicle · Vehicle convoy

1   Introduction

The advancement in information communication technology has paved way for connected cars known as the internet of vehicles (IoV). The IoV is defined as the exchange information between the vehicle and its environment through different communication media [1]. The IoV allows for vehicles to communicate with human drivers, other vehicles, road infrastructures, road users and provide several services such as the fleet management systems in real-time [2]. The ability to exchange such useful information is changing the way vehicle operates and making it possible to have intelligent transport systems and co-operative intelligent transport systems [3].

The communication is achieved via distributed networks or vehicular ad hoc networks (VANETs). The VANET is a mobile ad hoc network in which nodes (vehicle) are clustered based on correlated spatial distribution and relative velocity using clustering algorithms [4]. These vehicles are incorporated with wireless transceivers that allows to communicate with their environment via the internet or line of sight communication technologies. There are different types of IoV network communication. They include Intra-Vehicle [5], Vehicle-to-Vehicle (V2V) [6], Vehicle-to-Infrastructure (V2I) [7], Vehicle-to-Cloud (V2C), and Vehicle-to-Pedestrian (V2P) [8]. While the V2V involves exchange of information between automobiles the other forms of communication involve the exchange of information between the vehicle and its surroundings.

The V2V network enables vehicles to exchange information such as speed, position of surrounding vehicles, direction of travel, acceleration, braking and loss of stability [1]. Such information can be used to increase the safety by providing different level of services to vehicle such as warning, brake or steering control. The use of IoV has been proposed to reduce collision and maintain stability of moving vehicles at the roundabout in [9].

In this research work, we focus on the use of V2V to solve the problems of safety and comfort of passengers in a two-vehicle convoy system. This is because the scope of work is limited to exchange of information between automobiles. Two-vehicle convoy system is a configuration of vehicles in a convoy in such a manner whereby the control vehicle is using the information (spacing, speed, and acceleration) of the neighboring (two ahead) vehicles. The two vehicles ahead are called the predecessor and lead vehicles where the spring and damper techniques have been utilized. One of the two control policies (constant inter-vehicular spacing or variable inter-vehicular spacing) need to be adopted for safety. Musa et al. [10] used the concept of spring-mass-damper system to present an idea of vehicle convoys' topology using the control policy of variable time-headway.

String stability is the most significant aspect that needs to be consider in such convoy's configuration to achieve stable movement. The term string stability is the ability of the vehicles in a convoy to move without amplifying the oscillation of the convoy leader off string. The convoy system is said to be stable if the oscillation of the lead-vehicle dies as it propagates along the convoys' stream [11]. Finally, the passenger's comfort should also not to be compromise. Passengers comfort can be determined by the jerk of the controlled vehicle, which is the third derivative of the control vehicles displacement. To achieve this, the control vehicle needs the speed, spacing and acceleration of the neighboring vehicles. This information is exchanged between the vehicles involved in the convoy using two types of communication technologies which are the dedicated short-range communications (DSRC) and the 5GLTE cellular technology. There are challenges associated with the reliability of the 5GLTE wireless communication in high mobility networks. The application of machine learning and deep learning can be used to address some of the challenges such as complexity in channel estimation, allocation of resources and network topology.

In this chapter we present the concept of V2V communication and the State-Feedback Controller (SFC) to achieve the safety and comfort in a two-vehicle convoy system. The V2V would allow the exchange of information between the controlled vehicle and the neighboring vehicle. The State-Feedback Controller (SFC) would be used via the pole-zeros approach to control the movement of the controlled vehicle to achieve the desire parameters.

The rest of this chapter is structured as follows. Section 2 covers the system architecture of the of IoV and V2V application. Sections 3 and 4 explains the mathematical modeling of the proposed two look-ahead vehicle convoy and vehicle dynamic, respectively. Section 5 presents the results obtained and discussion. Section 6 provides the conclusion and further work.

2 System Architecture

In this section, we discuss the various components of the IoV and the architecture of the two-vehicle look-ahead convoy.

2.1 IoV Components

The internet of things consists of four major components which (1) IoT devices; (2) communication technology; (3) Internet; and (4) data storage and processing [12]. Similar to the IoT, the IoV consist of four major components. One is the node in this case the vehicle becomes the node. The second is the communication technology. The third is the internet that allows for remote connection between the vehicle and environment and finally the fourth is the data storage and processing for control management and intelligent services. The illustration of the IoV (V2V) is shown in Fig. 1.

The four components are described as follows.

The Node. This consist of sensors and actuators for sensing and control of the vehicle via the On-board Unit (OBU) and user interfaces. The sensors help to detect events, driving patterns and environmental situations. The OBU are designed to have various all kinds of interconnectivity for transmission of information between the vehicle and other entities. The user interface can be either auditive, visual or haptic. The user interfaces provide useful application and services to the drivers and passengers such as timely notifications of lane changes, emergency braking, traffic situations and avoidable obstacles or head on collisions.

The Communication Technology. There are various communication technologies that have been designed to support vehicular communications. Some of the technology's supports are either short-range or long-range communication. The two

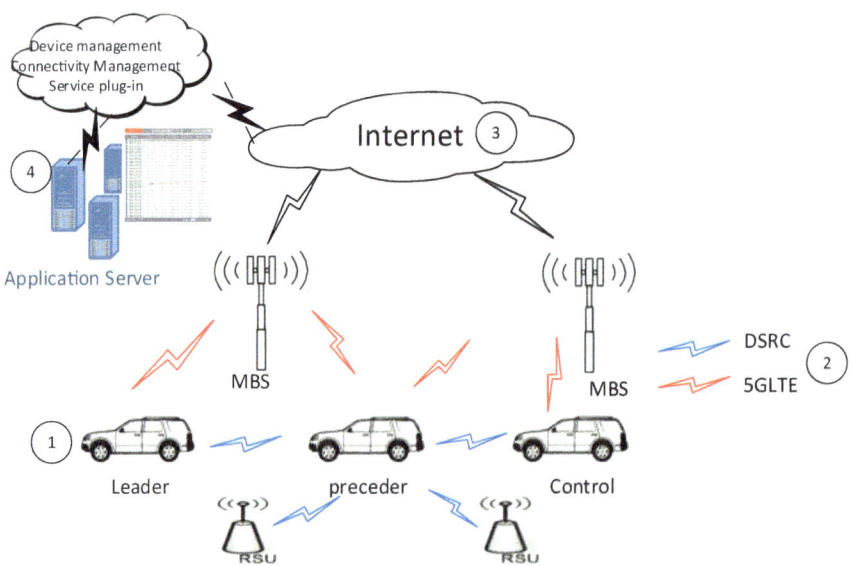

Fig. 1 Internet of Vehicle for V2V two-vehicle look-ahead convoy

major communication technologies for V2V communication. One is the dedicated short-range communications (DSRC) [13] and the second is the 5GLTE cellular technology. The DSRC is an established technology for V2V and operates in 75 MHz of spectrum (5.85–5.925) GHz. It supports a transmission range of 300 m and a latency of 25 ms. The 5GLTE is expected to deliver 1 ms latency and operate at cellular frequency with much wider bandwidth, better penetration and lesser interference [14]. Other forms of communication technology include the GSM, Wi-FI, and Bluetooth.

The Internet. The internet is a platform where paths are provided to carry and exchange data and network information between multiple subnetworks [12]. The connection of IoT devices to the Internet enables data to be available anywhere and anytime provided there is adequate security both at the network and physical level. Several network protocols have been designed to handle network communication for IoV both at the physical and network layer. Some of the protocols are IEEE 802.11p directional medium access control (DMAC) and vehicular cooperative media access control.

The Data Storage and Processing. This include the data acquisition, transmission, storage and computing of big data that is transferred for automotive telematics, autonomous vehicles and the dynamic traffic monitoring among many others. It also include statistical and artificial intelligence tools for processing of data needed for decision making in different scenarios such as traffic congestion, dangerous road conditions, lane and convoy coordination. The machine learning and deep leaning are subset of artificial intelligence needed in high mobility networks that exhibit strong dynamics in the areas such as wireless channels, network topologies and traffic. They enable machines to learn from large amounts of data in order to perform certain task without explicitly programming it. The challenge of complexity in channel estimation, resource allocation, link scheduling and routing for the vehicular communication in 5GLTE can be handled using the Artificial intelligence.

2.2 IoV Architecture for Two-Vehicle Look-Ahead Convoy

In this section we describe the operation of the two-vehicle look-ahead convoy using IoV five-layer architecture. The IoV is expected to enhance the exchange of information of the dynamics of the individual vehicles involved in the convoy. This information is sent to the controllers in the vehicles. The IoV architecture consists of five layers. We describe the five layers as follows.

The First Layer. The first layer is the data acquisition layer. This consist of sensors and actuators for sensing and control of the vehicle via the On-board Unit (OBU). The OBU is equipped with multiple wireless communication interface. This enables the vehicle to exchange information with other vehicles, the RSU and the internet. Data acquisition of the speed and position are obtained from the vehicle via inbuilt

Fig. 2 Packet structure

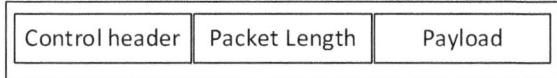

Control header	Packet Length	Payload

sensors. This data can be compressed into a packet as shown in Fig. 2. The control header consists of the control field for communication, the packet length hold length of the packet and the payload contains the data being sent.

The packets are sent with unique ID for each vehicle to the controlled vehicle. The OBU receives the packet and the data is sent to the state feedback controller (SFC). Based on the information, the SFC can adjust the movement of the controlled vehicle using the model described in Sect. 3.

The Second Layer. The second layer is the communication technology. It consists of the radio frequency technology required for communication between the OBU and the roadside unit (RSU) or mobile base station (MBS) required for the 5GLTE. It is assumed that all vehicles are connected to the network using the 5GLTE technology at the same time the vehicles can communicate with the RSU using the DSRC.

The Third Layer. The third layer is the edge network. It handles the distributed intelligence and fog computing needed for the V2V communication. Traffic efficiency and safety servers are installed in the edge network in order to reduce amount of load in terms of data for the core network and the reduce latency time.

The Fourth Layer. The fourth layer is the core network. The core network handles the security, multicast, IP routing and the mobile packet core needed for the V2V communication. It also coordinates communication between the edge networks.

The Fifth Layer. The last layer is the Data center and cloud. This layer handles the data processing and application layers. The information of each vehicle can be stored in the cloud. Also, the function of the SFC can be computed in the cloud and sent to the controlled vehicle. Figure 3 shows the IoV architecture for Internet of Vehicle.

2.3   Platform Used for Implementation of the Model

The MATLAB platform was utilized in which the Similink blocks were used to model the mathematical equation that represents the behavior of the single vehicle. Simplified vehicle dynamics was used to represent the single vehicle system. The modeled single vehicle was cascaded to form the topology of the two-vehicle look-ahead using the bases of the mass-spring damper as shown in Fig. 4. The real time three dimension (3D) visualization of the topology behavior was translated using MapleSim 2019. MapleSim is object oriented software, which permits the modeling of complex model within fraction of minutes. The MapleSim platform provides a simple and flexible ways to model and visualized the true behavior of the model at a glance. It also allowed instant changes of the model parameters and the effect

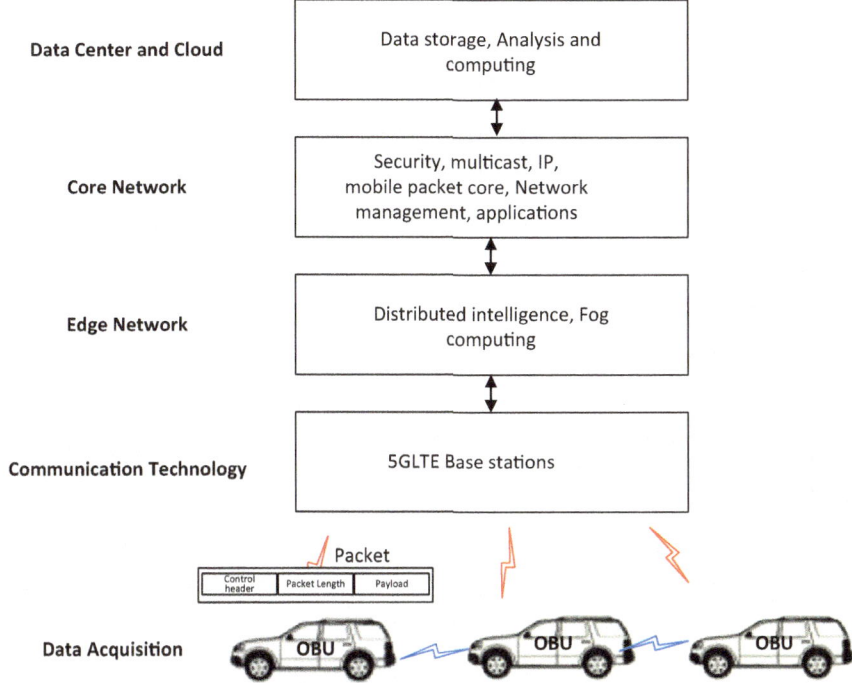

Fig. 3 IoV architecture for Internet of Vehicle

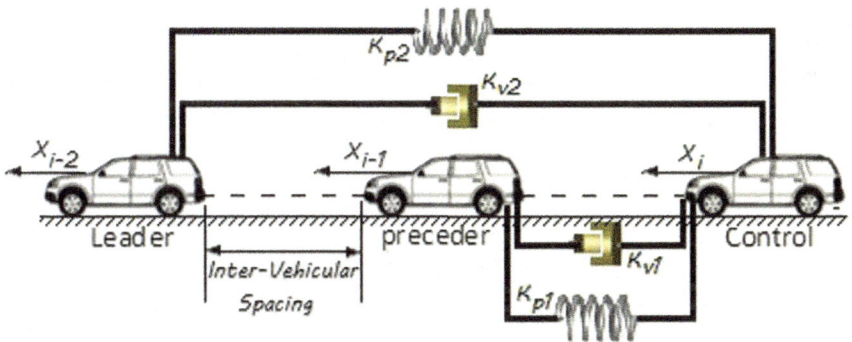

Fig. 4 A Conventional two-vehicle look-ahead topology

of the changes will immediately show on the graphical simulation and on the 3D visualization.

3 Modeling of the Two Look-Ahead Vehicle Convoy Strategy

In this section, the concept of the two-vehicle convoy system would be developed from the 2nd Newton's law and expressed using the mass-spring-damper as shown in Fig. 4 starting from the basics as.

Where, x_i, x_{i-1} and x_{i-2} are the positions of the controlling vehicle, preceding and lead vehicle respectively.

$$m\dot{x}_i = \sum_{n=1}^{2} \{kp_n(x_{i-n} - x_i) + kv_n(\dot{x}_{i-n} + \dot{x}_i)\} \tag{1}$$

Equation (1) gave the spring-damper-representation of the two-vehicle look-ahead model. Where by m represents the mass of the vehicle in kg, k_p is the spring constant, kv is the damper constant, derivative of x_i and derivative of x_{i-n} when n stands for 1 and 2 and are the speed of the controlled, predecessor and lead vehicle respectively. Taking the Laplace transformation of Eq. (1) at unit mass yield Eq. (2):

$$X_i = \frac{\sum_{n=1}^{2} \{(kp_n s + kv_n)X_{i-n}\}}{s\left(s + \sum_{n=1}^{2} kv_n\right) + \sum_{n=1}^{2} kp_n} \tag{2}$$

The transfer function in Eq. (2) depends on the vehicle following spacing policy adopted. Spacing policy is a predefine rule, which enable the controlled vehicle to regulate it distance with respect to the neighboring vehicles in the convoy. Controller should be implemented to regulate the vehicle speed according to the designed spacing policy [15, 16]. The time headway policy is used in this chapter it varies according to the speed of the neighboring vehicles. To achieve such spacing regulation, the time headway is introduced to Eq. (2) as:

$$X_i = \frac{\sum_{n=1}^{2} \{(kp_n s + kv_n)X_{i-n}\}}{s^2 + s\left\{\sum_{n=1}^{2} kv_n + h(kp_1 + 2kp_2)\right\} + \sum_{n=1}^{2} kp_n} \tag{3}$$

For stability to occur the pole-zero cancellation is utilized by introducing the constraints Eq. (4) [17]:

$$\frac{kp_1}{kv_1} = \frac{kp_2}{kv_2} = (kp_1 + 2kp_2)h \tag{4}$$

Under the constraints in Eq. (4), Equation (3) can now be reduce to single pole, which will lead to string stable convoy as achieved in Eq. (5):

$$X_i = \frac{\sum_{n=1}^{2} k v_n \cdot x_{i-n}}{s + \sum_{n=1}^{2} k v_n} \tag{5}$$

The control signal u_i is directly proportional to the applied force f on the vehicle, which was modelled as initial mass [15–18] as shown in Eq. (6).

$$\ddot{x}_i = f(\dot{x}_i, u_i) \tag{6}$$

Hence the control signal can be represented by (7):

$$\ddot{x}_i = u_i \tag{7}$$

where $\ddot{x}_i$ is the acceleration of the controlled vehicle. The control signal can be obtained from Eq. (3) as:

$$u_i = \sum_{n=1}^{2} k p_n (x_{i-n} - x_i - nh\dot{x}_i) + \sum_{n=1}^{2} k v_n (\dot{x}_{i-n} + \dot{x}_i) \tag{8}$$

4 Vehicle Dynamic

The vehicle internal dynamics only is considered in the model for simplicity. The simplified vehicle model used a lag function to represent the model. The concept of the lag function comes to existence since the actual vehicle's acceleration can only be obtained after certain time delay τ as in Eq. (9):

$$\tau\dot{x} + \ddot{x} = u \tag{9}$$

Now the time delay can be included in the convoy vehicle model of Eq. (3) as shown:

$$X_i = \frac{\sum_{n=1}^{2} (k p_n s + k v_n) x_{i-n} + (2s + k p_2) x_{i-n}}{s^2(\tau s + 1) + \left\{ \left(\sum_{n=1}^{2} k v_n + \sum_{n=1}^{2} n k p_n \right) h \right\} s + \sum_{n=1}^{2} k p_n} \tag{10}$$

With the introduction of the vehicle dynamics in Eq. (10) the command signal produced will now be used to drive the vehicle dynamic as shown in Fig. 5.

Taking the value of $\tau = 1$ (delay) and $h = 1$s (headway) the best acceptable acceleration of 2 ms^{-2} [10] is achieved when $k p_1 = k v_1 = 0.28$ and $k p_2 = k v_2 = 0.36$ from the transfer function as:

Fig. 5 Simplified vehicle
dynamic

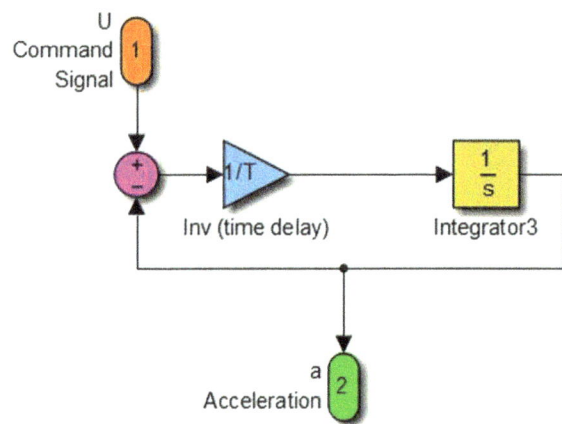

$$X_i = \frac{(s+1)(0.28x_{i-1} + 0.36x_{i-2})}{s^3 + s^2 + 1.64s + 0.64} \quad (11)$$

Certain trail an error where carried out with different values of spring and damper
constants to come up with the best values of, kp's and kv's which can be justified
from Figs. 6 to 7.

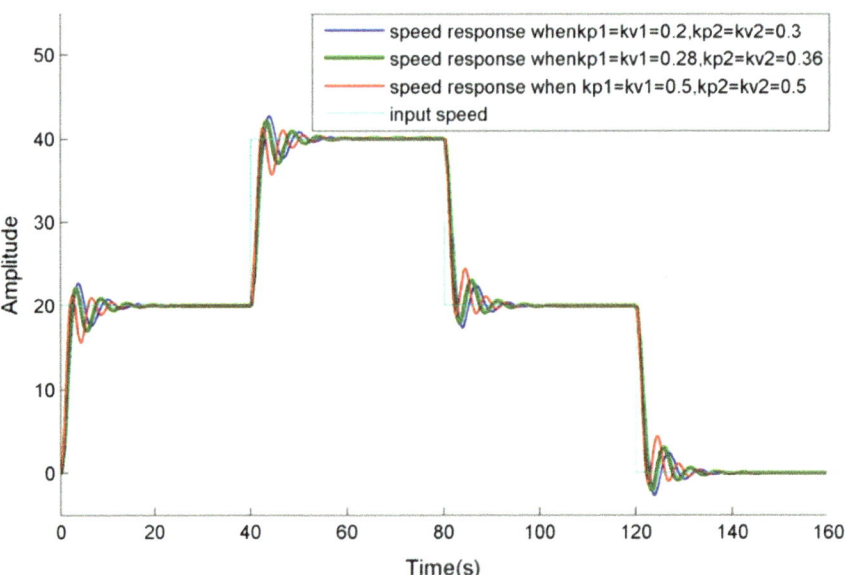

Fig. 6 Speed responses with different constants (kp's and kv's) values

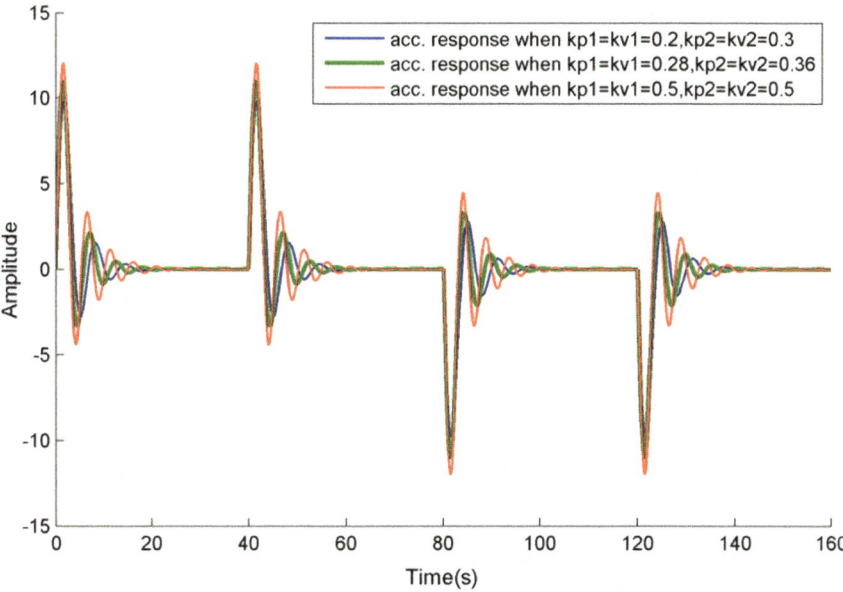

Fig. 7 Acceleration response with different constants (k_p's and k_v's) values

The best spring's and damper's gain selection can be seen in Fig. 6 using trial an error. Several gains are applied through tuning and the gain that gives the best performance was adopted throughout the convoy's model.

Figure 7 shows acceleration response using different gains, the selected gain values gave smooth, acceptable and low acceleration values of less than 2 ms^{-2}. The low the acceleration the low the jerk. Hence, the more the passenger's comfort.

4.1 SFC Design Using Pole-Placement Approach

The SFC was designed using the pole-placement approach to enable the transfer of all the poles of closed-loop independent to each other [19].

Pole Placement Topology

The state space equation for the plant can general be represented as:

$$\left. \begin{array}{l} \dot{x} = Ax + Bu \\ y = Cx + Du \end{array} \right\} \tag{12}$$

Fig. 8 Plant state-space

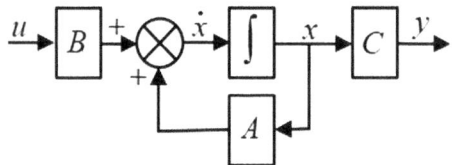

Fig. 8 Plant state-space

where A is the state transition matrix, B is the input matrix, C is the output matrix, x is the state vector, y is the output and $\dot{x}$ is the input. The state-space map graph is as shown in Fig. 8

In order to achieve the close-loop pole values from the plant state-variable feedback as shown in Fig. 9 the gain need to be adjusted to a suitable value.

Hence Eq. (12) can now be represented as:

$$\dot{x} = Ax + B(-kx + r) \tag{13}$$

In order to design a SFC with pole placement method, the state-space representation of the plant should be developed. In this chapter, the system transfer function is converted to state-space equation as illustrated in [10]. The plant representation in term of input and output is given as in Fig. 10.

Based on the transfer function Eq. (11) of the plant model, the following assumptions are made which resulted to Fig. 11:

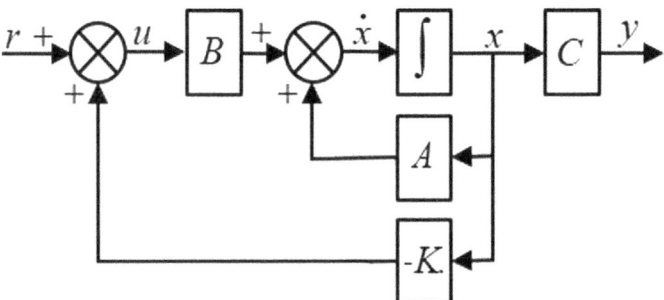

Fig. 9 Plant state-variable feedback

Fig. 10 Block diagram of the Plant state-variable feedback for the propose convoy

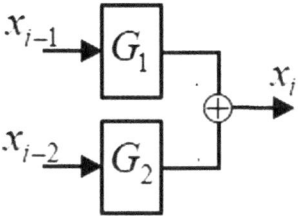

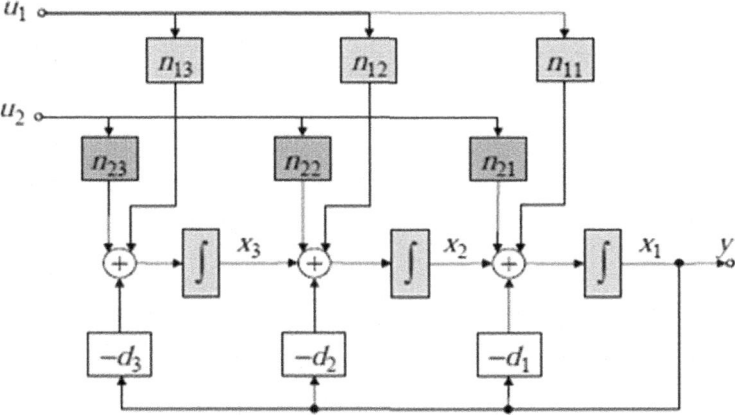

Fig. 11 Block diagram of the state variables, inputs, and output

$$x_{i-1} = u_1 \tag{14}$$

$$x_{i-2} = u_2 \tag{15}$$

Figure 10 is a Multi Input Single Output (MISO) system, therefore:

$$G(s) = \begin{bmatrix} G_1 & G_2 \end{bmatrix} \tag{16}$$

and

$$G(s) = \begin{bmatrix} \frac{b_1(s)}{a_1(s)} & \frac{b_m(s)}{a_m(s)} \end{bmatrix} \tag{17}$$

where $G(s)$ can also be express as:

$$G(s) = \frac{[n_1(s)\ldots n_m]}{d(s)} \tag{18}$$

$$G(s) = \frac{\left[n_{11}s^{n-1} + n_{12}s^{n-2} + \cdots + n_{1n}\cdots n_{m1}s^{n-1} + n_{m2}s^{n-2} + \cdots + n_{mn}\right]}{s^n + d_1 s^{n-1} + \cdots + d_n} \tag{19}$$

The numerator polynomial is defined as (20):

$$n_1(s) = \frac{b_1(s)d(s)}{a_1(s)} \tag{20}$$

The inputs and the output variables can be represented by Fig. 10.

In general, the MISO canonical form matrix is shown as Eqs. (21)–(22):

$$
A = \begin{bmatrix}
-d1 & 1 & 0 & \cdots & 0 \\
-d2 & 0 & 1 & \cdots & 0 \\
\vdots & \vdots & \vdots & \ddots & \vdots \\
-d_{n-1} & 0 & 0 & \cdots & 1 \\
-dn & 0 & 0 & 0 & 0
\end{bmatrix}
\tag{21}
$$

$$
B = \begin{bmatrix}
b_{11} & \cdots & b_{n-1} \\
b_{12} & \cdots & b_{n-2} \\
\vdots & \ddots & \vdots \\
b_{1(n-1)} & \cdots & b_{m(n-1)} \\
b_{1n} & \cdots & b_{mn}
\end{bmatrix}
\begin{bmatrix}
u_1 \\
\vdots \\
u_m
\end{bmatrix}
\tag{22}
$$

$$
C = \begin{bmatrix} 1 & 0 & \cdots & 0 \end{bmatrix}
\tag{23}
$$

Since the number n of states is equal to the degree of denominator $d(s)$, then the characteristics equation will be as Eq. (24):

$$
s^n + d_1 s^{n-1} + \cdots + d_{n-1} s + d_n = 0
\tag{24}
$$

According to Eqs. (21)–(23) the system transfer function can be written as Eq. (24) to Eq. (26) respectively:

$$
\dot{x} = \underbrace{\begin{bmatrix}
-1 & 1 & 1 \\
-1.64 & 0 & 1 \\
-0.64 & 0 & 0
\end{bmatrix}}_{A} x + \underbrace{\begin{bmatrix}
0 & 0 \\
0.28 & 0.36 \\
0.28 & 0.36
\end{bmatrix}}_{B} \underbrace{\begin{bmatrix}
u_1 \\
u_2
\end{bmatrix}}_{u}
\tag{25}
$$

$$
y = \underbrace{\begin{bmatrix} 1 & 0 & 0 \end{bmatrix}}_{C} x
\tag{26}
$$

4.2 Design Procedure of the SFC via Pole Placement Technique

Feeding back each state variable to the two inputs u_1 and u_2 gives Eq. (27):

$$
u = -k_{mn} x
\tag{27}
$$

where k is the feedback gains express as:

$$k_{mn} = \begin{bmatrix} k_{11} & \cdots & k_{1n} \\ \vdots & \ddots & \vdots \\ k_{m1} & \cdots & k_{mn} \end{bmatrix} \tag{28}$$

Hence the feedback gains matrix is can also be written as Eq. (29):

$$k = \begin{bmatrix} k_{11} & k_{12} & k_{13} \\ k_{21} & k_{23} & k_{23} \end{bmatrix} \tag{29}$$

Using Eqs. (11) and (25) the system matrix $A - BK$ for the closed loop will be:

$$A - BK = \begin{bmatrix} -1 & 1 & 0 \\ -(7*k_{11})/25 - (9*k_{21})/25 - 41/25 & -(7*k_{12})/25 - (9*k_{22})/25 & 1 - (9*k_{23})/25 - (7*k_{13})/25 \\ -(7*k_{11})/25 - (9*k_{21})/25 - 16/25 & -(7*k_{12})/25 - (9*k_{22})/25 & -(7*k_{13})/25 - (9*k_{23})/25 \end{bmatrix} \tag{30}$$

Now closed loop system characteristic equation will be as shown in Eq. (31):

$$\det[SI - (A - BK)] = \sum_{n=1,3} \left(\frac{7*k_{1n}}{25} + \frac{9*k_{2n}}{25} \right) s$$

$$+ \left\{ \left(\sum_{n=2}^{3} \left(\frac{7*k_{1n}}{25} + \frac{9*k_{2n}}{25} \right) + 1 \right) + s \right\} s^2$$

$$+ \frac{1}{25} \{ (s(41 + 14k_{12} + 18k_{22})) + 16 \}$$

$$+ \frac{1}{25} \sum_{n=1}^{3} (7*k_{1n} + 9*k_{2n}) = 0 \tag{31}$$

The desired-characteristic equation is in the form:

$$s^n + d_{n-1}s^{n-1} + d_{n-2}s^{n-2} + \cdots + d_2s + d_1s + d = 0 \tag{32}$$

where, d_i is are the desired coefficients.

For better transient response overshoot of $\leq 4.3\%$ and 5.3 s settling time must be produced. The characteristic equation for the dominant poles is gives as Eq. (33) using the design parameters ξ as 0.693 and w_n as 1.082:

$$s^2 + 1.5s + 1.1709 = 0 \tag{33}$$

where dominant poles location is at $-0.75 \pm j0.78$, the third pole is ≈ 10 times the distance from the imaginary axis. while at -0.55 the closed-loop system desired characteristic equation was found to be:

$$s^2 + d_{n-1}s^{n-1} + d_{n-2}s^{n-2} + \cdots + d_2 s + d_1 s = 0 \tag{34}$$

Comparing Eq. (31) with Eq. (34) and matching their coefficients, the k_{mn}'s could be evaluated as follows:

$$k_{11} = -1.4081, \quad k_{12} = 0.4737, \quad k_{13} = 0.9397,$$
$$k_{21} = -1.8104, \quad k_{22} = 0.6091 \text{ and } k_{23} = 1.2082 \tag{35}$$

The MATLAB coding used for the controller design and implementation on the system is:

```
g=[tf([0.28 0.28],[1 1 1.64 0.64]) tf([0.36 0.36],[1 1 1.64 0.64])]
sys2=tf2ss(num,d/'en)
a=[0 1 0;0 0 1;0 -171 -101.71]
b=[0 0;0.28 0.36;0.28 0.36]
c=[1 0 0;0 1 0;0 0 1]
d=[0 0;0 0;0 0]
plant=ss(a,b,c,d)
p=[-0.55 -75-0.78j -0.75+0.78j]
k1=place(a2,b2,p1)
cl_sys=feedback(plant,k1)
step(cl_sys)
```

Encouraging values of k_{mn}'s are obtained while running the MATLAB codes, which are the same as the computed.

4.3 Controllability System Test

The aspect of controllability C_M in system is widely established [20, 21]. According to Sinha [22] for an nth-order plant whose state equation is as in Eq. (12) then the totally controllability can generally be expressed as Eq. (36):

$$C_M \left[B \vdots AB \vdots A^2 B \ldots A^{n-1} B \right] \tag{36}$$

Similarly, the 3rd order system controllability is as shown:

$$C_M = \left[B : AB \ A^2 B \right] \tag{37}$$

$$C_M = \begin{bmatrix} 0 & 0 & \vdots & 0.28 & 0.36 & \vdots & 0 & 0 \\ 0.28 & 0.36 & \vdots & 0.28 & 0.36 & \vdots & -0.4592 & -0.5904 \\ 0.28 & 0.36 & \vdots & 0 & 0 & \vdots & -0.1792 & -0.2304 \end{bmatrix} \qquad (38)$$

Whereby the rank of C_M is equal to the number of linearly independent rows or columns. The rank can also be found by finding the highest-order square sub matrix that is non-singular. The determinant of the submatrix, which includes for example 2nd, 3rd and 4th column of C_M is not zero then the 3 by 3 matrix is non-singular, hence the rank of C_M is 3. It can be concluded that the system is controllable since the rank of C_M is equal to the system's order. Thus, the poles of the system can be placed using state-variable feedback design, which can be prove as:

```
A=[-1 1 0; -1.64 0 1; -0.64 0 0]
B = [ 0  0; 0.28 0.36; 0.28 0.36]
Cm=ctrb(A,B)
Rank=rank(Cm)
Ans=3
```

thus, the system is controllable.

5   Result and Discussion

Figure 12 shows the controlled vehicle's output speed with and without the SFC controller. The speed of the controlled vehicle with the SFC gave a smooth path with a precise tracking that has no oscillation as against that without SFC.

Figure 13 shows the acceleration response where the controlled vehicle with SFC gave the most acceptable acceleration of 2 ms^{-2}, while the acceleration of the controlled vehicle without SFC gave a value more than the accepted acceleration. Hence the acceleration with SFC is the accepted value [23].

Figure 14 shows the achievement recorded in terms of jerk in the controlled vehicle's jerk. A zero-jerk value (0 ms^{-3}) was recorded in the controlled vehicles jerk as seen in Fig. 14. This signifies maximum comfort in the controlled vehicle during the convoy movement.

6   Conclusion and Future Work

In this chapter, the use of IoV and state feedback controller for two-vehicle look-ahead convoy system was presented. The architecture of the system was discussed

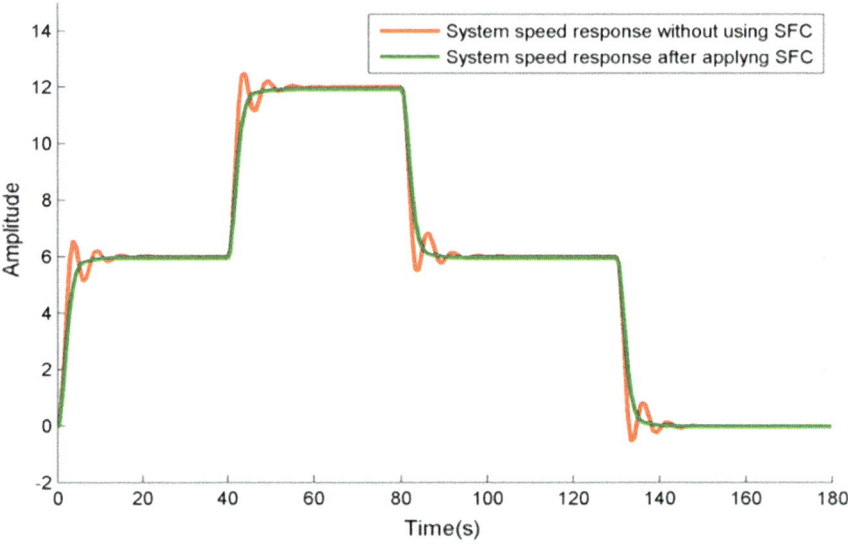

Fig. 12 Following vehicle's speed response with and without SFC

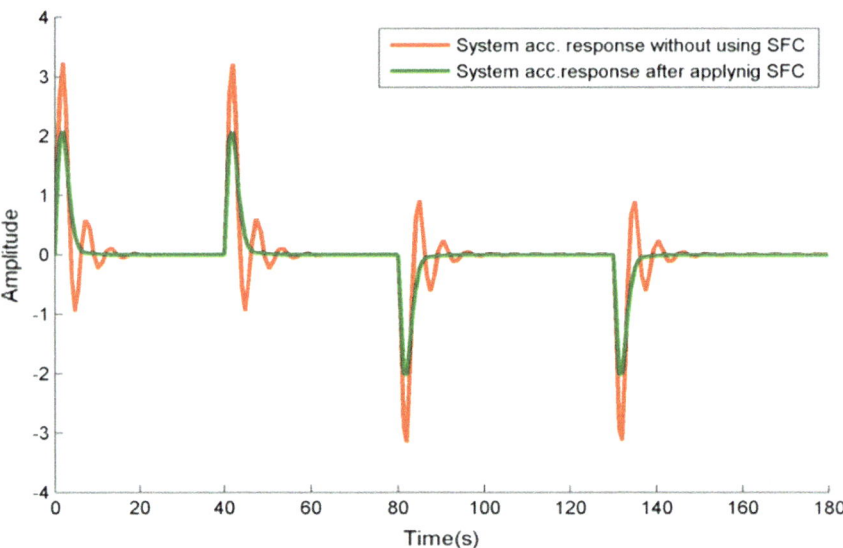

Fig. 13 Following vehicle's acceleration response with and without SFC controller

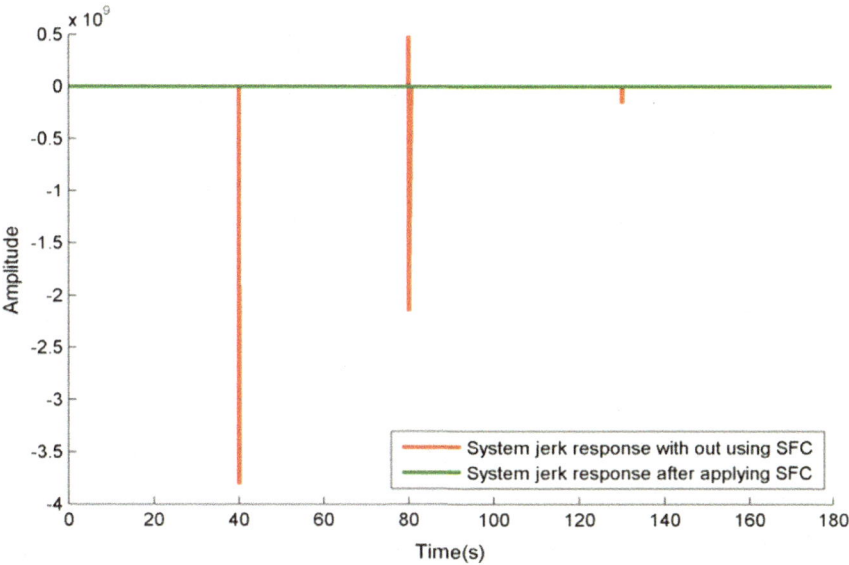

Fig. 14 Following vehicle's jerk response with and without SFC controller

and best system response was derived. The best system response was achieved with the design parameters of 0.28 for kp_1 and kv_1 and 0.36 for kp_2 and kv_2, respectively, which can be seen in Figs. 6 and 7. The state feedback controller design using pole placement technique for the two-vehicle look-ahead convoy system was designed and implemented. The overall performance of the vehicle convoy was analyzed in term of speed, acceleration and jerk and a comparative test was conducted with respect to the convoy with SFC against that without SFC. An improvement was achieved in the convoy behavior with SFC, such improvements included perfect tracking of speed by the controlled vehicle without amplifying the lead-vehicle oscillation off stream as seen in Fig. 12. Hence, stable string was achieved. Acceptable acceleration of $2 \ \text{ms}^{-2}$ was achieved with the SFC whereas non-accepted acceleration was observed in that without SFC as seen in Fig. 13. Passengers' comfort is achieved with no jerk associated with the controlled vehicle as seen in Fig. 14. In future work we shall explore how the SFC as a service can be deployed on the network server.

References

1. Contreras-Castillo, J., Zeadally, S., Guerrero-Ibañez, J.A.: Internet of vehicles: architecture, protocols, and security. IEEE Internet Things J. **5**(5), 3701–3709 (2018)
2. Mahmood, Z.: Connected vehicles in the IoV: concepts, technologies and architectures. In: Connected Vehicles in the Internet of Things, pp. 3–18. Springer, Cham, Switzerland (2020)
3. Sjoberg, K., Andres, P., Buburuzan, T., Brakemeier, A.: Cooperative intelligent transport systems in Europe: current deployment status and outlook. IEEE Veh. Technol. Mag. **12**(2), 89–97 (2017)

4. Cooper, C., Franklin, D., Ros, M., Safaei, F., Abolhasan, M.: A comparative survey of VANET clustering techniques. In: IEEE Communications Surveys & Tutorials, vol. 19, no. 1, pp. 657–681, Firstquarter (2017). https://doi.org/10.1109/comst.2016.2611524
5. Chen, M., Tian, Y., Fortino, G., Zhang, J., Humar, I.: Cognitive internet of vehicles. Comput. Commun. Sci. Direct **120**, 58–70 (2018)
6. Dey, K.C., Rayamajhi, A., Chowdhury, M., Bhavsar, P., Martin, J.: Vehicle-to-Vehicle (V2V) and Vehicle-to-Infrastructure (V2I) communication in a heterogeneous wireless network–performance evaluation. Transp. Res. Part C: Emerg. Technol. Sci. Direct **68**, 168–184 (2016)
7. Ubiergo, G.A., Jin, W.L.: Mobility and environment improvement of signalized networks through Vehicle-to-Infrastructure (V2I) communications. Transp. Res. Part C: Emerg. Technol. Sci. Direct **68**, 70–82 (2016)
8. Anaya, J.J., Merdrignac, P., Shagdar, O., Nashashibi, F., Naranjo, J.E.: Vehicle to pedestrian communications for protection of vulnerable road users. In: IEEE Intelligent Vehicles Symposium Proceedings, pp. 1037–1042, Dearborn, MI, USA (2014)
9. Ding, N., Meng, X., Xia, W., Wu, D., Xu, L., Chen, B.: Multivehicle coordinated lane change strategy in the roundabout under internet of vehicles based on game theory and cognitive computing. IEEE Trans. Industr. Inf. **16**(8), 5435–5443 (2020)
10. Musa, M.J., Sudin, S., Mohamed, Z., Nawawi, S.W.: Novel information flow topology for vehicle convoy control. In: Mohamed Ali, M., Wahid, H., Mohd Subha, N., Sahlan, S., Yunus, M., Wahap, A. (eds.) Modeling, Design and Simulation of Systems. Communications in Computer and Information Science, vol. 751, pp. 323–335. Springer Nature, Gateway East, Singapore (2017)
11. Musa, M.J., Sudin, S., Mohamed, Z.: Aerodynamic disturbance on vehicle's dynamic parameters. Journal of Engineering Science and Technology, School of Engineering. Taylor's University, Malaysia, vol. 13, no. 1, pp. 69–82 (2018)
12. Elijah, O., Rahman, T.A., Orikumhi, I., Leow, C.Y., Hindia, M.N.: An overview of Internet of Things (IoT) and data analytics in agriculture: benefits and challenges. IEEE Internet Things J. **5**(5), 3758–3773 (2018)
13. Khan, U.A., Lee, S.S.: Multi-layer problems and solutions in VANETs: a review. Electronics **8**(2), 204 (2019)
14. Zhang, H., Dong, Y., Cheng, J., Hossain, M.J., Leung, V.C.: Front hauling for 5G LTE-U ultra-dense cloud small cell networks. IEEE Wirel. Commun. **23**(6), 48–53 (2016)
15. Sudin, S., Cook, P.A.: Two-vehicle look-ahead convoy control systems. In: IEEE 56th Vehicular Technology Conference, vol. 5, pp. 2935–2939, Milan, Italy (2004)
16. Musa, M.J., Sudin, S., Mohamed, S., Sha'aban, Y.A., Usman, A.D., Hassan, A.U.: An improved topology model for two-vehicle look-ahead and rear-vehicle convoy control. In: IEEE 3rd International Conference on Electro-Technology for National Development (NIGERCON), pp. 548–553, Owerri, Imo, Nigeria (2017)
17. Hassan, A.U., Sudin, S.: Road vehicle following control strategy using model reference adaptive control method stability approach. Jurnal Teknologi (Sci. Eng.) **72**(1), 111–117 (2015)
18. Hassan, A.U., Sudin, S.: Controller gain tuning for road vehicle control system using model reference adaptive control method. Int. J. Res. Eng. Sci. (IJRES) **1**(4), 1–10 (2013)
19. Ashish, T.: Modern Control Design with MATLAB and Simulink. Indian Institute of Technology, Kanpur, India. Wiley (2002)
20. Ruths, J., Ruths, D.: Control profiles of complex networks. Science **343**(6177), 1373–1376 (2014)
21. Shields, R., Pearson, J.: Structural controllability of multiinput linear systems. IEEE Trans. Autom. Control **21**(2), 203–212 (1976)
22. Sinha, P.K.: Multivariable Control: An Introduction. Marcel Dekker, Inc., New York, NY, USA (1984)
23. Musa, M.J., Sudin, S., Mohamed, Z.: Effect of drag force on individual vehicle of heterogeneous convoy. In: IGCESH2016, Universiti Teknologi Malaysia, pp. 161–163, Johor Bahru, Malaysia (2016)

Vehicle Following Control with Improved Information Flow Using Two-Vehicle-Look-Ahead-Plus-Rear-Vehicle Topology

Mu'azu Jibrin Musa⬤, Shahdan Sudin, Zaharuddin Mohamed,
Abubakar Abisetu Oremeyi, Yahaya Otuoze Salihu⬤,
and Gambo Danasabe⬤

Abstract There exists an inherent trade-off among vehicle following performance indices (position, speed, acceleration and jerk) of any vehicle following system. The use of unrealistic information flow topology (IFT) affects the overall performance of a convoy. This chapter proposes an improved IFT of the two-vehicle look-ahead-plus-rear-vehicle control, which aimed to mitigate the trade-off with a wider range of control regions and to provide acceptable performance simultaneously. The proposed improved topology has explored the single vehicle's dynamic equations and derived the external uncertainties which are modeled together as a unit. The vehicle model is then integrated into the control strategy in order to improve the performance of the convoy. Changes in parameters of the improved convoy are compared with the most widely used conventional convoy topologies (one-vehicle look-ahead and the two-vehicle look-ahead). The results showed that the proposed following vehicle control topology has improved performance of an increase in the inter-vehicular spacing by 19.45 and 18.20%, reduce in both acceleration and jerk by 20.28, 15.17,

M. J. Musa (✉) · S. Sudin · Z. Mohamed
Division of Control and Mechatronics Engineering, Universiti Teknologi Malaysia, Johor Bahru, Malaysia
e-mail: mmjibrin@utm.my

S. Sudin
e-mail: shahdan@utm.my

Z. Mohamed
e-mail: zahar@utm.my

A. A. Oremeyi
School of Engineering, Kogi State Polytechnic Lokoja, Itakpe Campus, Nigeria
e-mail: abubakarabisetu@kogipolytechnic.edu.ng

Y. O. Salihu
Department of Computer Engineering, Kaduna Polytechnic, Zaria, Nigeria
e-mail: s.otuoze@kadunapolytechnic.edu.my

G. Danasabe
Department of Electrical and Electronics Engineering Technology, School of Engineering Technology, Nuhu Bamalli Polytechnic, Zaria, Kaduna, Nigeria
e-mail: danasabegambo@gmail.com

H. Chiroma et al. (eds.), *Machine Learning and Data Mining for Emerging Trend in Cyber Dynamics*, https://doi.org/10.1007/978-3-030-66288-2_11

25.09 and 6.25% as against the conventional, respectively. This signifies that the proposed vehicle follower will always remain in between the rear-vehicle and the predecessor-vehicle for safety, and it also gave more ride comfort due to the achieved low acceleration and jerk in the following vehicle.

Keywords Following topology · Passenger's comfort · Road vehicle · Vehicle dynamics · Vehicle safety

1 Introduction

Rise in population in urban areas resulted in high demand for highway travel. So building of more highways to meet with this challenge is never the solution to this growth in traffic density. It was projected that transportation of goods alone will almost double by 2020 as compared to 2012. The traffic problem is not only a main problem of the metropolis, but it is also common in small urban and rural areas [1].

The deployment of self-driving vehicles on the highway has the potential of playing important role in intelligent traffic systems by minimizing the issues associated with traffic congestion, facilitating people's safety, reducing energy wastage, maximizing ride comfort and cutting down fuel consumption [2]. Several vehicle convoy models and controllers were proposed in the literature [3, 4]. Vehicle control strategies need vehicles in the same convoy to move at a stable agreement in speed while maintaining the desired inter-vehicular spacing with respect to the neighboring vehicles within the convoy. Furthermore, it has to ensure stable string which is the ability of the controlled vehicle to move along the convoy without amplifying the oscillation of the leading vehicle upstream and also to provide minimum jerk in the control vehicle [2].

To achieve string stability, desired inter-vehicular spacing and ride comfort, the vehicle convoy must comply with either of the control policies, the variable spacing or the constant spacing. It is of importance to know that ride comfort is the third-order differential of the displacement of the vehicle, which is called as the jerk of the vehicle. In the variable spacing policy, the inter-vehicle spacing is large (a function of velocity), which is applicable for low traffic density conditions. This technique facilitates string stability using onboard information. This implies that vehicles do not rely largely on information from other vehicles. While constant spacing policy depends mostly on inter-vehicle communication, this policy facilitates string stability with little spacing and it is generally applicable in high traffic density conditions [2]. To achieve desired spacing, the time headway would play a significant role [5] in inter-vehicular spacing and to avoid collision with the vehicles of the convoy. The constant time headway (CTH) describes the desired inter-vehicular spacing is proportional to the control vehicle's speed, and the constant of proportionality from the CTH policy is referred to as the time headway (h) [6, 7]. To achieve the passenger's comfort, the control vehicle's jerk must be minimized to not more than one-third of the vehicle's acceleration (not more than 5 ms^{-3}) [8, 9]. The smaller the vehicle's jerk, the more comfortable are the passengers in the vehicle [10].

String stability is mostly achieved in situations where errors (spacing and information flow) are not amplified within the convoy as vehicles move. For the perfect cancellation of such errors, the errors must have the same sign so as to avoid collision within the convoy [11–13]. The concept of vehicle convoy refers to a string of vehicles that aims to keep a specified, but not necessarily constant inter-vehicle distance with respect to either of the two policies discussed above.

Researchers have been working in the area of vehicle convoy topology control with the aim to come up with the optimum topology, which will give an improved controlled vehicle communication, string stability, chattering free and maximize passengers comfort. Dunbar and Caveney [14] modeled the limited range (LR) following topology using the distributed receding horizon control of vehicle platoons. The LR model was tested to monitor the performance indices (stability and string stability), hence it was observed that the platoons suffered from chattering phenomena and slinky effect which are of importance to the optimum performance. Moreover, the communication among the vehicles in the platoons will never be possible on high speed due to the use of sensors on board of limited range; as a result of this, vehicles must be close together before the topology will be effective. Hence the need for an improved topology will cater for the said lapses. The art of topology experienced a remarkable improvement when vehicle following are modeled to be able to communicate using the unidirectional (UD) following [15–17] and the bidirectional (BD) following [16]. Though the vehicles in both the UD and BD are using distributive controllers, in this formation vehicles can only communicate with the immediate vehicle, and by doing that the information on other vehicle in the formation are not known. This leads to unnecessary poor communication and high jerk that affects the performance of the topology, hence an improvement is necessary.

An enhancement was achieved by [16] when bidirectional leader (BDL) following and two predecessor leader (TPL) following were implemented, whereby bidirectional communication between the leader and the follower occurs, and the two followers take orders from the lead vehicle, respectively [18]. The shortcomings of this research work are due to the fact that the formation does not allow any autonomous control of the vehicles but that the vehicles to be only following the lead vehicle. Consequently, any inherent instability in the lead vehicle will be amplified within the string and causes slinky effect and passengers' discomfort. Therefore, the need for improvement in the topologies is necessary.

The most recent topology is the two-look-ahead following by [19, 20], which permits the controlled vehicle to be able to receive information from both the predecessor and the leader. This topology leads to new knowledge in the area of convoy system. Cook and Sudin [20] recorded stable string and acceptable speed with good jerk but the topology does not fully mimic human driving habit, and it uses distributive control, whereby each vehicle has a built-in controller. Hence an improved topology is needed to achieve the full human driving habit and robust convoy that can lead to stable string, comfort and autonomous control of desire vehicle within the convoy.

This chapter introduces vehicle following control using an improved information flow topology (IFT) for vehicle convoy, where the controlling vehicle is expected to be controlled at consensual speed and to maintain desired space with the independent vehicles, and also to greatly reduce jerk. The proposed following vehicle control topology ensures the information flow from the leader, predecessor and the rear vehicle to the controlled vehicle, where the control vehicle utilizes the information received to adjust in speed and position in the convoy. A dynamic model for the proposed following vehicle is implemented to facilitate realistic, free slinky-effect, high passenger's comfort and safe spacing. The proposed following vehicle would utilize the IFT of the two-vehicle look-ahead plus rear-vehicle and is then compared with the conventional (one-vehicle and two-vehicle look-ahead) convoy to ascertain its dynamic parameters (relative position, speed, acceleration and jerk) performance.

Section 2 of this chapter describes the mathematical modeling of single-vehicle dynamics. The analysis of the proposed following vehicle convoy dynamics model is shown in Sect. 3. Section 4 provides a discussion on the results obtained. Finally, Sect. 5 concludes and gives possible future work.

2   Single-Vehicle External Dynamics

From the Newton's Second Law of Momentum, the acting force can be mathematically expressed as Eq. (1), which is stated as the rate of change in momentum of a body is directly proportional to the applied force in the direction of the applied force [21, 22];

$$\overrightarrow{F} = \frac{d}{dt}\left(m\overrightarrow{v}\right) \tag{1}$$

where $\overrightarrow{F}$ is the force acting on the object at time t in a specific direction, m is the mass of the object and $\overrightarrow{v}$ is the object's speed.

The fundamental law also stands for both translational and rotational motion. In the translational motion, the summation of all external forces acting on the object in a specific direction is equal to the product of its mass and acceleration in the same direction at a fixed mass [23];

$$\sum_{i=1}^{i=\infty} \overrightarrow{F}_{xi} = m\overrightarrow{a}_x \tag{2}$$

where $\overrightarrow{a}_x$ is the acceleration in the forward direction x and $\overrightarrow{F}_{xi}$ is the ith force acting on the object in the same direction as x.

For a vehicle moving in the horizontal direction only, the external forces considered include the aerodynamic drag force, friction drag and the rolling resistance. Equation (3) gives the actual acceleration performance of the vehicle with respect to external disturbances.

$$ma_x = F_x - F_d - F_{fd} - F_{rr} \tag{3}$$

where F_d is the aerodynamic drag force, F_{fd} is the viscous friction drag force and F_{rr} is the rolling resistance force.

2.1 Aerodynamic Drag

The impact of aerodynamic forces produced on a vehicle comes from two major sources, namely, drag and viscous friction. The aerodynamic drag F_d from air resistance depends on the changes in the squared velocity value of the vehicle (v^2) [24] as given in Eq. (4).

$$F_d \, \alpha \, v^2 \tag{4}$$

The complete expression for the aerodynamic drag is given in (5) [24–26]:

$$F_d = \frac{1}{2}\left(C_d A \rho_a v^2\right) \tag{5}$$

where C_d is the non-dimensional drag coefficient, A is the frontal area of the vehicle as shown in Fig. 1 and ρ_a is the density of the ambient air (1.225 kg/m^3).

Equation (5) shows that the aerodynamic drag of a vehicle is determined by the size of the vehicle's frontal area A, while the drag coefficient C_d depends on the vehicle's shape for which the aerodynamic quantity is characterized. The vehicle's frontal area contributes to the effect of the aerodynamic on the vehicle, which can be seen from Eq. (5) [24, 25]. Hence, the aerodynamic drag can also be reduced by decreasing the frontal area of the vehicle where it has direct contact with the wind and also to give shape with easy free airflow [26].

The frontal of each individual car class has been drastically shrunk to its lowest limit according to the Europe car class type [27]. An agreeable formula for an estimate in the vehicle frontal area has been arrived by all manufacturers [27] as in Eq. (6);

$$A \approx 0.81 \times b_t \times h_t \tag{6}$$

where b_t and h_t are as shown in Fig. 1.

Table 1 presents some values for the frontal area of some selected car classes [25].

Reducing the aerodynamic drag through changes in the vehicle shape is determined experimentally from wind tunnel tests. Its definition comes from (5) as:

Fig. 1 Definition of the
frontal area A of a vehicle

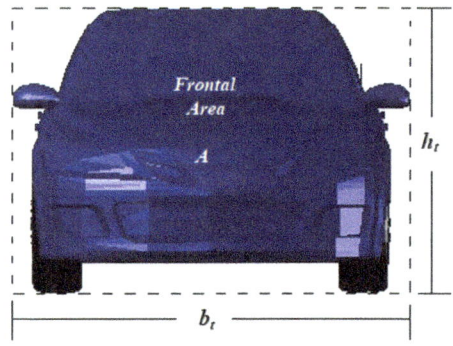

Table 1 Car classification and estimated area

Car class	Mini	Medium size	Upper medium size	Full size
Frontal area A (m^{-2})	1.8	1.9	2.0	2.1

$$C_d = \frac{F_d}{(\frac{1}{2}\rho_a v^2) \times A} \tag{7}$$

where $\frac{1}{2}\rho_a v^2$ is the dynamic pressure.

The drag coefficients of various bodies vary with each other [27, 28]. The typical modern car achieves a drag coefficient of between 0.30 and 0.35 [29]. A four-wheel-drive vehicle, with larger, flatter shape, typically achieves a drag coefficient of between 0.35 and 0.45. Less-powerful engines with the same maximum speed will be obtained from lower F_d [25].

2.2 Viscous Friction Drag

Viscous friction is another form of aerodynamic drag. The viscous friction drag happens when two things (surface of object and airflow) rub together as the object is moving through it. The viscous friction drag can be estimated as in Eq. (7):

$$F_{fd} = \frac{1}{2}(C_{df} \times \rho_f \times b_t \times l \times v^2) \tag{8}$$

where C_{df} is the non-dimensional friction drag coefficient and l is the characteristic length, the chord width of an airfoil.

The friction drag coefficient C_{df} depends on the type of airflow through the vehicle's body. The airflow is defined by the Reynolds number (Re) as in Eq. (9):

Table 2 Classification of friction drag coefficient under three Reynolds number conditions

Type of flow	Laminar flow $(\text{Re} < 5 \times 10^5)$	Turbulent flow $(5 \times 10^5 < \text{Re} < 10^7)$	Even higher Re $(\text{Re} > 10^7)$
C_{df} value	$\dfrac{2.656}{\sqrt{\text{Re}}}$	$\dfrac{0.148}{\sqrt[5]{\text{Re}}}$	$\dfrac{0.91}{(\log \text{Re})^{2.58}}$

$$\text{Re} = \frac{vl}{\upsilon} \qquad (9)$$

where υ is the kinematic viscosity that depends on pressure and temperature, which is expressed as in Eq. (10) [25]:

$$\upsilon = \frac{\mu}{\rho_f} \qquad (10)$$

For incompressible fluids at standard sea level the values for μ and υ are 1.7894 $\times 10^{-5}$ Nsm^{-2} and 1.4607 $\times 10^{-5}$ m^2 s^{-1}, respectively [25]. The dynamic viscosity of the fluid is represented by μ and ρ_f represents the fluid density.

To simplify our analysis the frictional drag coefficient C_{df} is classified under three conditions of Re, as tabulated in Table 2.

2.3 Rolling Resistance Force

Rolling resistance F_{rr} is the force that acts on the tire while in contact with the surface, which resists the motion of the tire. When the tire moves on the road, it is always deformed at the bottom. The energy spent on such a deformation process on the tire while rolling can be translated into a frictional force, which is called the rolling resistance [30].

The rolling resistance force can be expressed mathematically as in Eq. (11):

$$F_{rr} = C_{rr} \times mg \cos\theta \qquad (11)$$

where C_{rr} is the rolling resistance coefficient, g is the acceleration of free fall due to gravity and θ is the slope angle. The typical value for the rolling resistance coefficient is between 0.01 and 0.02 inclusive [31, 32].

2.4 Simplified Vehicle Dynamics

The simplified vehicle dynamic model is used in the simulation of this chapter because it provides an acceptable actual speed as compared with the desired speed [15, 33–35]. In the simplified model, the internal dynamics is represented as a lag function according to Liu et al. [33], in which the vehicle acceleration is obtained after some time delay τ from the given Eq. (12). Liu et al. [33] come up with a value of 0.2 s as τ to represent the vehicle's characteristics of the propulsion systems, which include the engine, transmission, wheels and any other internal controllers.

$$\tau \dot{a} + a = u \tag{12}$$

where τ is the time delay constant, $\dot{a}$ is the vehicle jerk, a is the vehicle acceleration and u is the command signal of acceleration.

Re-arranging Eq. (12) yield expression for the vehicle's jerk as:

$$\dot{a} = \frac{1}{\tau}(u - a) \tag{13}$$

The simplified model of Eq. (13), which considers only the internal dynamics of the vehicle, can be represented by the equivalent block diagram as in Fig. 2.

Next, is to include the effect of external dynamics; the modification will be in Fig. 2 to achieve Fig. 3. By integrating the acceleration performance, the speed and position of the vehicle can be obtained, respectively. Figure 3 gives the representation of the actual acceleration a_{act} performance considered, which is the difference between the acceleration obtained from the vehicle propulsive force a_{pro} and that from the external drag forces a_{dra}. This can be represented mathematically as in Eq. (14):

$$a_{\text{act}} = a_{\text{pro}} - a_{\text{dra}} \tag{14}$$

Fig. 2 Simplified vehicle model without external dynamics

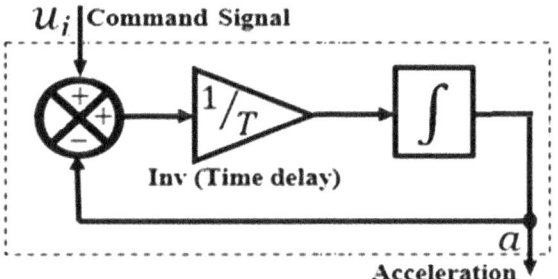

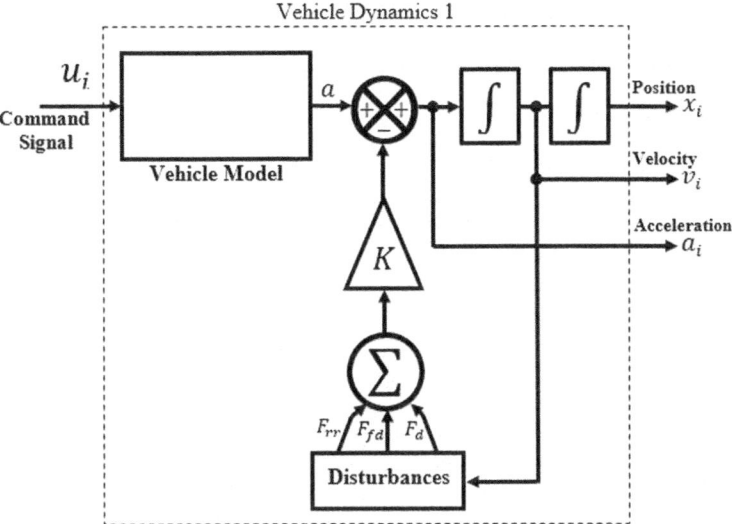

Fig. 3 Overall simplified vehicle dynamics

Reference [25] suggested typical values of a modern vehicle as follows: A is 2.0 m^2, m is 1000 kg, C_{rr} is 0.015, b is 1.4 m and l is 3.0 m for moderate vehicle [25]. These values will be used in the overall simplified vehicle dynamics in Fig. 3 and incorporated into the following vehicle topology.

3 Following Vehicle Convoy Dynamics

In the conventional unidirectional and two-vehicle look-ahead control scenarios, each rigid body mass is assumed to be connected only to its immediate predecessor or two-ahead, respectively. By so doing, the communication range in the conventional vehicle following topology would be limited to only one-vehicle or two-ahead, respectively. This limits the information received by the following vehicle and may lead to chattering, high jerk, passenger's discomfort and unstable string [24].

The proposed topology controls the following vehicle in a convoy using the information of both the preceding vehicle, the vehicle in front of the preceding vehicle and the vehicle at the back of the following vehicle called the rear-vehicle. Figure 4 shows the proposed topology in mass-spring-damper form. The improved convoy operation is considered in a longitudinal dimension where no lane change is permitted. Equation (15) presents the proposed topology derived from Fig. 4 using mass-spring-damper. In the conventional unidirectional and two-vehicle look-ahead control scenarios, each rigid body mass is assumed to be connected only to its immediate predecessor or two-ahead, respectively. By so doing, the communication range in the conventional vehicle following topology would be limited to only one-vehicle

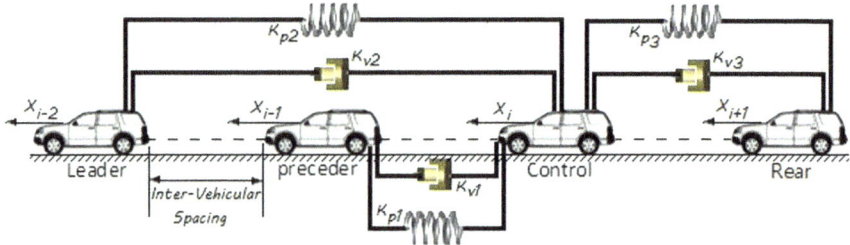

Fig. 4 Representation of the proposed control strategy

or two-ahead, respectively. This limits the information received by the following vehicle and may lead to chattering, high jerk, passenger's discomfort and unstable string [24].

The proposed topology is to control the following vehicle in a convoy using the information of both the preceding vehicle, the vehicle in front of the preceding vehicle and the vehicle at the back of the following vehicle called the rear-vehicle. Figure 4 shows the proposed topology in mass-spring-damper form. The improved convoy operation is considered in a longitudinal dimension where no lane change is permitted. Equation (15) presents the proposed topology derived from Fig. 4 using mass-spring-damper.

$$m\ddot{x}_i = K_{p1}(x_{i-1} - x_i) + K_{v1}(\dot{x}_{i-1} - \dot{x}_i) + K_{p2}(x_{i-2} - x_i)$$
$$+ K_{v2}(\dot{x}_{i-2} - \dot{x}_i) + K_{p3}(x_{i+1} - x_i) + K_{v3}(\dot{x}_{i+1} - \dot{x}_i) \tag{15}$$

The vehicle mass is represented by m, x_i, x_{i-1}, x_{i-2} and x_{i+1} stands for the instantaneous positions of the i, $(i-1)$, $(i-2)$ and $(i+1)$th vehicle, respectively, along the X-axis. $\dot{x}_i$, $\dot{x}_{i-1}$, $\dot{x}_{i-2}$ and $\dot{x}_{i+1}$ are the corresponding velocities of the vehicles, $\ddot{x}_i$ is the acceleration of the ith vehicle, K_{p1}, K_{p2}, K_{p3}, K_{v1}, K_{v2} and K_{v3} are the spring and damper constants, respectively.

The control signal (u_i) has a direct influence on the force applied to the vehicle of mass m [36].

$$\ddot{x}_i = f(\dot{x}_i, u_i) \tag{16}$$

The simplified model can be expressed as in Eq. (17), where u_i is the signal received to accelerate or decelerate the following vehicle. The amount of the controlled vehicle's acceleration is the same as the magnitude of the control signal [36]:

$$\ddot{x}_i = u_i \tag{17}$$

Hence, because the following (ith) vehicle has no influence on the independent vehicles [36] $((i-1), (i-2)$ and $(i+1)$th) of the same mass, Eq. (18) is obtained from Eq. (15) as:

$$u_i = K_{p1}(x_{i-1} - x_i) + K_{p2}(x_{i-2} - x_i) + K_{p3}(x_{i+1} - x_i)$$
$$+ \ K_{v1}(\dot{x}_{i-1} - \dot{x}_i) + K_{v2}(\dot{x}_{i-2} - \dot{x}_i) + K_{v3}(\dot{x}_{i+1} - \dot{x}_i) \tag{18}$$

Using a fixed spacing policy, the control strategy of the proposed model was presented in Eq. (18). It was evident in [37–41] that the slinky-effect is associated with the fixed spacing policy. Moreover, different inter-vehicular spacing is required for different convoy speed [36]. This is to avoid collision and to give enough time for the controlling vehicle to adjust for sudden changes in the speed of the neighboring vehicles in the convoy. Due to this control challenging problem, a promising policy is necessary; hence the choice of the speed-dependent policy was employed in the proposed model. An improved control signal was achieved by incorporating the constant time headway policy as shown in Eq. (19).

$$u_i = K_{p1}(x_{i-1} - x_i - h\dot{x}_i) + K_{p2}(x_{i-2} - x_i - 2h\dot{x}_i) + K_{p3}(x_{i+1} - x_i - h\dot{x}_i)$$
$$+ \ K_{v1}(\dot{x}_{i-1} - \dot{x}_i) + K_{v2}(\dot{x}_{i-2} - \dot{x}_i) + K_{v3}(\dot{x}_{i+1} - \dot{x}_i) \tag{19}$$

The speed-dependent spacing is represented by $h\dot{x}_i$ from Eq. (19). The product of the headway in s and the controlled vehicle's speed are combined to produce displacement, which is inserted in the spring's component as shown in Eq. (19). The product was used in the spring component of the rear and predecessor-vehicle and double of the product was used in the lead-vehicles spring component. The policy used is to create space between the vehicles and to discourage the means of collision. The convoy speed is directly proportional to the vehicle's inter-vehicular spacing [10]. At constant speed, it is assumed that the distance between the leader vehicles (x_{i-2}) and the third (controlling) vehicle (x_i) will be double than that between the fourth (rear) vehicle (t) and the controlling vehicle, or from the second (predecessor) vehicle (x_{i-1}) to the controlling vehicle [10].

Since the following vehicle topology is dependent on the information received from the $(i-1)$, $(i-2)$th vehicle look-ahead and the one-rear $(i+1)$th vehicle, the control signal coming to the controlled vehicle can be re-written from Eq. (19) as [42]:

$$u_i = \sum_{\Psi=1}^{n=2} \left(K_{p\Psi}(x_{i-\Psi} - x_i - \Psi h\dot{x}_i) + K_{v\Psi}(h\dot{x}_{i-\Psi} - \dot{x}_i) \right)$$
$$+ \sum_{\Psi=1} \left(+K_{p3}(x_{i+\Psi} - x_i - \Psi h\dot{x}_i) + K_{v3}(\dot{x}_{i+\Psi} - \dot{x}_i) \right) \tag{20}$$

For a more compact form, the control signal can be best represented as follows:

$$u_i = \sum_{\Psi=1}^{n=2} \left(K_{p\Psi}(\delta_{im1} - \Psi h\dot{x}_i) + K_{v\Psi}(\dot{x}_{i-\Psi} - \dot{x}_i) \right)$$

$$+ \sum_{\psi=1} \left(+K_{p3}(\delta_{im2} - \psi h \dot{x}_i) + K_{v3}(\dot{x}_{i+\psi} - \dot{x}_i) \right) \tag{21}$$

whereby δ_{im1} and δ_{im2} can be written as the following expressions:

$$\delta_{im1} = \sum_{\psi=1}^{2} x_{i+\psi} - x_i - \psi L \tag{22}$$

$$\delta_{im2} = \sum_{\psi=1} x_{i+\psi} - x_i - \psi L \tag{23}$$

$$\varepsilon_i = x_{i-1} - x_i - L \tag{24}$$

where ε is the inter-vehicular spacing, L is the length of the vehicle including desired spacing. L is the same for each vehicle since the homogeneous type of convoy is assumed, though it may not always be satisfied in practice where the heterogeneous type of vehicle convoy also exists.

Taking the Laplace transform of Eq. (20) gives:

$$s^2 X_i = K_{p1}(X_{i-1} - X_i - hsX_i) + K_{p2}(X_{i-2} - X_i - 2hsX_i) + K_{p3}(X_{i+1} - X_i - hsX_i)$$
$$+ K_{v1}s(X_{i-1} - X_i) + K_{v2}s(X_{i-2} - X_i) + K_{v3}s(X_{i+1} - X_i) \tag{25}$$

Re-arranging for X_i from Eq. (25) gives,

$$X_i = \frac{(K_{v1}s + K_{p1})X_{i-1} + (K_{v2}s + K_{p2})X_{i-2} + (K_{v3}s + K_{p3})X_{i+1}}{s^2 + (K_{v1} + K_{v2} + K_{v3} + (K_{p1} + 2K_{p2} + K_{p3})h)s + K_{p1} + K_{p2} + K_{p3}} \tag{26}$$

Equation (30) gives the reducing form of Eq. (26) to a single-pole system as shown in the following steps:

$$X_i = \frac{K_{v1}\left(s + \frac{K_{p1}}{K_{v1}}\right)X_{i-1} + K_{v2}\left(s + \frac{K_{p2}}{K_{v2}}\right)X_{i-2} + K_{v3}\left(s + \frac{K_{p3}}{K_{v3}}\right)X_{i+1}}{s^2 + (K_{p1} + 2K_{p2} + K_{p3})hs + K_{v1}\left(s + \frac{K_{p1}}{K_{v1}}\right) + K_{v2}\left(s + \frac{K_{p2}}{K_{v2}}\right) + K_{v3}\left(s + \frac{K_{p3}}{K_{v3}}\right)} \tag{27}$$

Hence:

$$X_i = \frac{K_{v1}\left(s + \frac{K_{p1}}{K_{v1}}\right)X_{i-1} + K_{v2}\left(s + \frac{K_{p2}}{K_{v2}}\right)X_{i-2} + K_{v3}\left(s + \frac{K_{p3}}{K_{v3}}\right)X_{i+1}}{s + (s + (K_{p1} + 2K_{p2} + K_{p3})h)s + K_{v1}\left(s + \frac{K_{p1}}{K_{v1}}\right) + K_{v2}\left(s + \frac{K_{p2}}{K_{v2}}\right) + K_{v3}\left(s + \frac{K_{p3}}{K_{v3}}\right)} \tag{28}$$

The pole-zero cancellation technique was used to minimize the complexity of the control law and to guarantee string stability [43]. To achieve a single-pole system (linear equation) Eq. (28) was reduced by incorporating the following constraint:

$$\frac{K_{p1}}{K_{v1}} = \frac{K_{p2}}{K_{v2}} = \frac{K_{p3}}{K_{v3}} = \left(K_{p1} + 2K_{p2} + K_{p3}\right)h. \tag{29}$$

This results in the simplification of Eqs. (28)–(30) as:

$$X_i = \frac{K_{V1}X_{i-1} + K_{V2}X_{i-2} + K_{V3}X_{i+1}}{s + K_{V1} + K_{V2} + K_{V3}} \tag{30}$$

From Eq. (30) (first-order) it can be seen that K_{v1}–K_{v3} are both positive, which indicates that the poles are to the left-hand side of the s-plane. Hence the proposed mathematical model for the convoy of Eq. (19) is string stable with respect to the constraint of Eq. (29). This implies that the system response of the convoy depends on h and $K'_p s$ as seen from Eq. (31).

$$K_{V1} + K_{V2} + K_{V3} = \frac{K_{p1} + K_{p2} + K_{p3}}{\left(K_{p1} + 2K_{p2} + K_{p3}\right)h} \tag{31}$$

From Eqs. (27) to (31) steps the model was reduced to a single-pole. Eq. (30) could be generalized for an arbitrary number of vehicles ahead and a single rear-vehicle as in Eq. (32) where $k_v = \gamma_v$.

$$X_i = \frac{\gamma_{v3}X_{i+1} + \sum_{m=1}^{n}\left(\gamma_{vm}X_{i-m}\right)}{s + \sum_{z=1}^{n}\gamma_{zm}} \tag{32}$$

Now the transfer function $G_m(s)$ is given in Eq. (33).

$$G_m(s) = \frac{\sum_{m=1}^{n=3}\left(K_{vm}s + K_{pm}\right)}{s^2\left(\sum_{\phi=1}^{m=2}\left(\left(K_{v\phi} + \phi h K_{pr}\right)s + K_{p\phi}\right)\right) + \sum_{\phi=3}\left(\left(K_{v\phi} + h K_{p\phi}\right)s + K_{p\phi}\right)} \tag{33}$$

Equation (32) can be greatly simplified by choosing the gain parameters to produce pole-zero cancellations in the transfer function. This can be done by rearranging Eq. (29) for the two-vehicle ahead and rear-vehicle inclusive as follows:

$$\sum_{\phi=1}^{m=3} K_{v\phi} = \frac{1}{h} \tag{34}$$

This simplifies the transfer function Eq. (33) to the form of a simple lag as before:

$$G_m(s) = \frac{K_{vm}}{s + \sum_{\phi=1}^{n=3} K_{v\phi}} \tag{35}$$

Hence, the convoy's string stability can be verified as achieved using the definition in Eq. (36) [20].

$$\rho = \sum_{\phi=1}^{n=3} K_{v\phi} \tag{36}$$

so that:

$$|X_{in}(j\omega)| \leq \frac{\rho}{|\rho + j\omega|} \mathop{\max}_{1 \leq m \leq n} |X_{out}(j\omega)| \tag{37}$$

The expression in Eq. (37) gives attenuation at all frequencies. The requirement of pole-zero cancellation is of course not necessary in principle, but its absence makes the analysis of string stability much more complicated.

4 Turning of Gains and Simulation

Both the model equation with the stated constraints of Eqs. (19) and (29) are used for the simulation of the vehicle following system. The simulation of only the controlled vehicle is presented, which was developed in Simulink as in Fig. 5.

The time headway (h) is set to a unit second, as suggested by Zhao et al. [35]. One h was used between the controlled vehicle and the predecessor or rear, while $2h$ was used between the controlled vehicle and the lead-vehicle due to their distance apart [35]. The constants k_{pn} and k_{vn} where $n = 1, 2$ and 3 are so chosen for stable operation of the control strategy when connected to the vehicle dynamics.

In order to ensure that the pole-zero cancellation of Eq. (26) occurs, which will produce Eq. (30), the appropriate tuning of the gain parameters must be made correctly by complying with Eq. (29). From the control law in Eq. (19), increasing the values of those gains k_p's and k_v's will increase the response of the command signal u_i, to changes in vehicles' position and speed.

Increasing the proportional gains, k_p's will speed up the system response. If the gains (k_p's) were kept increasing, a point will be reached where the system will overshoot the changes. When k_p is increased and k_v is kept constant in Eq. (29), a faster response could be achieved, but high-frequency oscillations are expected as the pole-zero cancellation occurs further away from the origin in s-plane. Increasing the differential gains ($k_v's$) will increase the noise to the system because the differential gains are associated with noise and high-frequency oscillations. When k_v is increased and k_p is kept constant in Eq. (29), the pole-zero cancellation occurs near to the origin of s-plane, which will dominate the system response if the exact cancellation is not

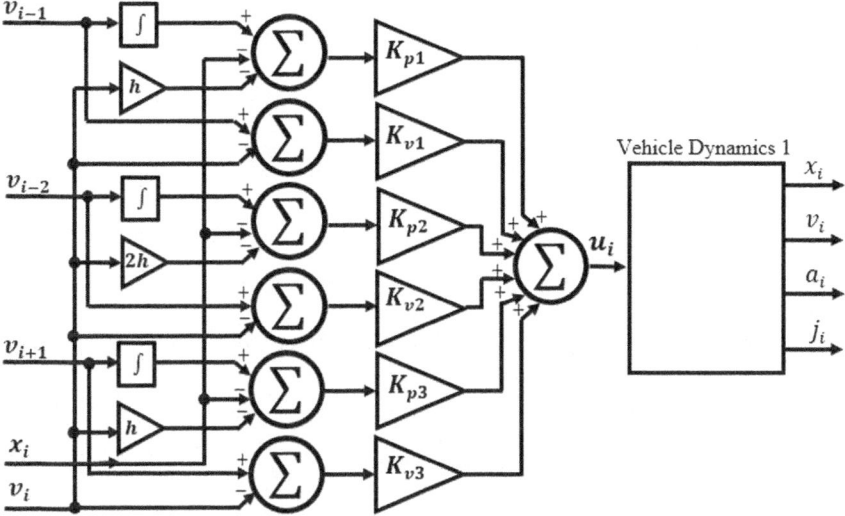

Fig. 5 Simulation model for one vehicle

properly achieved. Therefore, those gains must be kept low in order to avoid the above problems, but the gains must not be too low to prevent a slow response to the system.

Furthermore, it is of importance in this chapter to obtain not only the system control but a balanced response from the controller to changes in both the speed and position of the controlling vehicle. To achieve the said balance response while reducing the measured undesirable effects, the constant k_p is set as equal to k_v for the rear and the respective look-ahead vehicle information, i.e., $k_{v1} = k_{p1}$, $k_{v2} = k_{p2}$ and $k_{v3} = k_{p3}$. The constraints of Eq. (29) give:

$$\frac{K_{p1}}{K_{v1}} = \frac{K_{p2}}{K_{v2}} = \frac{K_{p3}}{K_{v3}} = \left(K_{p1} + 2K_{p2} + K_{p3}\right)h = 1 \tag{38}$$

In this case, the pole-zero cancellation occurs at $s = -1$ on the s-plane. The speed pattern used in this chapter is a deliberate design that gives the human-driven habit of accelerating, decelerating and maintaining a constant speed.

Several gains were used at Eqs. (39)–(41):

$$\frac{k_p}{k_v} \succ 1 \tag{39}$$

where gains $k_{p1} = k_{p2} = k_{p2} = 1$; $k_{v1} = k_{v2} = k_{v2} = 0.4$.

$$\frac{k_p}{k_v} \prec 1 \tag{40}$$

with gains $k_{p1} = k_{p2} = k_{p2} = 0.1$; $k_{v1} = k_{v2} = k_{v2} = 0.44$

$$\frac{K_p}{K_v} = 1 \tag{41}$$

where gains $K_{p1} = 0.36$, $K_{p2} = 0.88$, $K_{p3} = 0.053$, $K_{v1} = 0.36$, $K_{v2} = 0.88$ and $K_{v3} = 0.053$.

Hence the need for the correct gains tuning is justified. Rearranging Eq. (38) for each ratio of K_p / K_v yields the following:

$$K_{v1} = \frac{K_{p1}}{\left(K_{p1} + 2K_{p2} + K_{p3}\right)h} \tag{42}$$

$$K_{v2} = \frac{K_{p2}}{\left(K_{p1} + 2K_{p2} + K_{p3}\right)h} \tag{43}$$

$$K_{v3} = \frac{K_{p3}}{\left(K_{p1} + 2K_{p2} + K_{p3}\right)h} \tag{44}$$

Therefore, with $h = 1$ s, k_{p1}, k_{p2} and k_{p3} are so selected to satisfy Eq. (38), while k_{v1}, k_{v2} and k_{v3} are calculated from Eq. (42) to Eq. (44), respectively.

The gains that permit speed changes with the maximum acceleration of fewer than 2 ms^{-2} [44] and the maximum jerk of fewer than 5 ms^{-3} [8] are used. By comparing the gains [Eqs. (39)–(41)], it appears that the appropriate gain among the obtained gains is Eq. (41), hence Eq. (41) is further turned for a satisfactory performance using the gain values of Eq. (45):

$$K_{p1} = 0.11, \ K_{p2} = 0.38, \ K_{p3} = 0.13, \ K_{v1} = 0.11, \ K_{v2} = 0.38 \text{ and } K_{v3} = 0.13 \tag{45}$$

Figures 6, 7 and 8 show the responses with the gains $K_{p1} = 0.11$, $K_{p2} = 0.38$, $K_{p3} = 0.13$, $K_{v1} = 0.11$, $K_{v2} = 0.38$ and $K_{v3} = 0.13$.

The overall convoy configuration is presented in Fig. 9. The control Eq. (21) and vehicle dynamics Eq. (17) were used in Fig. 9. The overall vehicle convoy is designed and implemented to enable the utilization of human driving habits in the following vehicle. The inter-vehicular spacing together with the vehicle's length is kept at 5 m. The length of a normal car is about 4 m [45–47] which includes the initial spacing.

The headway h is taken to be 1 s [20]. This implies that the spacing of the front of each vehicle to the front of another vehicle is 5 m, assuming all vehicles are initially at rest. The gains k_{vn} and k_{pn} for the control law satisfy the required conditions where the corresponding gains for each vehicle are obtained in Eqs. (42)–(44).

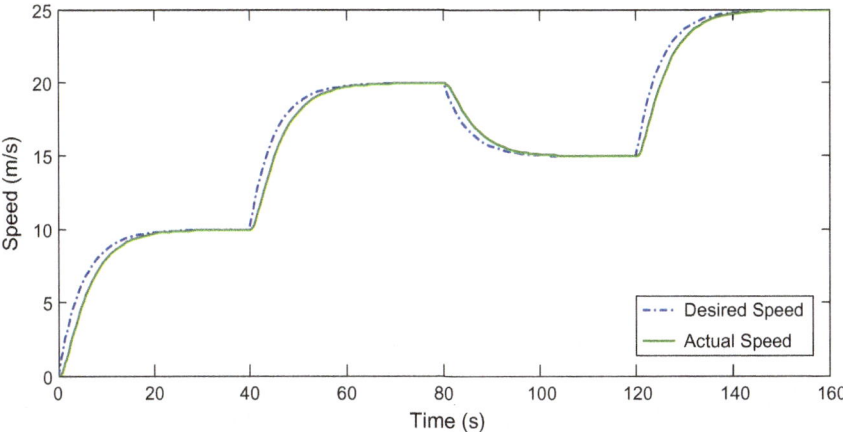

Fig. 6 Speed response of the vehicle when $\frac{K_p}{K_v} = 1$

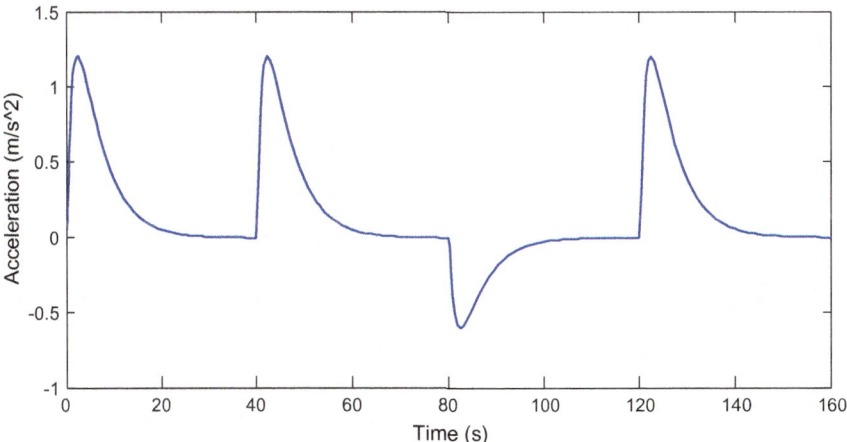

Fig. 7 Acceleration response of the vehicle when $\frac{K_p}{K_v} = 1$

The model can also be verified by doing a mathematical proof in terms of pole location and pole-zero cancellation under simple classical control theory. From Eq. (25) and substituting the constants obtained in Eq. (45) yields:

$$X_i = \frac{(0.11s + 0.11)X_{i-1} + (0.38s + 0.38)X_{i-2} + (0.13s + 0.13)X_{i+1}}{s^2 + 1.62s + 0.62} \quad (46)$$

Simplifying Eq. (46) gives:

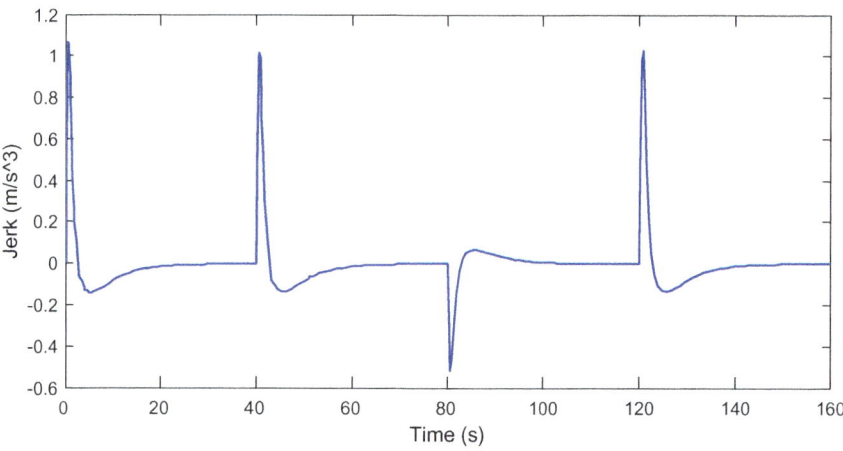

Fig. 8 Jerk response of the vehicle when $\frac{K_p}{K_v} = 1$

$$X_i = \frac{(s + 1)(0.11X_{i-1} + 0.38X_{i-2} + 0.13X_{i+1})}{(s + 1)(s + 0.62)} \tag{47}$$

This shows that the zero at –1 and pole at –1 cancel each other. Hence, leaving only the pole at –0.62 to be the effective pole of the model, that is:

$$X_i = \frac{0.11X_{i-1}}{(s + 0.62)} + \frac{0.38X_{i-2}}{(s + 0.62)} + \frac{0.13X_{i+1}}{(s + 0.62)} \tag{48}$$

In other words, the pole-zero cancellation occurs at $s = -1$ on s-plane and the model has an effective pole at $s = -0.62$ as shown in Figs. 10, 11 and 12.

5   Results and Discussion

The performance of the proposed improved information flow topology can be evaluated through simulation of the topologies' dynamic parameters (position, speed, acceleration and jerk) with a special interest in the following vehicle. The convoy is as designed in Fig. 4, where the following vehicle was controlled by the information (speed and velocity) received from the leading, preceding and the immediate rear-vehicle.

The lead-vehicle starts from rest and gradually rises to a speed of 10 ms^{-1} in 40 s, then accelerate to a velocity of 20 ms^{-1} in 80 s; it then decelerates to 15 ms^{-1} in further 40 s and finally accelerates to a speed of 25 ms^{-1} in 40 s more. The convoy maintains the steady speed trend of changes in velocity over the journey of 160 s with a smooth profile thereafter. The normal convoy operation of a single lane is shown in Figs. 13, 14, 15 and 16.

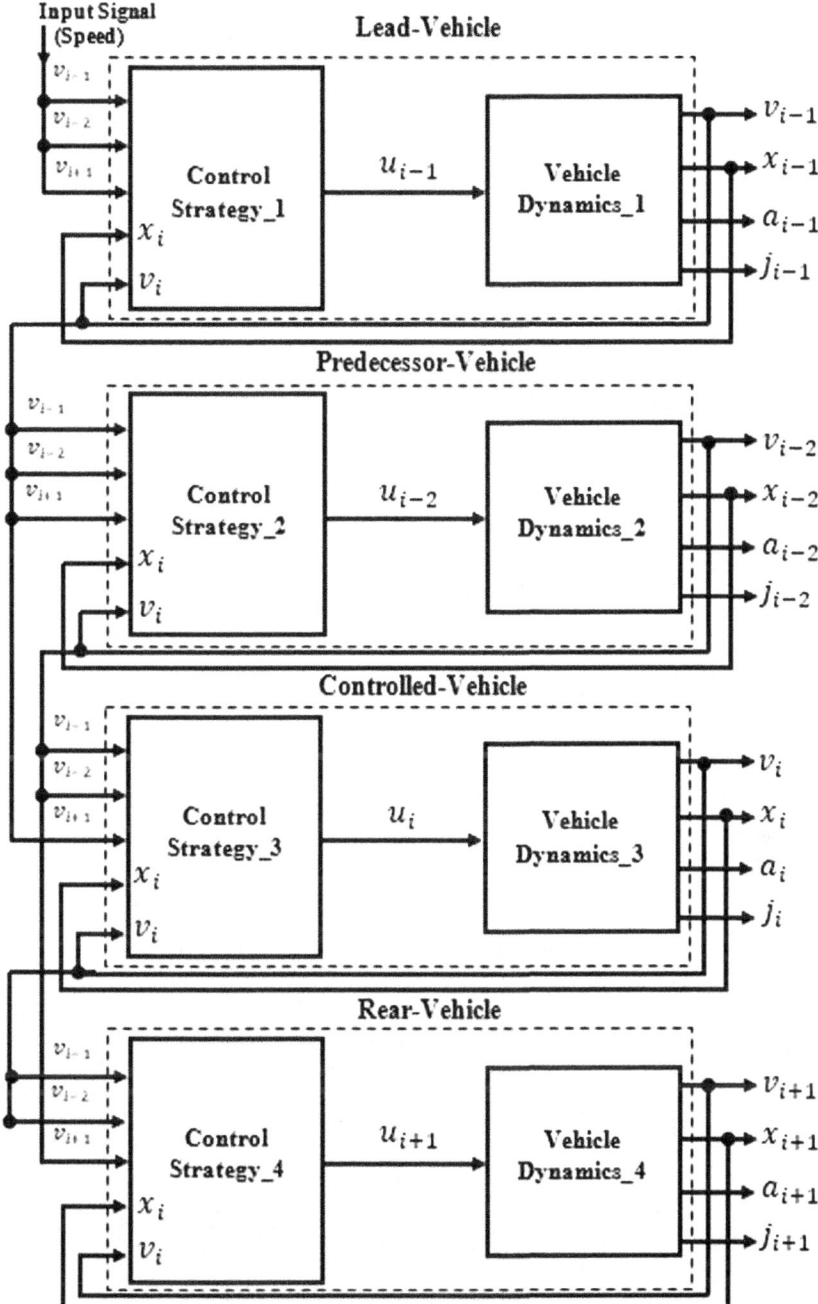

Fig. 9 The overall configuration of the two-ahead and rear-vehicle convoy control

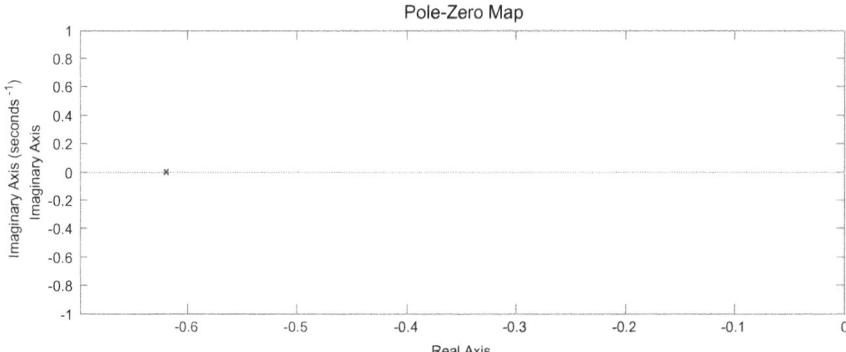

Fig. 10 Poles and zeros location with respect to X_{i-1}

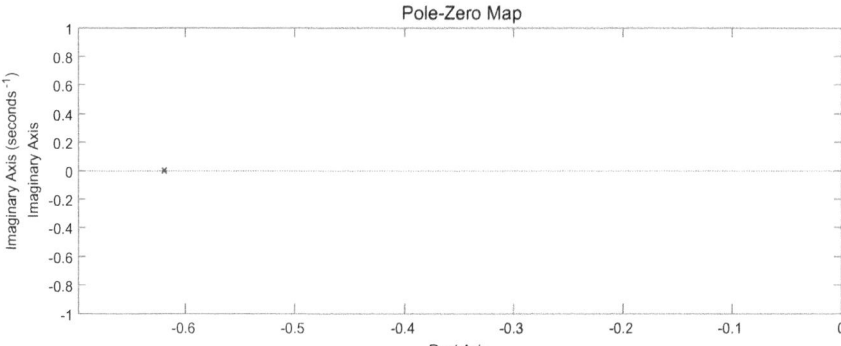

Fig. 11 Poles and zeros location with respect to X_{i-2}

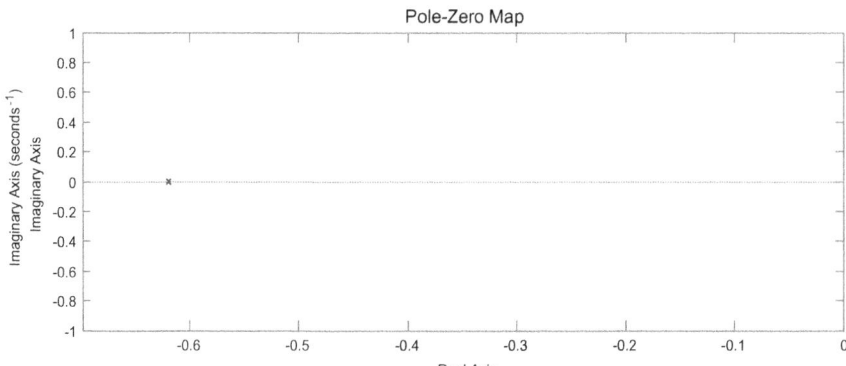

Fig. 12 Poles and zeros location with respect to X_{i+1}

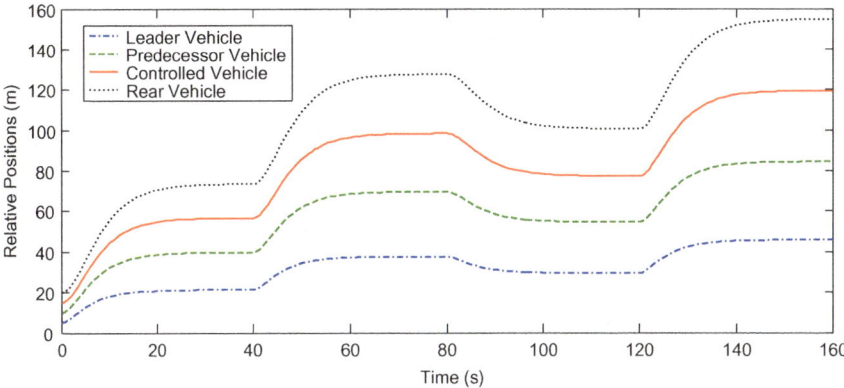

Fig. 13 Relative position of normal convoy operation of the improved topology

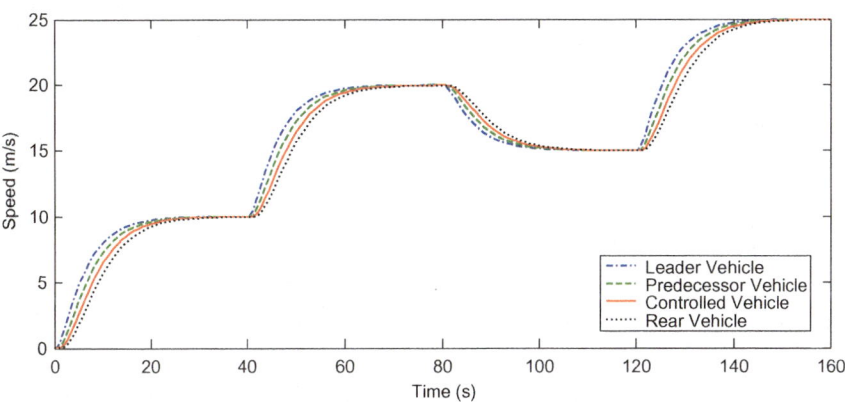

Fig. 14 Speed responses of normal convoy operation of the improved topology

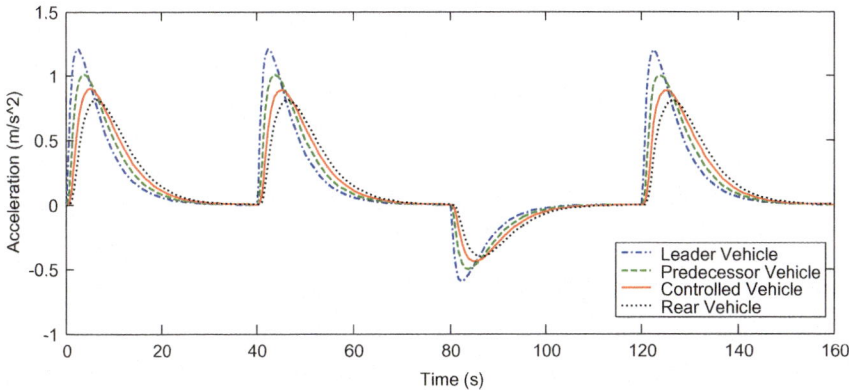

Fig. 15 Acceleration responses of normal convoy operation of the improved topology

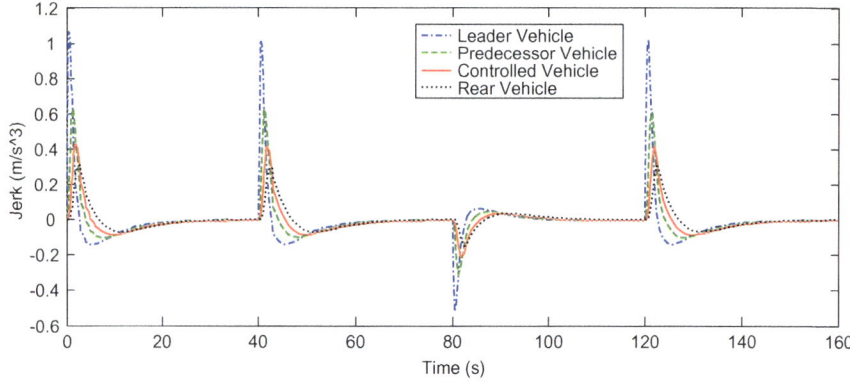

Fig. 16 Jerk responses of normal convoy operation of the improved topology

The convoy moves on the normal operation for the total period of 160 s starting from initial when the time is 0 s. The vehicles in the convoy follow all the changes in the speed of the leading vehicle and maintain close inter-vehicular spacing maneuvers throughout the journey. Figure 13 shows the relative position of each vehicle within the convoy. Figure 13 also shows that the inter-vehicular spacing within the convoy varies according to the convoy's changes in speed while the CTH is set to $h = 1$ s as explained earlier. For instance, at a convoy speed of 10 ms^{-1}, the aggregate inter-vehicular spacing is the combination of the desired spacing of 10 m and the initial set spacing of 5 m gives 15 m. Similarly, at a speed of 20 ms^{-1} the inter-vehicular spacing is 20 m, plus the initial spacing and it continues in this manner.

Figure 14 shows how the control of the proposed topology closely communicates in terms of vehicle speed and tracks the path of leader and predecessor vehicles with much cooperation with the rear-vehicle without collision. This shows the ability of the control vehicle to depend mainly on the acceleration, deceleration and constant speed of the neighboring vehicles.

The improved topology's acceleration was presented in Fig. 15, which shows a smooth and orderly mannered. The controlled vehicle's acceleration was maintained at 0.90 ms^{-2} which is below the maximum acceptable value of 2 ms^{-2} as stated by Rajamani and Shladover [44]. The controlling vehicle's acceleration is adjusted and remains to be between that of the predecessor and the rear-vehicle; this proves proper control of controlled vehicle's acceleration within the convoy.

Among other factors, the passenger's comfort also depends on the vehicle's jerk. The smaller the jerk, the more comfortable the vehicle will be. The jerk of the proposed topology in Fig. 16 is 0.437 ms^{-3}, which is low and far from the maximum rated jerk of 5 ms^{-3} [8, 10]. This value signifies that the control vehicle would be comfortable for passengers [48, 49].

5.1 Comparison of the One-Vehicle Look-Ahead and Two-Vehicle Look-Ahead Against the Proposed Topology

The performance of both the improved topology and conventional (one-look-ahead and two-look-ahead) topologies can be evaluated from the results obtained in Figs. 17, 18, 19, 20, 21, 22, 23, 24, 25, 26, 27 and 28. Both topologies were subjected to similar headway, vehicle profile and time duration of 160 s. Several variations in dynamics parameters were recorded; this is due to the use of ill-topology in the one-look-ahead and the two-look-ahead. Figures 17, 20, 23 and 26 of the one-look-ahead, Figs. 16, 21, 24 and 27 of the two-look-ahead are compared with Figs. 19, 22, 25 and 28 of the improved two-look-ahead and one rear-vehicle convoy topology, respectively, with

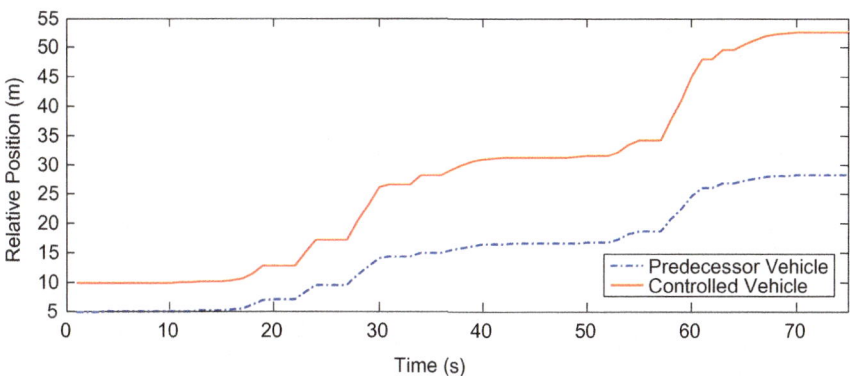

Fig. 17 Relative position responses of one-vehicle look-ahead control topology for t equal to 0–75 s

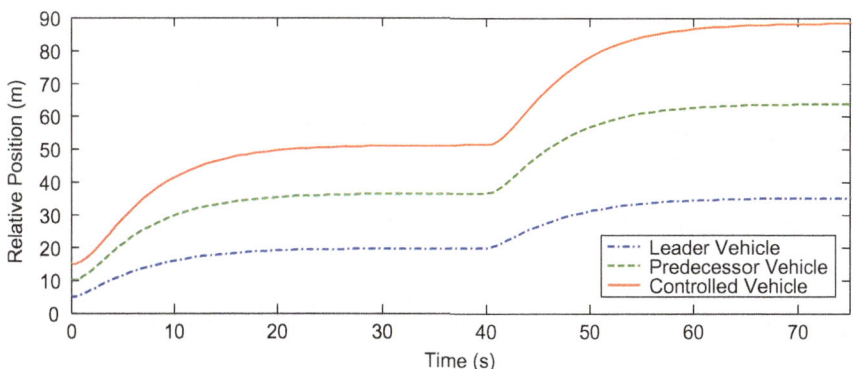

Fig. 18 Relative position responses from the two-vehicle look-ahead control topology for t equal to 0–75 s

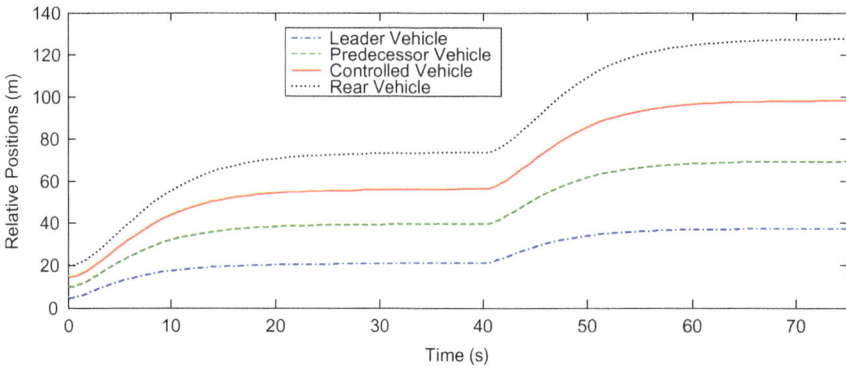

Fig. 19 Relative position responses from the two-vehicle look-ahead and one-rear-vehicle control topology for t equal to 0–75 s

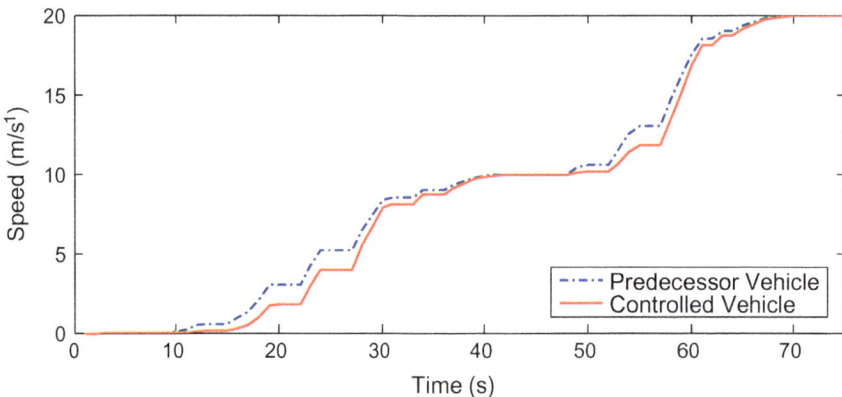

Fig. 20 Speed responses for the one-vehicle look-ahead control topology for t equal to 0–75 s

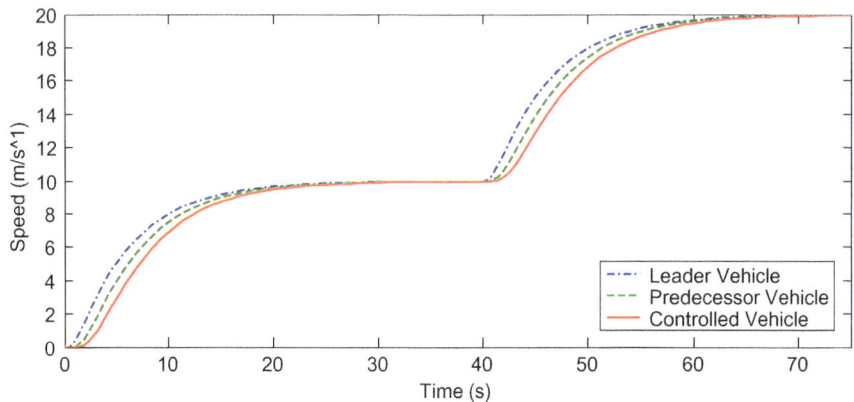

Fig. 21 Speed responses for the two-vehicle look-ahead control topology for t equal to 0–75 s

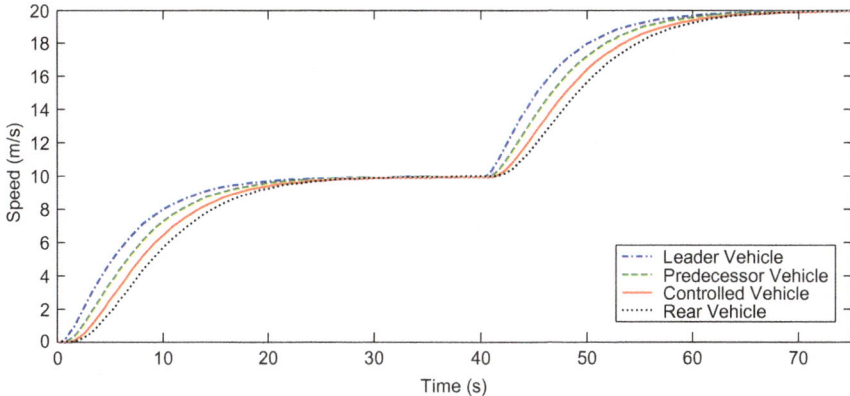

Fig. 22 Speed responses from two-vehicle look-ahead and one-rear-vehicle control topology for t equal to 0–75 s

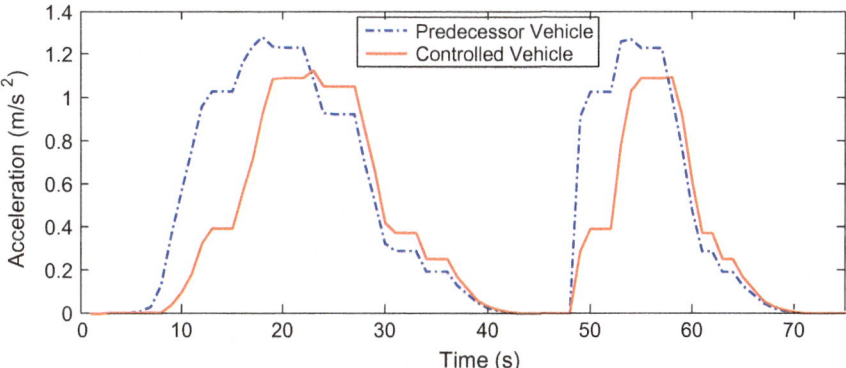

Fig. 23 Acceleration responses from the one-vehicle look-ahead control topology for t equal to 0–75 s

both the figures at selected time interest 0–75 s. The selected time interval of 0–75 s is due to the significant changes that occur within the period, which brings about the performance improvement in the topology.

Figure 19 shows the achievement of variable inter-vehicular spacing, as variation in spacing was achieved with respect to variation in the vehicle speed. By implication, for constant speed among the vehicles it will result in an equal inter-vehicular spacing. The inter-vehicular spacing gives an improvement in spacing and slinky free IFT than that of Fig. 17. This improved topology results in a smooth and free running of the vehicle over the test period of 160 s as shown in Fig. 13. Similarly, the policy used in the control-vehicle provides a wider inter-vehicular spacing of 35 m. The achieved wider spacing gives room for the control-vehicle to adjust its speed and position on any sudden changes in speed from the neighboring vehicles within the

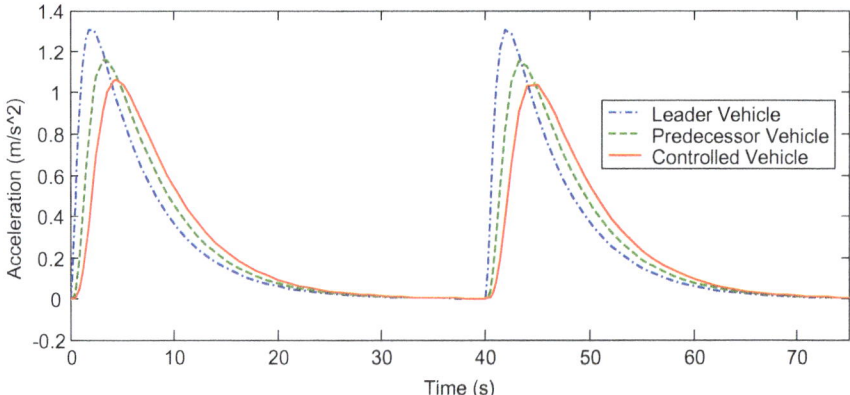

Fig. 24 Acceleration responses from the two-vehicle look-ahead control topology for t equal to 0–75 s

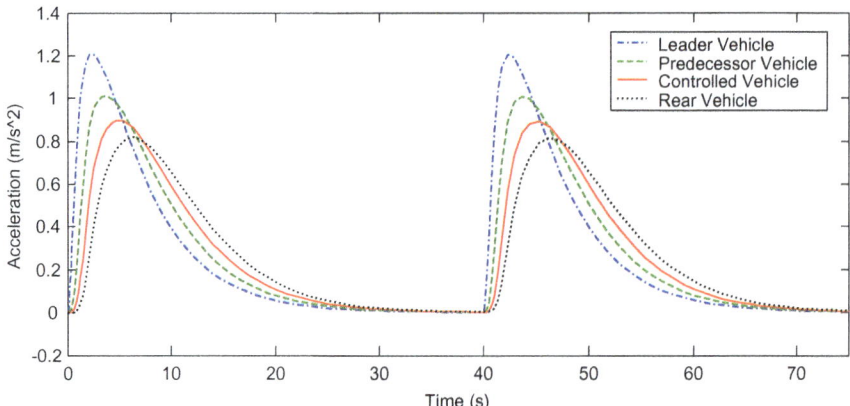

Fig. 25 Acceleration responses for the two-vehicle look-ahead and one-rear-vehicle control topology for t equal to 0–75 s

convoy. Figure 17 shows the spacing provided by the improved topology, which avoid collision as the wider spacing will allow the controlling vehicle to take decision due to communication delay.

Shorter inter-vehicular spacing of only 29.3 m was seen in Fig. 17 of the one-vehicle look-ahead topology as compared to the improved topology, even though they are both subjected to similar strategy of CHT. Moreover, chattering was seen for the first 75 s in the spacing provided by the conventional one-look-ahead. Hence, the one-look-ahead is porous to collision on high speed.

Inter-vehicular spacing of 35 m was achieved in the proposed topology as shown in Fig. 19. The improved spacing of the proposed topology is speed-dependent, hence it keeps increasing when the vehicles are on high speed and vice versa. Moreover, the

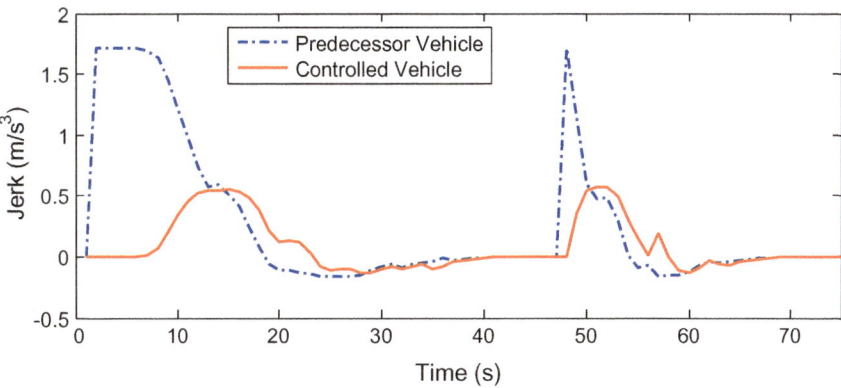

Fig. 26 Jerk responses from the one-vehicle look-ahead control topology for *t* equal to 0–75 s

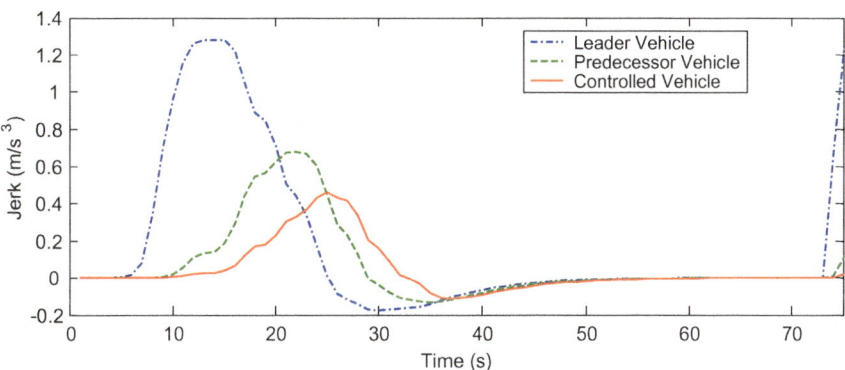

Fig. 27 Jerk responses from the two-vehicle look-ahead control topology for *t* equal to 0–75 s

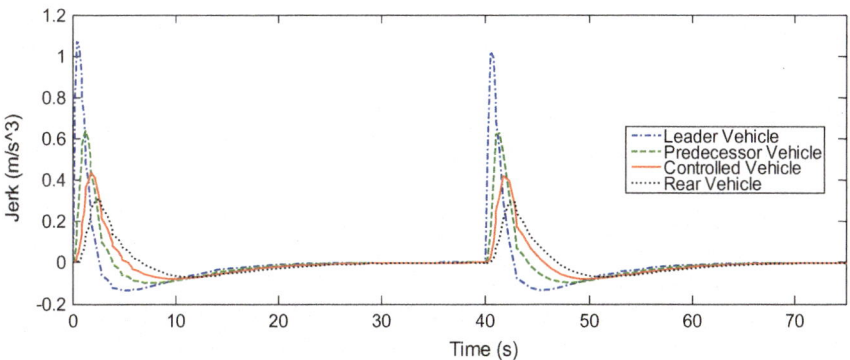

Fig. 28 Jerk responses from the two-vehicle look-ahead and one-rear control topology for *t* equal to 0–75 s

proposed topology gives smooth and free running of the convoy over the set period without overlapping or chattering, as also seen in Fig. 18 for the two-look-ahead topology. The inter-vehicular spacing of Fig. 18 is only 29.61 m, which is lower than that of the improved topology by 5.39 m. The larger inter-vehicular spacing of the proposed topology will avoid collision among and can safely react to sudden changes in the speed of the neighboring vehicles as compared to the two-look-ahead topology.

Figure 22 shows a precise control of the controlled vehicle's speed, who tracks the path of the neighboring vehicles by depending on the information received on their individual speeds without collusion as compared to Figs. 20 and 21 where opposite is the case. Elapses in speed were seen in the one-vehicle look-ahead of Fig. 20. Chattering effect can also be seen in the first 70 s within the journey. The elapses in speed and the chattering resulted in an overlap in speed within the convoy even at low speed of 20 ms^{-1}. This proves that the one-vehicle look-ahead topology is unrealistic.

Figure 22 reveals how the control vehicle of the proposed topology closely tracks the path of the leader, predecessor and rear-vehicle speeds without collusion. The speed of the two-vehicle look-ahead as in Fig. 21 shows an encouraging speed control with similar maneuvering as exhibited in the proposed topology Fig. 22, and all transitions were smooth throughout the journey of 160 s.

The four vehicles acceleration of the improved topology was shown in Fig. 25 within the period of interest of 0–75 s. The controlled vehicle's acceleration was maintained not to pass the threshold value of 0.9 ms^{-2}, which is less than the maximum acceptable value of 2 ms^{-2} [50]. The controlling vehicle's acceleration was adjusted and maintained between the predecessor and the rear-vehicle by using the information received from the neighboring vehicles in the convoy as compared to that of the conventional Fig. 23.

Unwanted oscillation was observed in Fig. 23, and acceleration of 1.129 ms^{-2} was recorded for controlling vehicle of this topology. Though the acceleration is within the acceptable range but it is not as low as that of the proposed topology Fig. 25. The most concern issue of this conventional topology is the chattering phenomena, which occurs at the beginning and last for 75 s. The difference in the acceleration values among the convoys is in favor to the proposed topology by 0.229 ms^{-2}. Hence the improved topology outperformed the one-look-ahead by precise acceleration control of 0.229 ms^{-2} less than that presented in Fig. 23. The less the controlled vehicles' acceleration, the lesser is the jerk.

The acceleration of all the four vehicles in the improved topology was shown in Fig. 25 within the period of interest of 0–75 s. It was found from Fig. 25 that the controlled vehicle's acceleration was maintained not to pass the maximum value of 0.9 ms^{-2}, which is less than the maximum acceptable value of 2 ms^{-2} [50]. The controlling vehicle's acceleration was adjusted and maintained between the predecessor and the rear-vehicle by utilizing the information received from the neighboring vehicles in the convoy. Also, an encouraging result was provided by the two-look-ahead convoy topology, where proper control in acceleration was recorded as well. The two-look-ahead provided an increment in the acceleration of 0.161 ms^{-2} ahead of the improved topology. This increment in the acceleration of the conventional topologies has consequence on the final controlling vehicle's jerks, which will jeopardize the overall comfortability of the vehicle users. Hence the improved topology outperformed the two-look-ahead by precise acceleration control of 0.161 ms^{-2} less than the one presented in Fig. 24.

To achieve passenger's comfortability, smaller jerk value is required; the smaller the jerk, the more is its comfortability. From Fig. 28 the controlled vehicle's jerk was found to be 0.430 ms^{-3}, which is by far below the maximum required jerk of 5 ms^{-3} [8]. Hence the 0.430 ms^{-3} shows that the controlled vehicle will be comfortable enough for passenger [48, 49] as that of Fig. 26.

Slow response and undesirable jerk of 0.547 ms^{-3} was seen from Fig. 26. The chattering in Fig. 26 indicated the presence of an oscillation as the vehicle is trying to settle to its final speed. The conventional topology of the one-vehicle look-ahead has a higher jerk value as against the proposed topology with 0.144 ms^{-3} difference. This difference in jerk in addition to the oscillation would lead to passenger's discomfort in the conventional convoy.

Similarly, the said jerk of 0.430 ms^{-3} proved to be the set as compared to that of the two-look-ahead as seen in Fig. 27 with a maximum jerk of value of 0.46 ms^{-3}. Moreover, the simulation result of Fig. 27 shows a fast response with minor oscillation at 10–35 s, which was inherited from the acceleration of Fig. 24. This difference in jerk in addition to the short period oscillation would lead to passenger's discomfort in the conventional two-look-ahead convoy topology.

It can be justified by the performance comparison between the improved topology and the conventional one-vehicle look-ahead that the improved topology proved to be of higher performance in terms of all the dynamics parameters.

Table 3 provides a summary of the performance comparison of the improved and the one-vehicle look-ahead convoy topology.

Table 3 Comparison of the improved and one-vehicle look-ahead convoy topology

Dynamics parameters	Position (m)	Speed (ms^{-1})	Acceleration (ms^{-2})	Jerk (ms^{-3})
Proposed	Good spacing of 35 m	Smooth and steady	0.900	0.430
One-vehicle look-ahead	Poor spacing of 29.3 m	Lapses in speed	1.129	0.574

Table 4 Comparison of the improved and two-vehicle look-ahead convoy topology

Dynamics parameters	Position (m)	Acceleration (ms^{-2})	Jerk (ms^{-3})
Proposed	Good spacing of 35 m	0.900	0.43
Two-look-ahead	Acceptable spacing of 29.61 m	1.061	0.46

An improvement of 5.7 m in spacing, smooth and steady speed, 0.229 ms^{-2} in acceleration, 0.144 ms^{-3} in jerk and no chattering phenomena were achieved in proposed improved topology over the conventional one-vehicle look-ahead convoy topology.

It can be justified from the performance comparison between the improved topology and the conventional two-vehicle look-ahead that both the two topologies proved satisfactory spacing, smooth and steady speed, while the improved topology outperforms the two-look-ahead in terms of acceleration and comfort. Table 4 provides a summary of the performance comparison of the improved and the two-vehicle look-ahead convoy topology.

An improvement of 5.39 m in spacing, 0.161 ms^{-2} in acceleration, 0.030 ms^{-3} in jerk and oscillation-free were achieved in the proposed improved topology over the conventional two-vehicle look-ahead convoy topology.

6   Conclusion and Further Work

This chapter aimed to propose vehicle control with improved information flow using two-vehicle look-ahead plus rear-vehicle topology. A resolution was achieved in producing a string stable following vehicle with driving comfort ahead of the conventional. The improved following vehicle can look at one vehicle behind and two vehicles ahead in stable speed and comfort. In conclusion, the improved following vehicle shows that the proposed improved IFT has been designed and implemented. It gave a wider operating range, effective communication and a more realistic vehicle convoy. A control strategy of the improved IFT has been designed and implemented for the following vehicle. Improved results were achieved against the two conventional via simulations since the proposed vehicle topology was string stable and smooth in any changes of vehicle speed due to its potentials in providing acceptable acceleration and rides comfort.

The proposed approach against the one-look-ahead and two-look-ahead achieved the following improvements: An increase in inter-vehicular spacing by 19.45% and 18.20%, respectively; smooth speed, chattering free and ride comfort; a reduction in acceleration by 20.28% and 15.17%, respectively; reduction in jerk by 25.09% and 06.25%, respectively. Hence, it is therefore important to give more emphasis not only on following vehicle convoy but also in the choice of IFT.

The reasons why the proposed topology performs better than the conventional approaches are due to the use of human-like driving habits, variable inter-vehicular spacing and considering unforeseen circumstances. The human driving habit is the only one that allowed the controlled vehicle to be able to look at the two vehicles ahead of it and the immediate rear vehicle. The proposed topology mimics how human drives by using direct visual to see the two vehicles in front while monitoring the immediate rear vehicle by using the mirrors of the vehicle. By doing so, a stable string was achieved unlike in the conventional topologies whereby only one or two vehicles in front were monitored; hence it leads to unstable string when any of the controlling vehicles suddenly match break. The variable inter-vehicular spacing used in the proposed vehicle topology allowed the controlling vehicle with enough time and space to react to the changes in the speed of the neighboring vehicles for safety. The conventional topologies used constant inter-vehicular which deprived the controlled vehicle with enough time and space to react to any changes in the speed of the neighboring vehicles. Finally, the proposed topology considers the effect of disturbances in terms of frictional force, aerodynamic resistance, rolling resistance and viscous force all in cooperated in the controlled vehicle model as against the conventional that considers only one effect (aerodynamic force). These make the proposed topology performs better than the conventional topologies, which suffered from lack of string stability, exposed to collusion and chattering phenomena.

This chapter opens a way to a new vehicle following control topology that needs to be further explored. Such thematic areas that need to be improved include behavior prediction and adoption, double lane, use of complete vehicle dynamic model and the cooperation of robust controller.

References

1. Rich, T.: 2012 Urban congestion trends operations: the key to reliable travel. U.S. Department of transportation, federal highway administration, US, pp. 1–5 (2013)
2. Lin, S., Zhang, Y., Yu, C., Hu, W., Chen, L., Chang, W.: A DC power-line communication based in-vehicle safety aided system for rear vehicles road safety. In: IEEE 1st Global Conference on Life Sciences and Technologies (LifeTech), pp. 137–138, Osaka, Japan (2019)
3. Ming, L., Haoyu, W., Zhengyu, W., Yifeng, J.: Analysis and warning of corner safety speed of curve based on vehicle structure modeling. In: IEEE 4th Information Technology, Networking, Electronic and Automation Control Conference (ITNEC), pp. 2389–2394, Chongqing, China (2020)
4. Wu, H., Si, Z., Li, Z.: Trajectory tracking control for four-wheel independent drive intelligent vehicle based on model predictive control. IEEE Access **8**, 73071–73081 (2020)
5. Feemster, M.: Using disturbance rejection to achieve inter-vehicle spacing within a convoy. IEEE Trans. Veh. Technol. **68**(6), 5331–5342 (2019)
6. Viel, C., Vautier, U., Wan, J., Jaulin, L.: Platooning control for heterogeneous sailboats based on constant time headway. IEEE Trans. Intell. Transp. Syst. **21**(7), 2078–2089 (2019)
7. Swaroop, D., Shyamprasad, K., Prabhakar, R.P.: Effects of V2V communication on time headway for autonomous vehicles. In: American Control Conference (ACC), pp. 1–14. Seattle, WA, USA (2018)
8. Batra, M., Maitland, A., McPhee, J., Azad, N.L.: Non-linear model predictive anti-jerk cruise control for electric vehicles with slip-based constraints. In: Annual American Control Conference (ACC), pp. 3915–3920, Milwaukee, WI (2018)

9. Turri, V., Kim, Y., Guanetti, J., Johansson, K.H., Borrelli, F.A.: Model predictive controller for non-cooperative eco-platooning. In: American Control Conference (ACC), pp. 2309–2314, Seattle, WA, USA (2017)
10. Musa, M.J., Sudin, S., Mohamed, Z., Nawawi, S.W.: Novel information flow topology for vehicle convoy control. In: Mohamed Ali, M., Wahid, H., Mohd Subha, N., Sahlan, S., Yunus, M., Wahap, A. (eds.) Modeling, Design and Simulation of Systems. Communications in Computer and Information Science, Springer Nature, vol. 751, pp. 323–335, Gateway East, Singapore (2017)
11. Lee, K., Kum, D.: collision avoidance/mitigation system: motion planning of autonomous vehicle via predictive occupancy map. IEEE Access 7, 52846–52857 (2019)
12. Arogeti, S., Ailon, A.: Formation control and string stability of a group of kinematic vehicles with front-steering wheels. In: 25th Mediterranean Conference on Control and Automation (MED), pp. 1011–1016, Valletta, Malta (2017)
13. Yousef, M., et al.: Dual-mode forward collision avoidance algorithm based on vehicle-to-vehicle (V2V) communication. In: IEEE 61st International Midwest Symposium on Circuits and Systems (MWSCAS), pp. 739–742, Windsor, ON, Canada (2018)
14. Dunbar, W., Caveney, D.: Distributed receding horizon control of vehicle platoons: stability and string stability. IEEE Trans. Autom. Control 57(3), 620–633 (2012)
15. Khan, M.S., Su, H., Tang, G.: Optimal tracking control of flight trajectory for unmanned aerial vehicles. In: IEEE 27th International Symposium on Industrial Electronics (ISIE), pp. 264–269, Cairns, QLD, Australia (2018)
16. Llatser, I., Festag, A., Fettweis, G.: Vehicular communication performance in convoys of automated vehicles. In: IEEE ICC Ad-hoc and Sensor Networking Symposium, pp. 1–6, Kuala Lumpur, Malaysia. (2016)
17. Ali, A., Garcia, G., Martinet, P.: Safe platooning in the event of communication loss using the flatbed tow truck model. In: IEEE 13th International Conference on Control, Automation, Robotics & Vision, (ICARCV), pp. 1644–1649, Singapore (2014)
18. Tsugawa, S., Kato, S., Aoki, K.: An automated truck platoon for energy saving. In: IEEE/RSJ International Conference on Intelligent Robots and Systems, pp. 4109–4114, San Francisco, CA, USA (2011)
19. Hassan, A.U., Sudin, S.: Road vehicle following control strategy using model reference adaptive control method stability approach. Jurnal Teknologi (Sci. Eng.) 72(1), 111–117 (2015)
20. Cook, P.A., Sudin, S.: Dynamics of convoy control systems. In: Proceedings of the 10th IEEE Mediterranean Conference on Control and Automation, pp. 9–12. WP7-2. Lisbon, Portugal (2002)
21. Fuse, H., Fujimoto, H.: Fundamental study on driving force control method for independent-four-wheel-drive electric vehicle considering tire slip angle. In: IECON 2018—44th Annual Conference of the IEEE Industrial Electronics Society, pp. 2062–2067, Washington, DC (2018)
22. Musa, M.J., Sudin, S., Mohamed, Z.: Aerodynamic disturbance on vehicle's dynamic parameters. J. Eng. Sci. Technol. School Eng. Taylor's Univ., Malaysia 13(1), 69–82 (2018)
23. Adam, N.M., Irawan, A.: Steering vehicle with force-based impedance control for inertia reduction. In: IEEE International Conference on Automatic Control and Intelligent Systems (I2CACIS), pp. 47–51, Shah Alam, Malaysia (2018)
24. Huo, Q., Mei, Y.: Study on aerodynamic drag characteristics of high-speed train. In: International Conference on Robots & Intelligent System (ICRIS), pp. 506–509, Changsha, China (2018)
25. Gan, W., Xiang, J., Ma, T., Zhang, Q., Bie, D.: Low drag design of radome for unmanned aerial vehicle. In: IEEE International Conference on Unmanned Systems (ICUS), pp. 18–22, Beijing, China (2017)
26. Muralidharan, V., Balakrishnan, A., Kumar, S.Y.: Design optimization of front and rear aerodynamic wings of a high-performance race car with modified airfoil structure. In: International Conference on Nascent Technologies in the Engineering Field (ICNTE), pp. 1–5, Navi Mumbai (2015)

27. Besselink, B., Johansson, K.H.: String stability and a delay-based spacing policy for vehicle platoons subject to disturbances. IEEE Trans. Autom. Control **62**(9), 4376–4391 (2017)
28. Kandula, J., Kumar, G.S., Bhasker, B.: Experimental analysis on drag coefficient reduction techniques. In: 4th International Symposium on Environmental Friendly Energies and Applications (EFEA), pp. 1–5, Belgrade, Serbia (2016)
29. Travis, W., Bevly, D.M.: Trajectory duplication using relative position information for automated ground vehicle convoys. In: IEEE/ION Position, Location and Navigation Symposium, pp. 1022–1032, Monterey, CA, USA (2008)
30. Trigui, O., Dube, Y., Kelouwani, S., Agbossou, K.: Comparative estimation of electric vehicle rolling resistance coefficient in winter conditions. In: IEEE Vehicle Power and Propulsion Conference (VPPC), pp. 1–6, Hangzhou, China (2016)
31. Seiler, P., Sengupta, R.: Analysis of communication losses in vehicle control problems. In: Proceedings of American Control Conference, IEEE, vol. 2, pp. 1491–1496, Arlington, VA, USA (2001)
32. Swaroop, D., Hedrick, J., Chien, C., Ioannou, P.A.: Comparison of spacing and headway control laws for automatically controlled vehicles. Veh. Syst. Dyn. **23**(1), 597–625 (1994)
33. Liu, K., Gong, J., Kurt, A., Chen, H., Ozguner, U.: Dynamic modeling and control of high-speed automated vehicles for lane change maneuver. IEEE Trans. Intell. Veh. **3**(3), 329–339 (2018)
34. Li, Y., He, C.: Connected autonomous vehicle platoon control considering vehicle dynamic information. In: 37th Chinese Control Conference (CCC), pp. 7834–7839, Wuhan, China (2018)
35. Zhao, Y., Yang, Z., Song, C., Xiong, D.: Vehicle dynamic model-based integrated navigation system for land vehicles. In: 25th Saint Petersburg International Conference on Integrated Navigation Systems (ICINS), pp. 1–4, St. Petersburg, Russia (2018)
36. Kieselbach, R.J.F.: Streamline cars in Germany. In: Aerodynamics in the Construction of Passenger Vehicles 1900–1945. Kohlhammer Ed. Auto and Verk, Stuttgart (1982a)
37. Uhlir, D., Sedlacek, P., Hosek, J.: Practial overview of commercial connected cars systems in Europe. In: 9th International Congress on Ultra-Modern Telecommunications and Control Systems and Workshops (ICUMT), pp. 436–444, Munich, Germany (2017)
38. Nie, G., Hao, Z., Zhao, F., Tian, F.: Fault tolerant control to ensure string stability of cooperative adaptive cruise control under communication interruption. In: IEEE 8th Data Driven Control and Learning Systems Conference (DDCLS), pp. 76–81, Dali, China (2019)
39. Gérard-Philippe, S., Henri, P.G.: Simple algorithms for solving steady-state frictional rolling contact problems in two and three dimensions. Int. J. Solids Struct. Sci. Direct **50**(6), 843–852 (2013)
40. Guo, X., Wang, J., Liao, F., Teo, R.S.H.: Distributed adaptive integrated-sliding-mode controller synthesis for string stability of vehicle platoons. IEEE Trans. Intell. Transp. Syst. **17**(9), 2419–2429 (2016)
41. Trigui, O., Mejri, E., Dube, Y., Kelouwani, S., Agbossou, H.: Energy efficient routing estimation in electric vehicle with online rolling resistance estimation. In: IEEE Vehicle Power and Propulsion Conference (VPPC), pp. 1–6, Belfort, France (2017)
42. Carbaugh, J., Godbole, D. N., Sengupta, R.: Safety and capacity analysis of automated and manual highway systems. In: Transportation Research Part C. Elsevier Science Ltd. UK, vol. 6, nos. 1 & 2), pp. 69–99 (1998)
43. Xie, Q., Filho, C.H.L., Clandfield, F.W., Kar, N.C.: Advanced vehicle dynamic model for EV emulation considering environment conditions. In: IEEE 30th Canadian Conference on Electrical and Computer Engineering (CCECE), pp. 1–4, Windsor, ON, Canada (2017)
44. Rajamani, R., Shladover, S. E.: An experimental comparative study of autonomous and co-operative vehicle-follower control systems. In: Transportation Research Part C. Elsevier Science Ltd. UK, vol. 9, no. 1, pp. 15–31 (2001)
45. Lee, J., Kim, D., Jeong, S., Kim, Y.: Range profile-based vehicle-length estimation using automotive FMCW radar. Electron. Lett. **55**(3), 151–153 (2019)

46. Vrábel, J., Jagelčák, J., Stopka, O., Kiktová, M., Caban, J.: Determination of the maximum permissible cargo length exceeding the articulated vehicle length in order to detect its critical rotation radius. In: XI International Science-Technical Conference Automotive Safety, pp. 1–7, Casta-Papiernicka, Slovakia (2018)

47. Avely, R.P., Wang, Y., Rutherford, G.S.: Length-based vehicle classification using images from uncalibrated video cameras. In: IEEE Intelligent Transportation Systems Conference, pp. 737–742, Washington, DC, WA, USA (2004)

48. Shet, R.A., Schewe, F.: Performance evaluation of cruise controls and their impact on passenger comfort in autonomous vehicle platoons. In: IEEE 89th Vehicular Technology Conference (VTC2019-Spring), pp. 1–7, Kuala Lumpur, Malaysia (2019)

49. Zuska, A., Więckowski, D.: The impact of unbalanced wheels and vehicle speed on driving comfort. In: XI International Science-Technical Conference Automotive Safety, pp. 1–6, Casta-Papiernicka, Slovakia (2018)

50. Zhu, H., Wang, J., Yang, Z.: Numerical analysis on effect of vehicle length on automotive aerodynamic drag. In: IET International Conference on Information Science and Control Engineering (ICISCE2012), pp. 1–4, Shenzhen, China (2012)

Extended Risk-Based Context-Aware Model for Dynamic Access Control in Bring Your Own Device Strategy

Shefiu Olusegun Ganiyu and Rasheed Gbenga Jimoh

Abstract The emergence of brings your own device (BYOD) strategy has brought considerable benefits to enterprises. However, secure access control to vital enterprise resources is one of the impedances to BYOD adoption. Thus, some researches were directed toward dynamic access control using concepts from risk evaluation, machine learning, or context-awareness. However, research efforts to harmonize the three concepts are yet to be established. Hence, this study proposed an Extended Security Risk Analysis Model (ExtSRAM) that combined the concepts to evolve a risk-based and context-aware model to mitigate access control challenges in BYOD. The proposed model comprised of three blocks, including static risk analysis, user contextual profiling, and risk computation. Furthermore, ExtSRAM utilized the Bayesian network to model user contextual profile and static enterprise risks. Again, the proposed model was formulated on six assumptions for it to be realistic for BYOD strategy. More so, a theoretical validation of ExtSRAM justified its soundness and completeness in estimating security risks for dynamic access control. Really, implementing ExtSRAM will proactively safeguard digital assets against unauthorized access. In doing so, an organization can strategically reposition its workforce for productivity while taking advantage of its investment in BYOD implementation.

Keywords BYOD · Risk-based access control · Context-aware access control · Risk evaluation model · Dynamic access control

S. O. Ganiyu (✉)
Department of Information and Media Technology, School of Information and Communication Technology, Federal University of Technology Minna, Minna, Nigeria
e-mail: shefiu.ganiyu@futminna.edu.ng

R. G. Jimoh
Department of Computer Science, Faculty of Information and Communication Sciences, University of Ilorin, Ilorin, Nigeria
e-mail: jimoh_rasheed@unilorin.edu.ng

H. Chiroma et al. (eds.), *Machine Learning and Data Mining for Emerging Trend in Cyber Dynamics*, https://doi.org/10.1007/978-3-030-66288-2_12

295

1 Introduction

Generally, several undertakings in current digital environments involve different categories of security risks, which can lead to varying degrees of threats to information assets. For instance, the security risks in Bring Your Own Device (BYOD) can be categorized into technological, organizational, implementational, human aspects, and policy (regulation) [54]. Therefore, these risks are either formally or casually determined before human and humanoid agents are allowed to perform certain actions on digital contents and facilities. In line with this, risk evaluation is a subprocess of risk assessment, which assigns qualitative or quantitative values to the likelihood of risk occurrence, the consequence after incidence, and the level of risk [34]. For completeness, it is expected that risk evaluation should appropriately consider the likelihood of risk in accordance with actions performed during an event [14].

Obviously, the adoption of BYOD strategy is rapidly becoming a global phenomenon. For example, some enterprises swiftly mandated their employees to work remotely through the *work from home policy* due to the COVID-19 pandemic [10, 16]. Meanwhile, some of these organizations that hurriedly keyed into BYOD strategy were hardly prepared for its security risk. Thus, the security risk in the emerging information technology (IT) landscape has continued to evolve [60], and the specific benefits of BYOD have been accompanied by security risks [4, 32, 43, 58]. For example, a survey conducted in the United Kingdom between 2017 and 2018 revealed that BYOD was embraced by both companies and charity organizations, in which 45 and 65% cases of BYOD security breaches were recorded respectively [53]. Specifically, classical risks pertaining to authorization is of serious concern to many enterprises [22]. Such risks are known to portend adverse impacts on the security of IT infrastructures resulting in performance degradation [52].

Furthermore, authorization is one of the issues encountered by BYOD implementers, which relates to allowing only the right employee to access only the right information, at right time, from the right location and through the right means [24, 45]. Especially, the recent upsurge in risks associated with BYOD adoption has called for more attention of security handlers [56]. Thus, the flexibility of user-oriented and infrastructure-centric features that are attributed to BYOD circumvents some traditional authorization controls mechanisms [2, 59]. Often, this flexibility contributes to unauthorized access problems in BYOD, which consequently leads to great financial loss, amongst other security challenges [27]. Unfortunately, this problem is yet to be sufficiently addressed by existing access control techniques, including those that assayed spatial and temporal properties of users for access control in pervasive environments, such as, BYOD and Internet of Things (IoT) [59]. Hence, there is s need for a dynamic and fine-grained access control that leverages risk or context-aware procedure to address the problem. Thus, a domain-specific risk estimation method must be developed [13], and integrated into the access control mechanism of BYOD to achieve such flexibility.

Therefore, this study presents two main contributions to the security of assets in emerging digital environments. First, it advances a unique approach that integrates context-awareness and machine learning into risk evaluation procedures to generate a dynamic architecture for access control in BYOD environment. Second, it develops a novel mathematical model for risk estimation that can serve as an add-on to existing static access control models, and reinvigorate them into fine-grained authorization mechanisms in other pervasive environments.

Primarily, the aim of this research is to extend a Security Risk Analysis Model (SRAM) developed by [23], with context-awareness features for dynamic and fine-grained authorization in BYOD strategy. More importantly, the rationale for selecting [23] was premised on our previous research [26]. The remaining sections of this chapter are organized as follows. Section 2 presents the background knowledge to foster understanding of the chapter. Section 3 presents the review of related works on risk evaluation models and context-aware access control models. Thereafter, Sect. 4 discusses the proposed model, whereas Sect. 5 presents a detailed process flow of the proposed model and Sect. 6 presents the theoretical evaluation of the proposed model. Lastly, Sect. 7 concludes the chapter.

2 Background

This section provides background information about dynamics approaches that can mitigate access control challenges in BYOD environment. Also, it presents the rudiment of a Bayesian network, which serves as a machine learning tool for the access control model presented in this study.

2.1 Dynamic Access Control in BYOD Strategy

Fundamentally, controlling access to vital information assets is one of the cardinal objectives of information security. To this end, access control models such as Access Control List (ACL), Role-Based Access Control (RBAC), Attribute Based-Access Control (ABAC), Policy-Based Access Control (PBAC), and Risk-Adaptive Access Control (RAdAC) have been developed to safeguard information assets from unauthorized access [21]. Unlike other models, RAdAC represents a significant paradigm shift and was primarily developed to utilize risk evaluation methods for access control decisions [47]. Thus, incorporating risk into access control did not only lead to a flexible and dynamic access control model [12], but it also contributed to the emergence of concepts like risk-based or risk-aware access controls [5, 19].

In addition, contextual information obtained from user's specific behavior, device attributes, and environmental factors have been utilized to form another class of access control mechanism [20, 24, 29, 61]. This class of control which is often referred to as context-based or context-aware access control offers fine-grained and dynamic authorization procedures to secure computing environments [31]. In order to further reinforce the performance of access control mechanisms, context-awareness has been combined with risk-awareness [1]. Furthermore, other researchers utilized the level of trust between interacting entities for authorization [6], combined trust with risk [18], or combined trust with opportunity to improve security posture [2]. Thus, context or trust is considered to be an additional parameter to the risk evaluation function, when it is included in risk-based access control [1, 2, 8].

The dynamism of contemporary computing strategies including cloud computing, IoT, and pervasive computing have paved the way for flexible access to information and technological infrastructures [5]. However, flexible access requires dynamic access control to mitigate security risks emanating from unauthorized entities [19, 42]. Hence, irrespective of the level of risk awareness built into the dynamic access control system, requests from each entity are expected to be treated on a case-by-case basis using historical data from previous access requests. Hence, internal security controls are usually deployed by security experts to mitigate every security risk in the requests for enterprise data [56].

Bring Your Own Device (BYOD) is a pervasive computing strategy that allows employees to use privately owned smart mobile devices to perform private and official tasks [41]. Hence, an employee can meet tight organization schedules with preferred devices and use any available network from comfortable locations at any time. This evolving strategy offers some benefits like an increase in employee productivity, improve work experience and reduction in IT budgets amongst others [8, 17, 44]. On the one hand, the adoption of BYOD has continued to permeate and reshape corporate business environments, non-profit organizations, and donor agencies. On the other hand, enrollees of the pervasive strategy are spreading beyond the organization workforce to customers, donors, ubiquitous business processes and much more.

2.2 Bayesian Network

The entire concept behind Bayesian network or probabilistic directed acyclic graph (DAG) model is based on Bayesian theorem. On its part, Bayesian theorem is a concept that converses the well-known conditional probability distribution. In probability distribution, an evidence is based on causes, whereas causes are based on evidence in Bayesian theorem. Thus, given two variables X and Y, the conditional probability of X, given that Y is true expressed in Eq. 1. The variable X normally represents a proposition or hypothesis, and Y represents an evidence or a new data.

$$p(X|Y) = \frac{p(X)p(Y|X)}{p(Y)} \tag{1}$$

Thus, provided that $p(Y) \neq 0$, then $p(X|Y)$ is the posterior probability of X considering a new evidence. Also, $p(X)$ is prior of X that shows a prior belief in X before new evidence Y, and $p(Y|X)$ is the likelihood function of evidence Y given that X has occurred. The main activities performed in Bayesian networking includes structure learning and parameter learning amongst others.

2.2.1 Structure Learning

Typically, Bayesian network comprises a network structure (or DAG) and a set of random variables $X = (x_1, x, x_3 \ldots . x_n)$ for learning the structure of the dataset by evaluating its posterior probabilities [49]. The DAG is represented as $g = (V, A)$, whereby V is vertex of each node in the graph, while A represents the direct and conditional dependencies between the variables of X. Also, a joint probability distribution is defined for variables of X, from network structure and the local probability distribution (P) at each node of the graph. In general, the joint probability distribution is expressed in Eq. 2 [28].

$$p(x) = \prod_{i=1}^{n} p(x_i | pa_i) \tag{2}$$

In which case, x_i is a node and its corresponding variable. Also, pa_i are the parent nodes of x_i in the graph, including the variables for the parent nodes. Hence, the entire ExtSRAM was built on Bayesian network and its integral artifacts like inferencing engine and conditional probability table (CPT).

2.2.2 Parameter Learning

The process of computing the probability of a desired variable (node) when given a Bayesian model is referred to as parameter learning. This probability must be computed because it is not directly obtainable from an arbitrary model. Thus, the probability of each node in the network is conditional upon its parent nodes. Although parameter learning is NP-hard problem, several methods have been developed to simplify the process of computing the CPT for a Bayesian network [60].

3 Related Work

In order to passably integrate context-awareness into risk evaluation procedure, this section reviewed previous works relating to core risk evaluation frameworks and models. Also, the section reexamined research efforts on the use of context-aware parameters for access control in ubiquitous computing.

3.1 Risk Evaluation Model

Alkussayer and Allen [3] proposed a dual-staged and hybridized model to simplify security analysis for software architecture using the analytic hierarchy process

(AHP). The study adopted a static approach to risk computation based on the risk factors that were relevant to the software domain. Nevertheless, this approach may be appropriate for relatively undynamic domains, but it will fall short of the security risks of a pervasive environment. In a similar study, Wei and Li [57] developed a security risk assessment model with formal safety assessment (FSA) and AHP to secure a typical information system. Similar to Alkussayer and Allen [3], the authors identified six risk factors that are peculiar to information systems without considering the roles of existing security countermeasures in the system.

Also, Sanchez [46] presented a risk and trust (ContextTrust) framework for a ubiquitous environment. The risk assessment module of the framework can model the contextual risk factors (and their dependencies) of participating mobile devices using logarithm random walk. However, the researcher did not consider the role of security countermeasures in the risk evaluation procedure and also assumed that all risk factors are characterized by logarithm random walk. Likewise, Wang and Jin [55] did not include security controls in the computation of risk scores in their proposed adaptive risk evaluation model for preservation of patient information in a health information system. The researchers applied the principle of need-to-know and Shannon entropy to ensure that medical doctors can access only necessary portion of patient's information during consultation. Similarly, Sharma et al. [50] developed task-based tisk-aware access control model for cloud-assisted eHealth. The researchers premised the module for risk evaluation on confidentiality, integrity, and availability (CIA). Analogous to Wang and Jin [55] and Sanchez [46], the risk evaluation module also excluded the effect of security control from the risk computation process.

Furthermore, Miura-ko and Bambos [40], Sato [48] presented risk analysis methods that incorporated the effectiveness of risk mitigation devices into risk computation. The two studies sufficiently justified the moderating effects of risk countermeasures in quantitative risk evaluation models. While the former represented countermeasures with a non-additive risk reduction matrix for generic risk computation, the latter employed a node-based evaluation of security controls that are implemented to independently safeguard computing infrastructures. On one hand, Sato [48] can be too basic for implementation in typical dynamic access control, on the other hand, Miura-ko and Bambos [40] considered only one active security control per risk evaluation session. The use of one countermeasure in risk evaluation is not always the case for secured operations in a pervasive strategy that requires a stack of countermeasures [7, 11]. Contrary to the approach adopted by Miura-ko and Bambos [40], Lo and Chen [37] developed a hybrid risk assessment procedure which measured mutuality among security countermeasures. Also, Lo and Chen [37] described the likelihood of threat occurrence as a function of vulnerabilities in security controls. Again, Lee et al. [36] presented a risk-adaptive access control framework that incorporated firewall provisioning as a countermeasure to accomplish zero-trust networking. Realistically, firewalls alone cannot mitigate threats in current pervasive networks.

In addition, Bojanc and Jerman-Blažič [9] proposed a monetary-inclined model, which primarily factored the role of security controls into a sequentially arranged four-phase risk management process. For adequate coverage and documentation, the authors classified security control into preventive, detective, or corrective measures. However, the model quantified security control from a financial perspective, also the model did not include the effect of security control stack. Furthermore, Aldini et al. [2] designed and validated an Opportunity-Enabled Risk Management (OPPRIM) methodology for regulating access to vital organization's resources in BYOD strategy. However, the risk evaluation module of OPPRIM did not explicitly outline the causal relationship among threats and the role of context-awareness parameters.

Similarly, Feng et al. [23] proposed a proactive security risk analysis model (SRAM) based on ant colony optimization (ACO) and Bayesian Network (BN). Mainly, the model considered casual relationships among risk factors and accounted for threats propagation path. In addition, Feng et al. [23] obtained information about model variables from experts and National Institute of Standards and Technology (NIST) guidelines on the standard for information systems. Remarkably, the risk evaluation component of the model computes risk scores from the probability of risk occurrence and severity after the occurrence. Also, SRAM carters for security countermeasures and allows new risk evidence to be added before computing risk value, thereby guaranteeing a dynamic risk analysis process. Implicitly, SRAM has some desirable elements for utilization in RAdAC, but it was not primarily designed for the purpose. Most especially, its direct application to access control in BYOD strategy without context awareness features cannot be guaranteed [31, 35].

3.2 Context-Aware Access Control Model

For some time now, research efforts that employed context data for secure access to crucial information assets have been conducted [15, 24, 29, 33, 39]. For instance, an in attempt to provide secured access control during military operations, Luo and Kang [38] presented risk-based mobile access control (RiMAC). The study employed contextual risk factors like location, authentications, threats, and timeouts to compute the risk of granting or denying access to information on battlefields. Nonetheless, RiMAC was specifically designed to function with onboard devices and a dedicated network, which cannot be imposed on BYOD stakeholders in civil business environments. Also, Kandala et al. [30] proposed an attribute-based framework that was superimposed on core components of RAdAC for dynamic and probabilistic risk determination. Majorly, the risk estimation function of the framework was based on request and access history. However, the framework only treated access control at coarse-level, because the computation of risk was not overtly defined for each access transaction.

Also, Atlam et al. [5] and Kang et al. [31] proposed a security framework for smart access control, which is based on historical and current context data that relate to user behavior patterns, and footprints of facilities usage, location, and time. More so, the authors suggested the use of Bayesian inference to provide dynamic and differential access to a resource in BYOD. Similarly, Ye et al. [59] developed a fine-grained access control called Attribute-tree Based Access Control (ATBAC) model, which was driven by users' contextual data for BYOD. Furthermore, Trnka and Cerny [51] extended the traditional RBAC with an additional security level, which is established on context-awareness components. However, Atlam et al. [5], Kang et al. [31], Ye et al. [59], and Trnka and Cerny [51] exclusively relied on contextual data to grant or deny a request by subjects without considering other elements of security, such as security countermeasures and threats.

Recently, Kang et al. [32] presented context-driven access control for BYOD called Poise, which combined the data plane programmability feature of software-defined network (SDN) with trust modeling concepts. The contextual data are extracted from network components like protocols, hardware, and data packets. Thus, rather than centralizing access control within enterprise servers, Poise employed the intelligent capabilities of SDN and high-tech topographies of user devices to decentralize access to vital organization resources. Nevertheless, Poise is hardware-centric, as a result, it did not explore the advances made in machine learning algorithms, risk attributes, and the contribution of existing security countermeasures when making access decisions.

3.3   Finding from Related Work

From the foregoing review, it was obvious that more research efforts are needed toward adaptive access control models that combine fundamental principles of risk evaluation, contributions of security countermeasures, machine learning algorithms, and contextual awareness. Such models should account for threats and their possible propagations, installed security countermeasures, especially with due attention to defense-in-depth. Similarly, in order to take advantage of the current and historical context variables for adaptive access control in pervasive environments like BYOD, the ensuing model should be driven by an appropriate machine learning algorithm.

4   Proposed ExtSRAM Model

Concisely, ExtSRAM is a risk-aware access control model that builds on the competency of SRAM, which was reviewed in Sect. 3.1 by incorporating context awareness into the latter. Thus, this section presents a bird's-eye view of ExtSRAM comprising of basic components such as risk factors, security countermeasures, and access control mechanism, together with their respective locations within a BYOD environment. Broadly, the high-level view of ExtSRAM can be segmented into two extensive areas namely; contextual risk factors and enterprise environment (enterprise information system).

4.1 Contextual Risk Factors

The contextual risk factor is a collection of dynamic factors that describes the behaviors of BYOD employee who uses a personal device to seek internal or remote access to the enterprise network. Thus, the risk factors are shown in Fig. 1, which represents the high-level view (architecture) of ExtSRAM, Generally, the factors are characterized by location (transiting, stationary), time, network, mobile device, and actions performed. These factors had been reported to be viable parameters for smart context mining and gaining insight into user behavior [31, 32, 36]. More so, personal attributes like keystrokes, mouse usage, mode of screen swipe, finger texture are

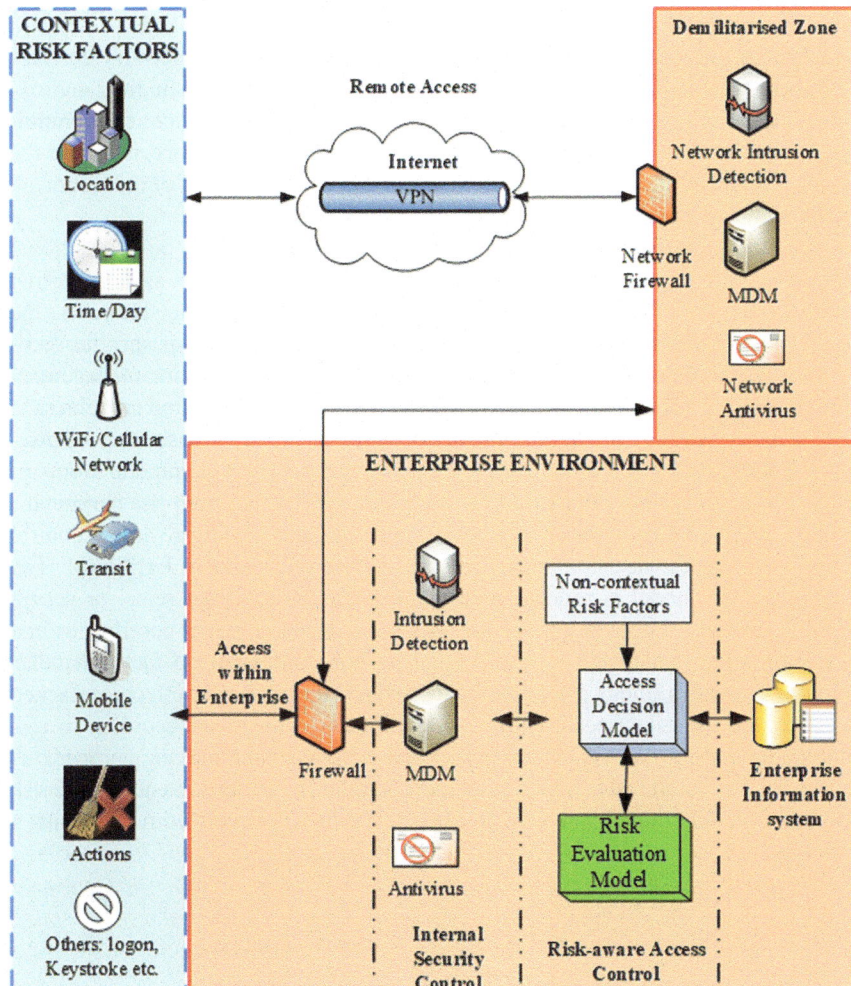

Fig. 1 High-level view of ExtSRAM model

other probable factors that can be captured as risk factors and utilized for contextual components of dynamic access control in pervasive environments. So, all necessary contextual information about a particular user will accompany each access request. Afterward, an access decision must be taken by the risk-aware access control, before a specific action can be performed by the user as expressed in the architecture.

4.2   Enterprise Environment

Generally, the enterprise environment comprises of non-contextual factors, enterprise systems, security countermeasures, and risk-aware access control. For clarity, Fig. 1 revealed the assemblage of IT infrastructures in a conventional enterprise environment that provide network connectivity and information delivery services via information systems to employees. Thus, static risk factors including environmental infrastructures, network, host/server computers, software, data, and communication facilities belong to enterprise environment [23]. Furthermore, it depicts the logical placement of countermeasures to ensure adequacy in terms of the technical, operational, and strategic security needs of the organization.

No doubt, providing unabridged security for BYOD still relies on conventional security techniques and controls like firewall, intrusion detection system, virtual private network (VPN), demilitarized zone (DMZ), antivirus, etc. However, the particular dimension of BYOD risks, necessitated the inclusion of specific security tools like mobile device management (MDM), mobile application management (MAM), mobile content management (MCM), virtualization, secured containers to mention but few [25]. In addition, the architecture further showcased the defense-in-depth initiative of modern security settings to support spatiotemporal forms of employee access to enterprise resources within or outside the enterprise perimeter.

Still, on the enterprise segment of the architecture, the risk-aware access control comprises of access decision model which is the central point of ExtSRAM. The access decision model is saddled with taking decision to either reject or accept incoming requests initiated by the user. Also, the decision model coordinates and formats the values relating to both contextual and non-contextual risk factors handles causal relationships among risk factors and considers the mitigating effects of stacked security controls. Likewise, it enforces the case-based decision according to risk value computed by risk evaluation model and other decision-making components such as risk threshold and heuristic judgement supplied by experts. Basically, the risk evaluation model computes risk value using the values acquired from the decision model.

5 ExtSRAM Process Flow

As described in Sect. 4, SRAM serves as the baseline model for ExtSRAM. Precisely, the last task performed by SRAM was to determine risk treatment plan for probable threat and vulnerability propagation paths in enterprise information system. However, this function is not required for adoption in dynamic risk-aware access control, so it is not included in the model presented in this study.

ExtSRAM adopts Bayesian network as its model building tool and functional block diagram was employed to illustrate its methodology (conceptual components). Thus, the entire ExtSRAM was built on Bayesian network and its integral artfacts like inferencing engine and CPT. Functional block diagram is modeling tool, which is mostly used in software engineering to represent the major functions in a model and their interactions. Fundamentally, a block diagram comprises some core components and their interactions Therefore, this study employed a functional block diagram as design tool to concretize the methodology in the extended model.

5.1 Assumptions on ExtSRAM Model

The proposed model is based on some assumptions to make its development realistic as well as to avoid replicating what has already been achieved in the baseline model.

1. Security risk propagation is assumed to be unidirectional, that is, from the subject (employee or mobile unit) to object (enterprise information systems).
2. Security risk is assessed to be a change in subject context, as well as, enterprise system variables.
3. There may be conditional dependencies among some contextual and enterprise system variables.
4. The proposed model covers the authorization aspect of access control.
5. The assumptions made during SRAM development are taken to be realistic.
6. The validity test conducted on SRAM are presumed to be reliable.

5.2 ExtSRAM Methodology

The entire ExtSRAM is vertically divided into four sections namely; variable definition and management, Bayesian model, Bayesian inference, and risk module as shown in Fig. 2. To begin with, the discovery and formal description of all model variables will be carried out in the variable definition and management section. In order to evolve these variables, relevant guidelines and standards for risk assessment in pervasive computing and domain experts will be consulted. In addition, the tasks scheduled for this topmost section include management of historical data, as well as, the development and use of ACO as described in the baseline model.

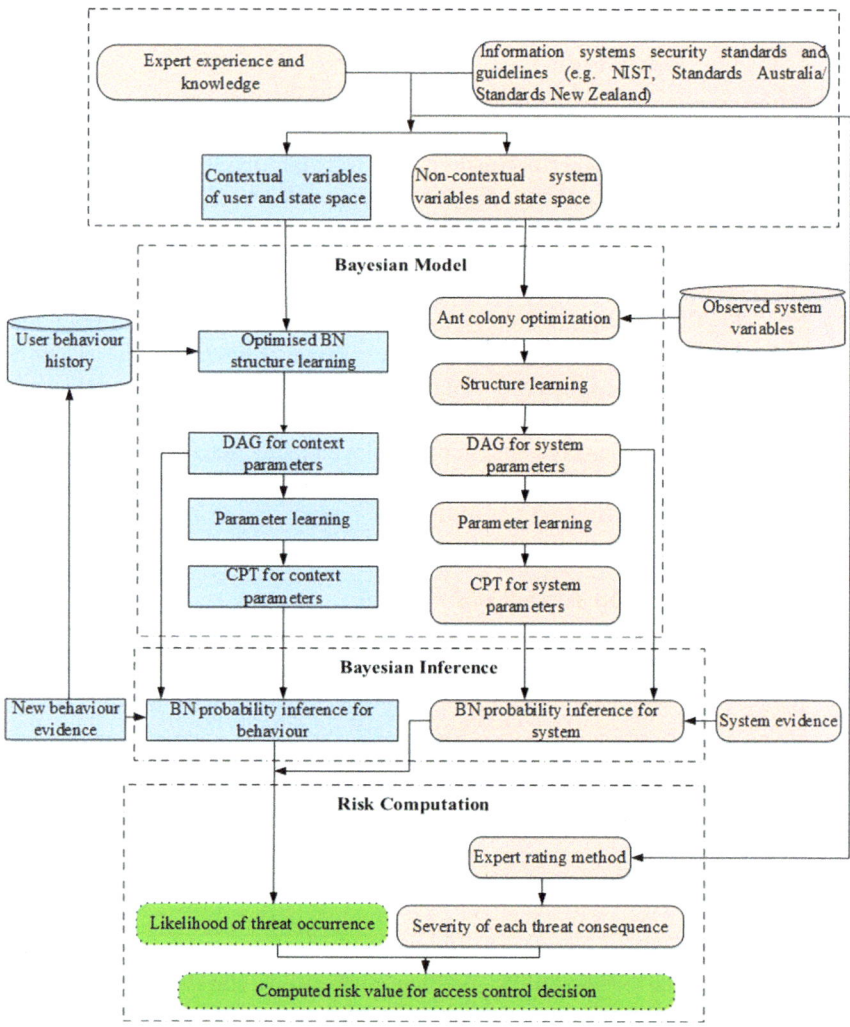

Fig. 2 Detailed ExtSRAM model

Next, a Bayesian model comprising of network structure and CPT will be developed separately from contextual variables and system variables for user profiling and enterprise system risk modeling, respectively. Thereafter, the current security posture of the system and user behavioral evidences will be utilized for BN inferencing. Lastly, the quantified risk value associated with a particular request will be evaluated by the risk module from two key parameters, namely, the likelihood of threat occurrence and severity of impact after occurrence.

Furthermore, the methodology was rendered with functional block diagram as shown in Fig. 2. Generally, the methodology consists of set blocks which are logically classified into three, namely, core, user profiling, and risk computation blocks.

5.2.1 Static Risk Analysis Blocks

The blocks with round edges are the components inherited from SRAM for handling static risk variables pertaining to exposure of enterprise system to BYOD security risks. Hence, some tasks including, identifying non-contextual risk factors by experts, harnessing experts' opinions on the method and scale for rating the severity of threats, evaluating security controls within enterprise environment, identifying threats and their causal relationships, maintaining the database of observed system variables, developing Bayesian oriented risk evaluation model and drawing inference with system evidence variables are undertaken in these blocks. A specific but comprehensive description of each block is available in [23].

5.2.2 User Profiling Blocks

The user profiling blocks are represented with square edges and are built on contextual variables of BYOD users. The collective target of these blocks is to estimate the probability of threat occurrence, arising from a request made by BYOD users. This probabilistic estimation is based on contextual parameters. Thus, to easily combine the results from profiling blocks and static risk analysis blocks, this study proposed the use of Bayesian model. Specifically, the Bayesian modeling is employed as a machine learning tool, to understand and draw inferences for the likelihood of threat from user behavior analytics with regards to interactions with the enterprise system. Therefore, Bayesian network and related CPTs would be formulated on contextual data. Also, inferences would be drawn from both current user profile and new contextual evidence that are extracted from the user's access request. Hence, some blocks that are similar in terms of functionalities, and which are located within the Bayesian model and Inference sections of the model can utilize the same techniques for static risk analysis and user profiling tasks. For brevity, understanding of the fundamental techniques for such blocks is assumed. Consequently, the following subsections explain the new blocks or those that significantly vary from corresponding static analysis blocks.

Contextual Variables of User

Primarily, the function of this block (contextual variables and state space) is to define and harness contextual variables and their particular state spaces that can uniquely differentiate users. The variables are expected from sources like experts' opinions and excerpts from information security standards (or guidelines) and academic publications. More so, the sources of contextual variables should provide adequate definitions for all the variables. Also, the definitions of variable are essential for predicting user's pattern of seeking access to information system, level of granularity for access control configuration by security administrators and user privacy issues concerning contextual data acquisition and storage. For example, in the case of spatiotemporal data, time as a risk factor might be defined as an hour of the day, rather than a day of the week to achieve a well-refined access control. Likewise, location coordinates obtained via Global Positioning System (GPS) can aid fine-grained access, better than the name assigned to a particular location. However, the right balance should be struck between fine-grained metrics for measuring sensitive contextual variables and their implications on user privacy.

Optimized Bayesian Network Structure Learning

The block determines the most appropriate Bayesian network to represent user contextual data that are stored in the user behavior repository. Within this block, the task of selecting an apt structure for subsequent DAG construction is optimized. This optimization can be achieved by setting the appropriate parameters of search and score algorithms that are employed for the Bayesian network development.

Probability Inference of Behavior

Mainly, the block can query the Bayesian model to estimate the probability that a specific action has been carried out by a certain user in the past with respect to other contextual variables. The values of the variables are mined from new evidences in the current user's request for access to an action with previous occurrences. Thus, the result from this block determines the overall likelihood of threat occurrence, which may either increase or decrease. Afterward, the estimated threat likelihood contributes to the overall risk value, which enterprise information system operating in BYOD strategy might be exposed to. Therefore, the risk value depends on the anomaly or similarity between the new evidence and the previous access profile of the user.

5.2.3 Risk Computation Blocks

These blocks are further subdivided into two separate blocks to handle threat likelihood and overall risk computation. These two blocks are represented with dashed round edges. Precisely, the likelihood of threat occurrence comprises of inferences drawn from user behavior and system threat profile for a particular user request. In order to compute the impending security risk in user's request, the risk computation block utilizes outputs from likelihood of threat occurrence and envisaged severity of each threat after occurrence. Functionally, Eq. 3 represents the main parameters for risk evaluation in ExtSRAM, which allows an organization to adopt quantitative or semi-quantitative implementation for the risk computation block.

$$R = f(\omega_c, \omega_s, l_c, l_s, I) \tag{3}$$

where R is the estimated risk of granting access for an action to be performed, I is severity of threat occurrence, l_c and l_s are derivates of Bayesian models, and they represent the likelihoods (probabilistic values) of threat occurrences through contextual risk factors and risk factors of enterprise information system, respectively. Similarly, ω_c means the contextual weight and ω_s represents the weight of enterprise information system. That is, if l_c and $l_s \in L$, then $L : [0, 1] \rightarrow [0, 1]$. In which case, L is the combined likelihood of threat occurrences from contextual and system risk factors. Also, the weights are arbitrary constants which can moderate the likelihood of threat when its value is not equal to 1. For example, the values assigned to the weights might be configured to differentiate between requests from BYOD users who are within enterprise network and those outside the network. For the simplest case, the parameters are modeled as shown in Eq. 4.

$$R = I\left(\frac{\omega_c l_c + \omega_s l_s}{2}\right) \tag{4}$$

Nevertheless, when multiple threats can accompany a request with intent to individually or collectively exploit an enterprise information system, then risk value can be evaluated as shown in Eq. 5. In such case, n is the number of threats, whereas l_{cj} and l_{sj} are likelihoods of threat occurrences through contextual and system risk factors respectively, for particular threat j.

$$R = \sum_{j=1}^{n} I_j\left(\frac{\omega_c l_{cj} + \omega_s l_{sj}}{2}\right) \tag{5}$$

6 Theoretical Validation of the Model

This section provides theoretical validation for ExtSRAM, which is premised on the soundness and completeness of the proposed model. The validation follows the earlier stated assumptions (5) and (6) in Sect. 5.1 that a reliable validation was conducted on SRAM.

6.1 Soundness of ExtSRAM

In order to validate soundness of the model, the authors postulated that the combined likelihood of threats and weights from both user contextual variables and enterprise information systems (static variables) will functionally satisfy the five conditions listed below.

1. $\forall l_c, l_s, \omega_c, \omega_s \in \mathbb{R}^+$
2. $0 < l_c, l_s, \omega_c, \omega_s \leq 1$
3. $0 < (l_c + l_s)/2 \leq 1$
4. $0 < (\omega_c + \omega_s)/2 \leq 1$
5. $0 < \omega_c l_c, \omega_s l_s \leq 1$

Note that, the values of l_c and l_s are probabilistic outcomes of the Bayesian models developed from static risk analysis and user profiling blocks, respectively. Thus, the combined values from these two major blocks would still satisfy basic probability conditions. With regards to the computed risk values, these are necessary conditions for the derivatives of Eq. (5) to be true representatives of real-life implementations.

In order to reinforce the soundness of Eq. (5), ten sets of random values were generated for the parameters in the equation. Each set of values is assumed to represent a known security threat that can be linked to an access request. The generated values for ω_c, ω_s, l_c and l_s were constrained to be within the range [0, 1] for the five conditions listed for the equation to be satisfied. However, the values for impact (I) of threat can be determined by the organization. For this demonstration, the value of impact is a random number between 1 and 10. The estimated risk value for each threat and the overall risk value for all the threats are shown in Table 1.

Table 1 Risk estimation for access request

Threat	l_s	l_c	w_s	w_c	$(w_s l_s + w_c l_c)/2$	Impact (I)	Estimated risk value
1	0.54	0.26	0.63	0.1	0.18	3	0.548764951
2	0.17	0.59	0.01	0.16	0.05	3	0.144674895
3	0.34	0.45	0.2	0.31	0.1	1	0.102939648
4	0.94	0.17	0.15	0.91	0.15	8	1.183142913
5	0.46	0.75	0.07	0.08	0.05	4	0.182067431
6	0.41	0.12	0.42	0.56	0.12	2	0.235882753
7	0.54	0.5	0.19	0.62	0.21	9	1.864801307
8	0.06	0.21	0.14	0.26	0.03	2	0.063811237
9	0.58	0.82	0.91	0.74	0.57	2	1.137792092
10	0.36	0.07	0.43	0.13	0.08	5	0.408285538
					Total estimated risk		**5.872162764**

Likelihood of threat from contextual factor (l_s), Likelihood of threat from enterprise information system (l_c), Enterprise information system weight (w_s), Contextual weight (w_c), Likelihood of threat ($(w_s l_s + w_c l_c)/2$)

As revealed in Table 1, all the computed values for $\frac{\omega_c l_c + \omega_s l_s}{2}$, which represent the likelihood of threat occurrence is within probabilistic range of [0, 1]. Moreover, the total estimated risk value will always depend on the values assigned to the impact of a known threat. Therefore, the result from the ExtSRAM will consistently quantify the risk of granting access to BYOD user's request.

6.2 Completeness of ExtSRAM

In order to completely describe the ExtSRAM, the contextual risk factors for user behavior profiling and the counterpart risk factors for static risk analysis will be drawn from domain experts, well-established standards, and guidelines. Therefore, the experts' input into the selection and definition of the factors will ensure the inclusion of all relevant contextual and non-contextual variables that are necessary for the development of a proposed model for BYOD security risks. In addition, the experts and relevant literatures will serve as a guide for itemizing the existing security countermeasures and their likely performance ratings. Similarly, such guidance would reveal current security threats relating to access control in BYOD further enrich the model.

7 Future Research Directions

Really, ExtSRAM as presented in this study is meant to provoke researchers' thoughts on the benefits and implications of combining static risk factors, contextual risk factors, countermeasures, and risk evaluation concepts for dynamic authorization in pervasive environments. Importantly, subsequent researches can explore abundant possibilities of implementing the model for specific BYOD domains like academic institutions, health management organizations, public commissions, and agencies. More so, attempts can be made to utilize other contemporary machine learning and deep learning algorithms to model the static risk analysis blocks and user profiling blocks. In addition, future researches can implement the model to mitigate access control challenges in emergent computing strategies including cloud computing, IoT and Fog computing. As part of future researches in this direction, the authors intend to develop and empirically validate the user profiling blocks of ExtSRAM.

8 Conclusion

In a word, BYOD remains an imperative IT strategy that continues to dynamically reshape enterprise work environments, improves productivity, and enhances employee workstyles. Unfortunately, as the pervasion of BYOD spreads with abundant benefits, so also, are the security consequences it leaves behind for organizations to surmount. Certainly, dynamic access control remains one viable countermeasure for securing access to vital organization resources against the security risks that accompanied BYOD adoption. Thus, this study developed ExtSRAM as a dynamic risk-based and context-aware paradigm that integrated three major components namely; user contextual data, static risk analysis (which incorporates threat propagation and existing countermeasures), and machine learning algorithm. Similarly, a novel mathematical model was developed for dynamic risk evaluation in ExtSRAM. For flexibility and dynamism, the novel model can serve as an add-on to static access control models for possible adoption in pervasive domains. As a case in point, subsequent implementation of ExtSRAM will certainly assist organizations to reap lofty benefits from BYOD while securing the critical resources from security risks. In another case, ExtSRAM opened another frontier for research activities on risk-aware access control for other pervasive environments.

References

1. Ahmed, A., Zhang, N.: An access control architecture for context-risk-aware access control: architectural design and performance Evaluation. In: Fourth International Conference on Emerging Security Information Systems and Technologies (SECURWARE), IEEE, Venis, pp. 251–260 (2010). https://doi.org/10.1109/SECURWARE.2010.48
2. Aldini, A., Carlos, J.S., Lafuente, B., Titi, X., Guislain, J.: Design and validation of a trust-based opportunity-enabled risk management system. Inf. Comput. Secur **25**(1–31) (2017)
3. Alkussayer, A., Allen, W.H.: Security risk analysis of software architecture based on AHP. In: 7th International Conference on Networked Computing, pp. 60–67. IEEE, Gyeongsangbuk-do, Korea (2011)
4. Alotaibi, B., Almagwashi, H.: A review of BYOD security challenges, solutions and policy best practices. In: 2018 1st International Conference on Computer Applications & Information Security (ICCAIS), pp. 1–6. IEEE (2018). https://doi.org/10.1109/CAIS.2018.8441967
5. Atlam, H.F., Alenezi, A., Walters, R.J., Wills, G.B., Daniel, J.: Developing an adaptive Risk-based access control model for the Internet of Things. In: International Conference on Internet of Things (iThings) and IEEE Green Computing and Communications (GreenCom) and IEEE Cyber, Physical and Social Computing (CPSCom) and IEEE Smart Data (SmartData), pp. 655–661. IEEE, Exeter, UK (2017). https://doi.org/10.1109/iThings-GreenCom-CPSCom-SmartData.2017.103
6. Bedi, P., Kaur, H., Gupta, B.: Trust-based access control for collaborative systems. J. Exp. Theor. Artif. Intell. **26**(1), 109–126 (2014). https://doi.org/10.1080/0952813X.2013.813973
7. Bijon, K.Z., Krishnan, R., Sandhu, R.: A framework for risk-aware role based access control. In: 2013 IEEE Conference on Communications and Network Security (CNS), pp. 462–469. IEEE, Washington (2013). https://doi.org/10.1109/CNS.2013.6682761
8. Blizzard, S.: Coming full circle: Are there benefits to BYOD? Comput. Fraud Secur. **2015**(2), 18–20 (2015). https://doi.org/10.1016/S1361-3723(15)30010-5
9. Bojanc, R., Jerman-Blažič, B.: A quantitative model for information security risk management. Eng. Manag. J. **25**(2), 25–37 (2013)
10. Bonate, P.L.: COVID-19: opportunity arises from a world health crisis. J. Pharmacokinet Pharmacodyn. **47**(2), 119–120 (2020). https://doi.org/10.1007/s10928-020-09681-5
11. Cabarcos PA (2011) Risk assessment for better identity management in pervasive environments. In: Fourth Annual Ph.D. Forum on Pervasive Computing and Communications, pp. 389–390. IEEE, Seattle, WA (2011)
12. Chen, L., Crampton, J.: Risk-aware role-based access control. In: Meadows, C., Fernandez-Gago, C. (eds.), Security and Trust Management, pp. 146–150. Springer, Berlin (2012). https://doi.org/10.1007/978-3-642-29963-6_11

13. Cherdantseva, Y., Burnap, P., Blyth, A., Eden, P., Jones, K., Soulsby, H., Stoddart, K.: A review of cyber security risk assessment methods for SCADA systems. Comput. Secur. **56**(2016), 1–27 (2016). https://doi.org/10.1016/j.cose.2015.09.009

14. CLUSIF: Risk management-concepts and methods. CLUSIF 2008/2009. Paris (2008).

15. Costabello, L., Villata, S., Delaforge, N., Gandon, F.: Linked data access goes mobile: context-aware authorization for graph stores. In: CEUR Workshop Proceedings, p. 937 (2012).

16. Djalante, R., Lassa, J., Setiamarga, D., Mahfud, C., Sudjatma, A., Indrawan, M., Haryanto, B., Sinapoy, M.S., Rafliana, I., Djalante, S., Gunawan, L.A., Anindito, R., Warsilah, H., Surtiari, I.G.A.: Review and analysis of current responses to COVID-19 in Indonesia: Period of January to March 2020. Prog. Disaster Sci. **6**, 100091 (2020). https://doi.org/10.1016/j.pdisas.2020.100091

17. Doargajudhur, M.S., Dell, P.: The effect of Bring Your Own Device (BYOD) adoption on work performance and motivation. J. Comput. Inf. Syst. **00**(00), 1–12 (2018). https://doi.org/10.1080/08874417.2018.1543001

18. Dorri Nogoorani, S., Jalili, R.: TIRIAC: a trust-driven risk-aware access control framework for Grid environments. Future Gener. Comput. Syst. **55**, 238–254 (2016). https://doi.org/10.1016/j.future.2015.03.003

19. dos Santos, D.R., Marinho, R., Schmitt, G.R., Westphall, C.M., Westphall, C.B.: A framework and risk assessment approaches for risk-based access control in the cloud. J. Netw. Comput. Appl. **74**, 86–97 (2016). https://doi.org/10.1016/j.jnca.2016.08.013

20. Emami, S.S., Amini, M., Zokaei, S.: A context-aware access control model for pervasive computing environments. In: The 2007 International Conference on Intelligent Pervasive Computing (IPC 2007), pp. 51–56. IEEE, Jeju City, South Korea (2007). https://doi.org/10.1109/IPC.2007.28

21. Fall, D., Blanc, G., Okuda, T., Kadobayashi, Y.: Toward quantified risk-adaptive access control for multi-tenant cloud computing. In: Proceedings of the 6th Joint Workshop on Information Security (JWIS 2011), pp. 1–14. JWIS, Kaohsiung (2011)

22. Fani, N., Von Solms, R., Gerber, M.: Governing information security within the context of "Bring Your Own Device in SMMEs." In: IST-Africa 2016 Conference Proceedings, pp. 1–11. Durban, South Africa (2016)

23. Feng, N., Wang, H.J., Li, M.: A security risk analysis model for information systems: causal celationships of risk factors and vulnerability propagation analysis. Inf. Sci. **256**(2014), 57–73 (2014). https://doi.org/10.1016/j.ins.2013.02.036

24. Fischer, G.: Context-aware systems: the 'right' information, at the 'right' time, in the 'right' place, in the 'right' way, to the 'right' person. In: Advanced Visual Interfaces International Working Conference, pp. 287–294. ACM, Capri Island (Naples), Italy (2012)

25. Ganiyu, S.O., Jimoh, R.G.: Characterising risk factors and countermeasures for risk evaluation of bring your own device strategy. Int. J. Inf. Secur. Sci. **7**(1), 49–59 (2018)

26. Ganiyu, S.O., Jimoh, R.G.: Comparative analysis of risk evaluation models for risk-aware access control in bring your own device environment. Int. J. Inf. Secur. Res. (IJISR) **8**(2), 810–820 (2018)

27. Hajdarevic, K., Allen, P., Spremic, M.: Proactive security metrics for bring your own device (BYOD) in ISO 27001 supported environments. In: 2016 24th Telecommunications Forum (TELFOR), pp. 1–4. IEEE, Belgrade, Serbia (2016)

28. Heckerman, D.: A tutorial on learning with Bayesian networks. Stud. Comput. Intell. **156**(January), 33–82 (2020). https://doi.org/10.1007/978-3-540-85066-3_3

29. Hong-yue, L., Miao-lei, D., Wei-dong, Y.: A context-aware fine-grained access control model. In: 2012 International Conference on Computer Science and Service System(SSS), pp. 1099–3656. IEEE, Nanjing, China (2012). https://doi.org/10.1109/CSSS.2012.278

30. Kandala, S., Sandhu, R., Bhamidipati, V.: An attribute based framework for risk-adaptive access control models. In: Proceedings of the 6th International Conference on Availability, Reliability and Security, pp. 236–241. IEEE, Vienna (2011). https://doi.org/10.1109/ARES.2011.41

31. Kang, D., Oh, J., Im, C.: Context based smart access control on BYOD environments. In: International Workshop on Information Security ApplicationsWISA 2014: Information Security Applications, pp. 165–176 (2015). https://doi.org/10.1007/978-3-319-15087-1

32. Kang, Q., Xue, L., Morrison, A., Tang, Y., Chen, A., Luo, X.: Programmable In-Network Security for Context-aware BYOD Policies. In 29th USENIX Security Symposium (SEC'20), pp. 1–21. Boston, MA (2020).

33. Kayes, A.S.M., Han, J., Rahayu, W., Islam, M.S., Colman, A.: A policy model and framework for context-aware access control to information resources. Comput. J. bxy065, 1–24 (2013)

34. Kiran, K.V.D., Mukkamala, S., Katragadda, A., Reddy, L.S.S.: Performance and analysis of risk assessment methodologies in information security. Int. J. Comput. Trends Technol. (IJCTT) **4**(10), 3685–3692 (2013)

35. Koh, E.B., Oh, J., Im, C.: A study on security threats and dynamic access control technology for BYOD, smart-work environment. In Proceedings of the International Multi Conference of Engineers and Computer Scientists 2014, vol. II. IMECS, Hong Kong (2014)

36. Lee, B., Vanickis, R., Rogelio, F., Jacob, P.: Situational awareness based risk-adapatable access control in enterprise networks. In: 2nd International Conference on Internet of Things, Big Data and Security (IoTBS), pp. 400–405. Porto, Portugal (2017). https://doi.org/10.5220/0006363404000405

37. Lo, C., Chen, W.: A hybrid information security risk assessment procedure considering inter-dependences between controls. Expert Syst. Appl. **39**(2012), 247–257 (2012). https://doi.org/10.1016/j.eswa.2011.07.015

38. Luo, J., Kang, M.: Risk based mobile access control (RiBMAC) policy framework. In: The 2011 Military Communications Conference, pp. 1448–1453. IEEE, Baltimore, Maryland, USA (2011)

39. Miettinen, M., Heuser, S., Sadeghi, A.: ConXsense –automated context classification for context-aware access control. In: 9th ACM Symposium on Information, Computer and Communications Security (ASIACCS 2014), pp. 293–304. Kyoto, Japan (2014)

40. Miura-ko, R.A., Bambo, N.: Dynamic risk mitigation in computing infrastructures. In: Third International Symposium on Information Assurance and Security, pp. 325–328. IEEE, Manchester (2007). https://doi.org/10.1109/IAS.2007.91

41. Morrison, A., Xue, L., Chen, A., Luo, X.: Enforcing context-aware BYOD policies with in-network security. In: 10th USENIX Workshop on Hot Topics in Cloud Computing (HotCloud'18). Boston, MA (2018)

42. O'Neill, M.: The Internet of Things: Do more devices mean more risks? Comput. Fraud Secur. **2014**(1) (2014). https://doi.org/10.1016/S1361-3723(14)70008-9

43. Redmond, J., Fong, B.: Crafting an effective Bring Your Own Device (BYOD) policy. Units Magazine (2018)

44. Rose, C.: BYOD: An examination of bring your own device in business. Rev. Bus. Inf. Syst. **17**(Second Quarter), 65–70 (2013)

45. Sadiku, M.N.O., Nelatury, S.R., Musa, S.M.: Bring your own device. J. Sci. Eng. Res. **4**(4), 163–165 (2017)

46. Sanchez, C.A.: A risk and trust security framework for the pervasive mobile environment. University of Oklahome (2013)

47. Santos, D.R., Westphall, C.M., Westphall, C.B.: A dynamic risk-based access control archi-tecture for cloud computing. In: Network Operations and Management Symposium, pp. 1–9. IEEE, Krakwo, Poland (2014). https://doi.org/10.1109/NOMS.2014.6838319

48. Sato, H.: A new formula of information security risk analysis that takes risk improvement factor into account. In: International Conference on Privacy, Security, Risk, and Trust, and IEEE International Conference on Social Computing, pp. 1243–1248. IEEE, Boston (2011). https://doi.org/10.1109/PASSAT/SocialCom.2011.44

49. Scutari, M.: An empirical-bayes score for discrete Bayesian networks. In: Proceedings of the Eighth International Conference on Probabilistic Graphical Models, vol. PMLR 52, pp. 438–448). Lugano (Switzerland) (2016)

50. Sharma, M., Bai, Y., Chung, S., Dai, L.: Using risk in access control for cloud-assisted eHealth. In: Proceedings of the 14th IEEE International Conference on High Performance Computing and Communications, HPCC-2012, pp. 1047–1052. IEEE, Liverpool, United Kingdom (2012). https://doi.org/10.1109/HPCC.2012.153

51. Trnka, M., Cerny, T.: On security level usage in context-aware role-based access control. In: The 31st ACM/SIGAPP Symposium on Applied Computing, pp. 1192–1195. ACM , Pisa, Italy (2016).

52. Ubene, O.E., Agim, U.R., Umo-Odiong, A.: The impact of Bring Your Own Device (BYOD) ON Information Technology (It) security and infrastructure in the Nigerian Insurance Sector. Am. J. Eng. Res. (AJER) **7**(5), 237–246 (2018)

53. Vaidya, R.: Cyber Security Breaches Survey 2018: Statistical Release. Portsmouth (2018)

54. Veljkovic, I., Budree, A.: Development of bring-your-own-device risk management model: a case study from a South African Organisation. Electron. J. Inf. Syst. **22**(1), 1–14 (2019)

55. Wang, Q., Jin, H.: Quantified risk-adaptive access control for patient privacy protection in health information systems. In: ASIACCS'11, pp. 406–410. ACM, Hong Kong (2011)

56. Weber, L., Rudman, R.J.: Addressing the incremental risks associated with adopting bring your own device. J. Econ. Financ. Sci. **11**(1), 1–7 (2018). https://doi.org/10.4102/jef.v11i1.169

57. Wei, Z., Li, M.: Information security risk assessment model base on FSA and AHP. In: Proceedings of the Ninth International Conference on Machine Learning and Cybernetics, pp. 11–14. IEEE, Qingdao (2010)

58. Yang, X., Wang, X., Yue, W.T., Sia, C.L., Luo, X.: Security policy opt-in decisions in Bring-Your-Own-Device (BYOD)–a persuasion and cognitive elaboration perspective. J. Organ. Comput. Electron. Commer. **29**(4), 274–293 (2019). https://doi.org/10.1080/10919392.2019. 1639913

59. Ye, D., Mei, Y., Shang, Y., Zhu, J., Ouyang, K.: Mobile crowd-sensing context aware based fine-grained access control mode. Multimed. Tools Appl. **75**(21), 13977–13993 (2015). https:// doi.org/10.1007/s11042-015-2693-3

60. Zhiwei, Y., Zhongyuan, J.: A survey on the evolution of risk evaluation for information systems security. Energy Procedia **17**(2012), 1288–1294 (2012). https://doi.org/10.1016/j.egypro.2012. 02.240

61. Zhu, Z., Xu, R.: A context-aware access control model for pervasive computing in enterprise environments. In 4th International Conference on Wireless Communications, Networking and Mobile Computing, WiCOM'08, pp. 1–6. IEEE, Dalian, China (2008). https://doi.org/10.1109/ IPC.2007.28

.